Information Security and Cryptography

More information about this series at http://www.springer.com/series/4752

Adrian Perrig · Pawel Szalachowski
Raphael M. Reischuk · Laurent Chuat

SCION: A Secure Internet Architecture

Adrian Perrig
Network Security Group
ETH Zürich
Zürich
Switzerland

Raphael M. Reischuk
Network Security Group
ETH Zürich
Zürich
Switzerland

Pawel Szalachowski
Network Security Group
ETH Zürich
Zürich
Switzerland

Laurent Chuat
Network Security Group
ETH Zürich
Zürich
Switzerland

ISSN 1619-7100 ISSN 2197-845X (electronic)
Information Security and Cryptography
ISBN 978-3-319-88374-8 ISBN 978-3-319-67080-5 (eBook)
https://doi.org/10.1007/978-3-319-67080-5

Printed on acid-free paper

This Springer imprint is published by Springer Nature
The registered company is Springer International Publishing AG
The registered company address is: Gewerbestrasse 11, 6330 Cham, Switzerland

To Miyoung,
Thank you for your unwavering support.
Love, forever!

Adrian

To Henio,
For all these sleepless nights.

Paweł

To my family and those
who supported me along my way.

Raphael

To Manon,
For your patience and encouragement.

Laurent

Contents

Foreword

VIRGIL GLIGOR (CARNEGIE MELLON UNIVERSITY)

Despite having worked with Adrian Perrig for a few years at Carnegie Mellon University's CyLab, where he embarked on the task of developing a secure architecture for the Internet, I had had no in-depth exposure to SCION until I attended a presentation he gave at Singapore Management University in late 2010. Entitled "SCI-FI: Secure Communication Infrastructure for a Future Internet," his talk described the early project that was to become SCION. The audience reaction was predictable and all too familiar: you can't change the Internet; its foundation is immutable!

But in fact it had been clear for a long time that the Internet design had to change, as security cracks had gradually been appearing in its foundation since its early days. By the mid-1980s, it was obvious that the denial-of-service problem was not effectively addressed by Internet protocols. By the mid-90s, it was clear that BGP was prone to cascading instability, and by the mid-2000s distributed denial of service had become a predictable Internet "feature." Other security issues arose, such as prefix hijacking, IP source address spoofing, and packet-content alteration. Even when cryptographic protocols, such as SSL/TLS, were finally applied in response to e-commerce pressure, their worldwide deployment was more an exception than the rule. Besides, the public-key infrastructure (PKI) supporting SSL/TLS continues to be extremely fragile. As the Internet has expanded in size and use, security problems have become increasingly severe: both organized crime and nation states have started to launch massive attacks for economic or political gain.

Despite repeated wake-up calls for Internet redesign, the response has generally been something of a "boiling frogs" reaction: the severity of the problems has continued to increase relentlessly, but perception of the enormous effort required to solve them has blocked, frustrated and foiled any impulse for redesign from ground up. Over the past decade, it has become clear that security is a fundamental problem of Internet design, but it remains a secondary concern. So against that background, the audience reaction to Adrian Perrig's 2010 SCI-FI presentation in Singapore was only to be expected.

Since my first exposure to SCION, I have been impressed with several of its innovative ideas and new properties. For instance, the concept of isolation domains provides control-plane protection and simplifies construction of PKI infrastructures due to the natural scoping of trust roots. (Although a concept similar to that of isolation domains was considered for the initial Internet design, the focus in that early phase was on getting the network to function at scale before introducing hierarchical decomposition mechanisms.) SCION's concepts of transparency and control, which weave through the entire architecture, result in many desirable properties, e.g., both high-performance and multipath communication for hosts. Also, cryptographically protected packet-carried forwarding state brings forwarding-path authorization without incurring any router-state cost. SCION's architecture integrates these concepts seamlessly into a coherent secure system.

This book offers a fascinating view of both the high-level concepts that drive SCION's design and its implementation, and it leads the reader to draw some surprising new conclusions.

Contrary to the common belief that security causes a loss of performance, several SCION operations are efficient despite performing cryptographic operations; e.g., SCION packet forwarding can be faster and require less energy than IP forwarding. This suggests that redesigning the Internet can be rewarding in more areas than security. I am not aware of any other project that has gone so deeply and broadly in redesigning an entire secure Internet architecture.

The SCION project contradicts another widely held opinion in demonstrating that deployment of a new Internet architecture at scale is in fact possible. This book illustrates the basic ingredients of deployment success: SCION has provided a multitude of incentives for ISPs and end domains, so that local deployment can already provide benefits to early adopters. The book also describes some of SCION's secret deployment sauce: keep the updates of the current routing infrastructure of both ISPs and end domains to a minimum, and reuse the existing intra-domain communication to the maximum extent. It should not be surprising that (e.g., Swiss) ISPs have already found it possible to deploy SCION routers in their core infrastructure and develop new services on it.

Contrary to another common belief, a single Internet architecture can enable integrated defenses against multiple types of attacks, as opposed to one which requires piecemeal solutions. In my opinion, the SCION architecture is unique in this sense, and this book illustrates the fact through the solutions it describes to long-standing problems. For example, SCION provides these unique properties:

- Global security without any global root of trust. This implies that a global "kill switch," an unavoidable feature of other secure network architectures, is not possible in SCION.
- Control-plane functions for secure path withdrawals and control messages. Although any network can always cryptographically sign messages in an

attempt to achieve secure operation, SCION secures the control plane in a very efficient way while enabling high-speed router operation.

- Global resource allocation without requiring per-flow or per-computation fairness mechanisms. This stands in contrast to the current Internet design, in which these mechanisms enable massive DDoS attacks by commercially available botnets. The book shows how SCION leverages its global resource allocation architecture to offer a range of DDoS countermeasures.
- Practical multipath architecture without having to rely on multiple communication media and heterogeneous routing interfaces; e.g., cellular or WiFi connection on cell phones. SCION is currently the only architecture I am aware of that provides general homogeneous multipath communication.
- A robust TLS PKI design with a very limited attack surface; i.e., several independent entities need to be compromised for an attack to be launched. In contrast, the current TLS PKI has a huge attack surface; e.g., if a single key is compromised of the thousand or more that are trusted to sign domain certificates, an adversary can compromise any TLS-protected channel.

So can the Internet be changed and secured from the ground up? This book provides a beacon of hope, proposing that the seemingly unsolvable problem of changing the Internet can in fact be solved. With the open-source SCION implementation and a readily available testbed, researchers can experiment on a firmer network foundation and develop solutions to today's pressing security problems. It is only through hands-on experiments on common platforms like SCION that we can build a new Internet, one that we can rely on with confidence. Let's embrace it!

Preface

ADRIAN PERRIG

The SCION project started in Summer 2009 at Carnegie Mellon University (CMU), when we began meeting weekly with Haowen Chan, Hsu-Chun Hsiao, and Xin Zhang to consider what a secure inter-domain Internet architecture would look like if we could start from a clean slate. The goal was to create an architecture that offered high availability and security for basic point-to-point communication — which other architectures that provide content-centric or mobility-centric properties could build upon.

The project was arduous, because for every approach we came up with, we saw at least two new problems. After several months of meetings, all we had was many pages filled with requirements that the architecture should meet, but no approach to satisfy even a major subset of the requirements. As time went on, the project seemed to be increasingly hopeless. But our perseverance paid off. In Summer of 2010 the basic ideas of beaconing and the creation of end-to-end paths through path-segment combination emerged. Although we would have been happy with any approach that satisfied half of the requirements, our basic approach appeared to meet most of our requirements. Delighted with our discovery, we accelerated the pace of the project. We were encouraged by the fact that our architecture could elegantly address every issue we came up with. We called it the Secure Communication Infrastructure for a Future Internet (SCI-FI).

In Fall 2010, Dave Andersen and Geoff Hasker joined the project and we started writing a paper. Many people took issue with the designation SCI-FI, so we went with Geoff Hasker's suggestion of SCION — despite its rather presumptuous meaning of "heir to the throne" — as an acronym for *scalability, control, and isolation on next-generation networks*. Our paper quickly took shape, and was accepted for publication at the IEEE Symposium on Security and Privacy in 2011. Oddly, the paper was placed in the "Secure Information Flow and Information Policies" session, which usually hosts papers of a different type. Unfazed, Xin Zhang gave a strong presentation and the work was well received.

Buoyed by the early promise of the project, we continued working on SCION and convinced the eXpressive Internet Architecture (XIA) team at CMU that

SCION was a worthwhile choice for host-to-host communication. So initially, SCION developed in the context of XIA, which helped support the early research.

The project developed along two major axes: research and implementation. The early research results leveraged SCION for DDoS defense [114] and anonymous communication [113]. To achieve source authentication and path validation, we designed OPT [132], and performed a formal verification of the protocol [263]. With the goal of producing a stronger public-key infrastructure (PKI) for SCION, the Accountable Key Infrastructure (AKI) was developed [133].

The initial implementation effort started with the help of several student projects. However, much of the progress was made when Soo-Bum Lee joined the project and completed a first SCION prototype in 2011, which we continuously improved throughout 2012.

In view of the opportunities offered by ETH Zurich, we built up a new research group around the SCION project in Switzerland. Pawel Szalachowski, a promising postdoctoral researcher from Poland, joined the group in March 2013 and became the core designer and developer of SCION. Under his guidance, the SCION prototype and testbed went through several generations of software and matured into the system that we currently deploy. Much progress was made when Stephen Shirley joined the group, as he improved numerous aspects of the system including design and implementation. Jason Lee deserves credit for his work on the multipath socket and the high-speed router (the latter project was in collaboration with Takayuki Sasaki who was visiting from NEC). More recently, Tobias Klausmann and Ercan Ucan joined the developer team, greatly improving SCION's infrastructure and deployment. All the hard work has paid off: in Summer 2016 we started a deployment of SCION routers in the production networks of Swisscom and SWITCH, two large ISPs in Switzerland, with several of their customers now engaging in test deployments.

On the research side, many newcomers joined the team at ETH, assisted by the postdoctoral researchers David Barrera, Raphael Reischuk, and Pawel Szalachowski. With SCION as the core focus of the research group, much progress was accomplished in many directions, such as PKIs [23, 52, 168, 169, 233–235], DDoS defense [22, 143], anonymous communication and privacy [49, 51, 153, 156], efficient forwarding [154], fault localization [21], energy analysis [50], high-speed duplicate detection [155], as well as public-policy and legal aspects [26, 194]. Besides the research contributions, Raphael Reischuk successfully contributed to outreach and promotion by designing the SCION logo and creating the SCION website, initiating a newsletter, and giving outreach presentations to help attract early adopters. Many PhD students contributed to SCION — for instance Sam Hitz has made several major contributions by suggesting Python as a base language (to speed up implementation and increase code clarity), implementing major parts of the (early) SCION core code, and designing and implementing the secure link revocation mechanism. Also many researchers contributed to the project, for instance Virgil Gligor, Yih-Chun Hu, and members of the XIA project team, who were involved in several research projects and contributed much feedback and many insights to the project.

Over the past eight years, numerous people helped on the project through research discussions, feedback on publications, setup and operation of SCION infrastructure, research projects, and more. We estimate that around 80 people have so far played a significant role in the project (about 30 people from our group, about 30 bachelor or master students have completed a semester project or thesis, and about 20 external collaborators and industry visitors who worked closely with us). We are very grateful for everyone's help, without which the project would not have reached its current status. When adding up the amount of time researchers and engineers worked on the SCION architecture, we arrive at approximately 75 person-years of endeavor that has been spent by the end of 2016. Consequently, much thought and deliberation have gone into the design decisions presented in this book.

When we started the project in 2009, it was mostly security researchers who agreed on the importance of re-designing the Internet from a security perspective [27]. However, many events that have occurred since have brought Internet security to the forefront of awareness: several cases of Internet censorship, the Snowden revelations, NSA backdoors (e.g., in Juniper routers, standardized cryptographic algorithms), Internet kill switches, IANA's stewardship transition to a multi-stakeholder governance, increasingly large DDoS attacks, attacked certification authorities, the emergence of quantum computers, etc. Today, Internet security and privacy is a common topic of conversation. In the IETF, the main body for standardizing Internet protocols, awareness of security concerns has greatly increased — with an IETF draft stating that pervasive monitoring by governments constitutes an attack [85]. These events have given impetus to the SCION project, as it matured during this period and provides solutions to the exact problems that have moved into public awareness. Consequently, the SCION architecture goals appear aligned with the public interests and we do not seem to be swimming against the mainstream goals.

Bob Kahn mentioned that simplicity and elegance were the main reasons why TCP/IP has lasted as long as it has. When a system is simple and elegant, it is easy to understand, implement, and maintain. Thus, simplicity and elegance are important goals in SCION, besides availability, security, scalability, and efficiency. In the entire architecture, we attempt to minimize complexity to achieve the desired properties, leveraging well-understood technologies. Unless they were in line with the approach we deemed best, we avoided the urge to use "trendy" technologies of the day, such as blockchain or doubly homomorphic encryption. We hope that the readers will also appreciate the results of our endeavors to produce a clean-slate re-design of a highly available point-to-point communication architecture, and that they will join us on our journey towards a secure Internet.

How to Read This Book

This book describes the essential components of the SCION future Internet architecture prototype (V1.0) including functional specifications of the SCION network elements (e.g., servers, routers, gateways), communication protocols among these elements, data structures, and configuration files. In particular, the book focuses on the specification of a working prototype and additional features that are not described in academic papers. We highlight contributions that we believe are particularly important and interesting with a diamond symbol.

The aim of this book is to provide an easy-to-follow introduction to SCION. To help the reader, it contains a glossary (Page 417) defining important terms and supplying background information. We indicate terms with a glossary entry as follows:

glossary term*

A gray bar in the margin indicates the presence of an example:

This is an example.

We also provide an index (Page 423), a list of abbreviations (Page 421), and answers to frequently asked questions (Page 409). A comprehensive example of SCION's operations is on Page 223 and illustrates the end-to-end communication between two hosts, including name resolution, path resolution, packet origination, and packet forwarding. The example provides references to detailed explanations of the underlying concepts and techniques, and thus serves as a good starting point for the more technically adept readers.

The book also aims to provide a comprehensive description of the main design features for achieving a secure Internet architecture. While many of the detailed design aspects are described in research papers, we have added relevant details where necessary to understand the important concepts. We have structured the book in such a way that the technical details gradually increase as it proceeds: starting with an overview and moving along to the format of configuration files at the end.

Additional SCION resources (research papers, talks, presentations, source code, and links to contributing efforts) are available on our web page:

`https://www.scion-architecture.net`

We also encourage interested readers to sign up to the SCION mailing list (through the above website). Furthermore, a discussion board for the SCION community takes questions and offers support regarding the development and deployment of SCION. As we encounter errors in the book, we will document them in an errata list on our web page.

Acknowledgments

Many people contributed toward this book. Special thanks go to Jeffrey Barnes for his excellent copy editing, and Ronan Nugent our editor at Springer who guided us through the publication process. We also thank the following individuals for providing valuable feedback that improved the content of this book (in alphabetical order):

David Basin	ETH Zurich
Jan Boogman	Swisscom AG
Srdjan Capkun	ETH Zurich
Alexander Gall	SWITCH
Virgil Gligor	Carnegie Mellon University (CMU)
David Hausheer	Technische Universität Darmstadt
Yih-Chun Hu	University of Illinois at Urbana-Champaign
Jill Jermyn	Columbia University
Burt Kaliski	Verisign, Inc.
Ayumu Kubota	KDDI Corporation
Jovan Kurbalija	Geneva Internet Platform
Heejo Lee	Korea University
Simon Leinen	SWITCH
René Merz	Magnetron Labs
Peter Müller	ETH Zurich
Radha Poovendran	University of Washington
Timothy Roscoe	ETH Zurich
Mark Ryan	University of Birmingham
Ankit Singla	ETH Zurich
Christoph Sprenger	ETH Zurich
Peter Steenkiste	Carnegie Mellon University (CMU)
Laurent Vanbever	ETH Zurich
David Watrin	Swisscom AG

The project was made possible by the generous support of the following organizations (in alphabetical order):

- CyLab at Carnegie Mellon University;
- ETH Zurich, which provided the majority of funding for the project;
- European Research Council, under the European Union's Seventh Framework Programme (FP7/2007-2013) / ERC grant agreement 617605;
- Infosec Global, through a contract;
- Institute for Information and Communications Technology Promotion (IITP), grant funded by the Korean government (MSIP) (No. R0190-16-2011, Development of Vulnerability Discovery Technologies for IoT Software Security);

- Intel Corp., which provided equipment;
- KDDI Corporation, through a gift;
- National Science Foundation (NSF), under awards CCF-0424422 and CNS-1040801;
- Swisscom AG, through a contract;
- Zurich Information Security and Privacy Center (ZISC), through gifts from Google, NEC, Open Systems, SIX, and ZKB.

Without these sources of support, the project would not have been possible. We would like to express our sincere gratitude to all who contributed.

Part I

Overview

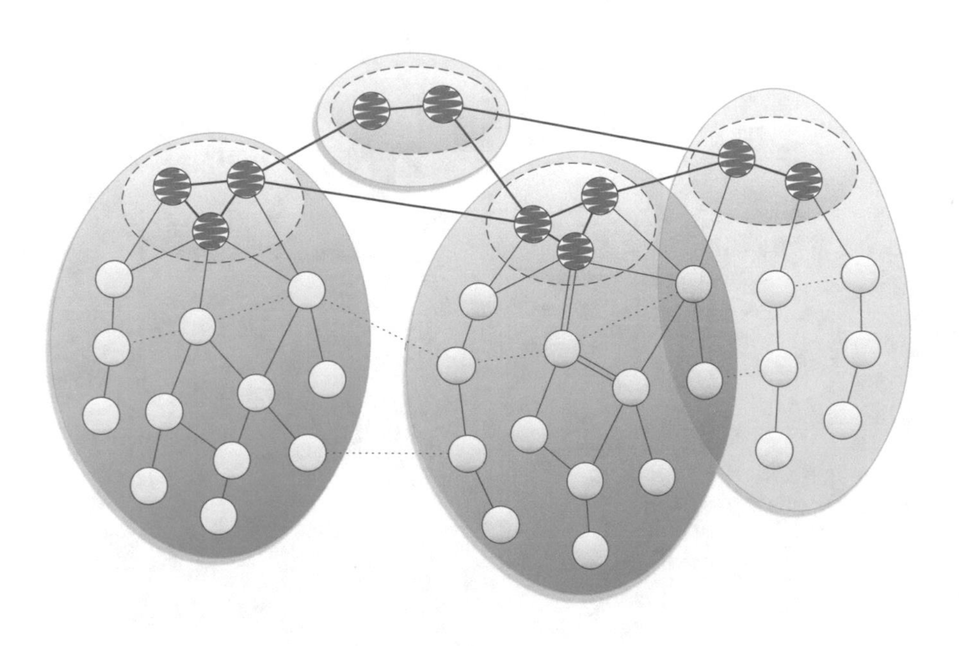

1 Introduction

David Barrera, Laurent Chuat, Adrian Perrig,
Raphael M. Reischuk, Pawel Szalachowski

The Internet has been successful beyond even the most optimistic expectations. It permeates and intertwines with almost all aspects of our modern society and economy. The success of the Internet has created a dependency on communication as many of the processes underpinning the foundations of modern society would grind to a halt should communication become unavailable. However, much to our dismay, the current state of safety and availability of the Internet is far from commensurate with its importance.

Although we cannot conclusively determine what the impact of a 1-minute, 1-hour, 1-day, or 1-week outage of Internet connectivity on our society would be, anecdotal evidence indicates that even short outages have a profound negative impact on governmental, economic, and societal operations. To make matters worse, the Internet has not primarily been designed for high availability in the face of malicious actions by adversaries. Recent patches to improve Internet security and availability have been constrained by the design of the current Internet architecture. A new Internet architecture should offer availability and security by design, provide incentives for deployment, and consider economic, political, and legal issues at the design stage.

In this book, we describe SCION, an inter-domain network architecture designed to address these issues by providing a fundamental building block: highly available point-to-point communication. We present SCION's goals, design, limitations, specifications, extensions, and the results of several years of research conducted since the initial publication [266]. But we start, as a motivation for our work, by reflecting on the current state of the Internet.

1.1 Today's Internet

Witnessing the fast advancement of Internet-based services, applications, and technologies, it might seem that the Internet is evolving at a rapid pace. In reality though, only parts of the protocol stack have changed significantly since the Internet's inception. The application and physical layers have adapted to

A. Perrig et al., *SCION: A Secure Internet Architecture*, Information Security and Cryptography, https://doi.org/10.1007/978-3-319-67080-5_1

new needs and trends, but the core protocols have remained mostly the same for decades. This situation has been referred to as the "Internet Hourglass", meaning that a handful of protocols form a thin — and seemingly irreplaceable — waist in the protocol stack, while both ends of the stack continue to increase in diversity. In this section, we start by discussing the two core technologies of the current Internet: the Internet Protocol (IP) [201] and the Border Gateway Protocol (BGP) [209].

Nobody could have predicted how impressively these protocols would stand the test of time, as they remained relatively static over the past 25 years. However, as the Internet continued to expand and needed to accommodate new uses, numerous issues of the architecture came to light. Since a comprehensive treatment of the Internet's problems would require an entire book, in this section we only present an overview of the salient issues that demonstrate the need for a new architecture.

1.1.1 The Internet Protocol (IP)

IP is one of the fundamental protocols of the Internet, as it enables the forwarding of packets between end hosts. Its first major version (IPv4) was specified in 1981 and extended by IPv6 in 1998 [64] (as of April 2017, IPv6 is estimated to be used by around 15% of hosts [102]). IP routes packets between a source and a destination along a single path that is opaque from the end host's perspective. To forward packets, end hosts (as well as routers) do not need a complete path, but only a table to determine the next hop solely based on the destination address. Neither senders nor receivers can influence the path that their packets take. This approach is simple, but it also comes with many drawbacks:

- **Lack of separation between routing and forwarding:** IP packet forwarding depends on forwarding tables in routers, which change dynamically over time. Hence, a working path can suddenly change in direction or even break after an update to forwarding tables.
- **Lack of transparency and control:** Being able to select and verify the path that packets take is desirable in many situations. End hosts might want to avoid packets being routed through adversarial or untrusted networks, or they might want to choose the most suitable path with regard to a specific metric (e.g., latency or bandwidth). Unfortunately, IP does not offer such an option. Although loose and strict source routing exist, these extensions are not commonly supported in today's networks. It is also not possible to simultaneously use multiple distinct paths towards the same destination — even though multipath communication can offer numerous beneficial properties; as we will demonstrate throughout the book.
- **Stateful routers:** IP routers maintain forwarding tables to determine the next hop of a received packet. This basic requirement has undesirable

consequences. Performing a route table lookup for every packet is a time-consuming operation. Therefore, high-performance networking equipment typically relies on ternary content-addressable memory (TCAM) hardware, which is expensive and energy-intensive. Moreover, the constantly growing size of forwarding tables poses a problem for routers, as the storage capacity of TCAM hardware is limited. Routers that keep state for network information can also suffer from denial-of-service (DoS) attacks that rely on the exhaustion of the router's state [219].

1.1.2 The Border Gateway Protocol (BGP)

BGP is the routing protocol that provides connectivity between independently operated networks or **autonomous systems (ASes)**⋆ such as Internet service providers (ISPs).[1] Each AS advertises its reachability information as a list of **IP prefixes**⋆ through a BGP update message. Such BGP updates accumulate the sequence of ASes through which they have passed, and they contain a list of *attributes* characterizing the advertised routes. There are two main types of business relationships between ASes: a *customer-provider* relationship (one AS pays another to forward traffic), and a *peering* relationship (two ASes agree that directly connecting to each other without payment is mutually beneficial). BGP lets ISPs perform traffic engineering and select routes based on *policies* that reflect these business relationships through an intricate decision process that is used to select the best route to a destination [45]. Unfortunately, BGP comes with a number of shortcomings:

- **Outages:** Since the **control plane**⋆ and the **data plane**⋆ are not clearly separated in today's Internet, forwarding may suddenly stop during route changes. By attacking routing, an adversary can thus prevent forwarding from functioning correctly. Furthermore, when BGP update messages are sent, the network may require up to tens of minutes to converge to a stable state [145], which can lead to outages. Studies have shown that a sudden degradation in user-perceived quality of VoIP calls is highly correlated with BGP updates [144].
- **Lack of fault isolation:** BGP is a globally distributed protocol, running amongst all BGP speakers in the entire Internet. BGP update messages are thus disseminated globally. Due to the lack of any routing hierarchy or isolation between different areas, a single faulty BGP speaker can affect routing in the entire world, as occurred in the AS 7007 incident, which disrupted global connectivity due to a single faulty router [176].
- **Lack of scalability:** The amount of work required to be performed by BGP is proportional to the number of destinations. Moreover, path changes are disseminated profusely and sometimes throughout the entire Internet. This reduces scalability and prevents BGPsec (a proposal for a

[1] The definition of words marked with a star can be found in the glossary starting on Page 417.

secured version of BGP) from frequently disseminating freshly signed routing updates.

- **Single path:** At the end of the BGP decision process used to determine how to reach a given destination, a single path is selected. Although some multipath protocols allow simultaneous use of multiple network interfaces, BGP does not provide path control to end hosts and does not allow use of multiple AS-level paths.

1.1.3 Lack of Authentication and Trust

Authentication is another important feature that the original Internet protocols lacked. The necessity of authenticating digital data is becoming increasingly prevalent, as adversaries exploit the absence of authentication to inject malicious information to attack the network.

Infrastructures to provide authentication have been added in an ad hoc manner: RPKI provides the roots of trust for the authentication of BGPsec messages; TLS allows browsers to authenticate web servers; and DNSSEC provides authentication for DNS. Nevertheless, the current situation is still unsatisfactory in many regards. For example, all these protocols are sensitive to the compromise of a single entity. BGPsec and DNSSEC both rely on a single or very small number of roots of trust, while TLS is based on an oligopolistic trust model in which any one of hundreds of authorities can issue a certificate for any domain name. The Internet Control Message Protocol (ICMP) does not even have an authenticated counterpart, thus allowing the injection of fake ICMP packets. The Internet also lacks a general infrastructure to enable two end hosts to establish a shared secret key for end-to-end encrypted and authenticated communication; the simplest mechanism today is to rely on trust-on-first-use (TOFU) approaches [250], which opportunistically send the public key unprotected to the other communicating party.

1.1.4 Attacks

In this section, we present a series of attacks against which the current Internet architecture offers little to no protection.

Prefix Hijacking

Due to a lack of fault isolation in BGP, numerous Internet outages are caused by a malicious or erroneous announcement of IP address space, a problem called prefix hijacking. Perhaps the most famous case of prefix hijacking happened in February 2008 when Pakistan's internal censorship attempt resulted in a global outage of YouTube that took close to two hours to resolve [211]. This was not the first nor the last such event.

A related attack is prefix redirection, where an adversary wants to eavesdrop on traffic towards a destination and hijacks its prefix to receive its packets, but also engineers BGP updates such that the packets finally do reach the intended destination. Renesys (now Dyn) documented such cases of prefix redirection, where the adversary managed to re-direct traffic to take a detour across another continent [63].

This problem is exacerbated by the fact that defining BGP routing policies is often a complicated, manual, and thus error-prone process. It can occur that a backup path is rejected by a routing policy, hence limiting possible recovery paths.

Spoofing and DDoS Attacks

ICMP can be employed to send error or diagnostic messages (used by tools such as *ping* or *traceroute*). Because ICMP packets are not authenticated, the source address can easily be spoofed, which can lead to distributed denial-of-service (DDoS) attacks [142], or be used to disconnect two BGP routers from each other [99]. Since regular IP packets are not authenticated either, they suffer from the same problem, i.e., the source IP address can be spoofed.

Distributed denial-of-service (DDoS) attacks have been widely used to prevent access to servers or network resources. For example, a large-scale attack against Estonia made much of the country's critical infrastructure inaccessible during one week in April 2007 [109], and recently a very large attack with an unprecedented amount of attack traffic — exceeding 1 Tbps — on Dyn's DNS infrastructure rendered numerous web sites unavailable [137, 198].

Forged TLS Certificates

Compromised trust roots have been used to create rogue TLS certificates [166, 167]. In a famous case, the government of Iran used forged certificates for Google and Yahoo services to perform man-in-the-middle attacks on its citizens; Iran is suspected to have mounted the attack on the DigiNotar **certification authority (CA)***, which signed these certificates [90,228]. CAs hold significant power in the TLS public-key infrastructure, as any trusted CA can produce a valid certificate for any domain name. However, browser and OS vendors hold even more power, as they control which CAs are trusted by default.

1.1.5 Transition to a New Architecture

Changing network protocols as fundamental as IP and BGP is not an easy task. But in the long run, as for any technology, evolution is inevitable. It is clear, however, that the current architecture cannot be replaced overnight. Consequently, we need to propose a set of models and tools to achieve a progressive transition towards the desired properties. By redesigning the entire architecture from a clean-slate perspective, we follow a holistic approach and aim at fixing a broad range of problems, exploiting the benefit of hindsight and leveraging the inventions made over the past decades.

1.2 Goals of a Secure Internet Architecture

In this section, we present high-level goals that an inter-domain point-to-point communication architecture should accomplish; we illustrate why these goals are important and how they can be achieved. We also briefly discuss non-goals, i.e., specific properties that we intentionally exclude from the design of our secure Internet architecture.

1.2.1 Availability in the Presence of Adversaries

Our overarching goal is the design of a point-to-point communication infrastructure that remains *highly available* even in the presence of distributed adversaries: as long as an attacker-free path between endpoints exists, that path should be discovered and used with guaranteed bandwidth between these endpoints.

Availability in the presence of adversaries is an exceedingly challenging property to achieve. An *on-path adversary* may drop, delay, or alter packets that it should forward, or inject additional packets into the network. The architecture hence needs to provide mechanisms to circumvent such malicious elements. An *off-path adversary* could launch hijack attacks to attract traffic to flow through network elements under its control, and then perform on-path attacks. Such traffic attraction can take various forms; for instance, an adversary could announce a desirable path to a destination by using forged paths or attractive network metrics. Conversely, an adversary could render paths not traversing its network less desirable (e.g., by inducing congestion). An adversary controlling a large botnet could also perform distributed denial-of-service (DDoS) attacks, congesting selected network links. Finally, an adversary could interfere with the discovery of legitimate paths (e.g., by flooding the control plane with bogus paths).

1.2.2 Transparency and Control

We aim to provide greater transparency and control for the forwarding paths of network packets, and the trust roots used for authentication.

Transparency and Control over Forwarding Paths

When the network offers path transparency, endpoints know (and can verify) the forwarding path taken by network packets. Applications that transmit sensitive data can benefit from this property, as it can be ensured that packets traverse certain Internet service providers (ISPs) and avoid others.

Taking transparency of network paths as a first property, we aim to additionally achieve path control, a stronger property that enables ASes to control the incoming **path segments*** through which they are reachable. Given path segments, senders can then create end-to-end paths. This seemingly benign requirement has several repercussions — beneficial but also fragile if implemented incorrectly. The beneficial aspects of path control for senders and receivers include the following:

- **Separation of control plane and data plane:** To enable path control, the control plane (which determines networking paths) needs to be separated from the data plane (which forwards packets according to the determined paths). The separation ensures that forwarding cannot retroactively be influenced by control-plane operations, e.g., routing changes. The separation contributes to enhanced availability.
- **Enabling of multipath communication:** Path control lets any sender select multiple paths to carry packets towards the destination. Multipath communication is a powerful mechanism to enhance availability [8].
- **Defending against network attacks:** If the packet's path is carried in its header (which is one way to achieve path control), then the destination can reverse the path to return its response to the sender, mitigating reflection attacks. Path control also enables circumvention of malicious network entities or congested network areas, providing a powerful mechanism against DoS and DDoS attacks.

The fragile aspects that need to be handled with care are the following:

- **Respecting ISPs' forwarding policies:** If senders have complete path control, they may violate ISPs' forwarding policies. We thus need to ensure that ISPs offer a set of policy-compliant paths which senders can choose from.
- **Preventing malicious path creation:** A malicious sender could exploit path control for attacks, for example by forming malicious forwarding paths such as loops that consume increased network resources.
- **Scalability of path control:** Source routing does not scale to interdomain networks, as a source would need to know the network topology to determine paths. To make path control scale, we ensure that sources select amongst a relatively small set of paths. We thus rely on source-selected paths and packet-carried forwarding state instead of full-fledged source routing.

- **Permitting traffic engineering:** Fine-grained path control would inhibit ISPs from operating and performing traffic engineering. We thus seek to provide end-host path control at the granularity of autonomous systems on the level of ingress/egress interfaces, allowing ISPs to fully control internal paths. ISPs can further perform traffic engineering based on per-path bandwidth allocations, which can be encoded in the forwarding information.

Transparency and Control over Trust Roots

Roots of trust are used for the verification of entities in today's Internet; for example, verification of a web server's public key in a TLS certificate, or verification of a Domain Name System (DNS) response in DNSSEC [13]. Transparency of trust roots provides the property that an end host or user can know the complete set of trust roots that it needs to rely upon for the validation of a certificate. Such enumeration of trust roots is complicated today because of intermediate certification authority (CA) certificates that are not explicitly listed but implicitly trusted, e.g., in the TLS public-key infrastructure (PKI). In fact, independent studies have counted over 300 roots of trust in the TLS PKI [1, 78], but because of the lack of transparency there may be additional ones these studies have missed. Providing control over trust roots enables *trust agility* [165], allowing *users* to select or exclude the roots of trust they want to rely upon.

1.2.3 Efficiency, Scalability, and Extensibility

Aside from the lack of availability and transparency, today's Internet also suffers from a number of stability deficiencies. For instance, the Border Gateway Protocol (BGP) encounters stability issues in cases of network fluctuations, where routing protocol convergence can require minutes [216]. A 2006 earthquake in Taiwan that severed several undersea communication cables caused Internet outages throughout Asia for several days [25]. Moreover, forwarding tables have reached the limits of their scalability due to IP prefix de-aggregation (i.e., announcement of more specific prefixes) and **multihoming**⋆ [117]. Unfortunately, extending the memory size of routing tables is challenging as the underlying **ternary content-addressable memory (TCAM)**⋆ hardware is expensive and power-hungry, consuming on the order of a third of the total power consumption of a router. Extending the routing-table memory would thus drastically increase the cost and power consumption of routers.

Security and high availability come at a cost, usually resulting in lower efficiency and potentially diminished scalability. High performance and scalability, however, are required for viability in the current economic environment. We therefore explicitly seek *high efficiency* as a goal, so that packet forwarding is at least as efficient (in terms of latency and throughput) as current IP forwarding, in the common cases. Moreover, we seek *improved scalability* compared to the current Internet, in particular with respect to BGP and the size of routing tables.

An approach to achieving efficiency and scalability is to avoid storing forwarding state on routers wherever possible. We thus aim to encode state into packet headers and to protect that state cryptographically, enabling simpler router architectures compared to today's IP routers. We observe that modern block ciphers such as AES can be computed faster than performing memory lookups. For example, on current PC platforms, computing AES requires on the order of 50 cycles while fetching a byte from main memory requires around 200 cycles [4]. Moreover, a modern block cipher can be implemented in hardware with a few tens of thousands of gates, which is sufficiently small to replicate it profusely, which in turn enables high parallelism — the high complexity of a high-speed memory system prevents such replication at the same scale. Besides higher efficiency, avoiding state on routers also prevents state-exhaustion attacks [219] and state inconsistencies across routers.

Our goal of efficiency and scalability is in line with the design rationale of end hosts assisting with network-layer functionality such as path selection. A selected path is communicated to the network by packet-carried forwarding information, which in turn removes the need for inter-domain routing tables at border routers. Consequently, end-host path selection results in a simpler forwarding plane and thus more efficient routers. Furthermore, end-host path selection is in line with the *end-to-end principle*, which states that a network functionality should be implemented by the entity that has the required information, and is thus in the best position to correctly implement the functionality [217]. Since the end host has the most information about its internal state, functions such as bit-error recovery, duplicate suppression, or delivery acknowledgments are most efficiently handled by the end host itself. Similarly, the end host has the knowledge of preferred or undesirable network paths and thus should be involved in path selection.

To future-proof SCION, we design the core architecture and codebase to be *extensible*, such that additional functionality can be easily built and deployed. SCION clients and routers should (without overhead or expensive protocol negotiations) discover the minimum common feature set supported by all intermediate nodes.

1.2.4 Support for Global but Heterogeneous Trust

Given the diverse nature of constituents in today's Internet with diverse legal jurisdictions and interests, an important challenge is how to scale the authentication of entities (e.g., autonomous systems for routing, name servers for DNS, or domains for TLS) to a global environment.

The trust roots of currently prevalent PKI models (monopoly and oligopoly) do not scale to a global environment because mutually distrusting entities cannot agree on a single trust root (monopoly model), and because the security of a plethora of trust roots is only as strong as its weakest link (oligopoly model).

We thus seek an architecture that supports *a global environment with heterogeneous trust.*

1.2.5 Deployability

Incentives for deployment are important to overcome the resistance to upgrading today's Internet. A multitude of features is necessary to offer the initial impulse: high availability even under control-plane and data-plane attacks (e.g., built-in DDoS defenses), path transparency and control, trust-root transparency and control, high efficiency, robustness to configuration errors, fast recovery from failures, high forwarding efficiency, multipath forwarding, and so on.

If early adopters cannot obtain sufficient benefits from migrating to a new network architecture, even initial deployment is unlikely to be successful. So ideally, already the first deploying ISP should gain a competitive advantage through the ability to sell a service that is desirable even for the initial customers.

Migration to the new architecture should require minimal added complexity to the existing infrastructure. Deployment should be possible by re-utilizing the internal infrastructure of an ISP, and only require installation or upgrade of a few border routers. Moreover, configuration of the new architecture should be similar to that of the existing architecture, such as in the configuration of BGP policies, minimizing the amount of additional personnel training.

Economic and business incentives are also of critical importance. ISPs should be able to define new business models and sell new services. Users should derive a business advantage from the new architecture, for example by obtaining properties similar to a leased line at a smaller cost. Migration cost should be minimal, requiring only the deployment of low-cost routers. Finally, a new architecture should not disrupt current Internet business models, but maintain the current Internet topology and business relationships (e.g., support peering).

1.2.6 Non-goals

We deliberately exclude certain properties and goals that could be added as additional functionality later on. For example, we do not consider multicast or efficient content dissemination as part of the basic communication infrastructure, as we recognize the significant complexity these features would add. Also, these features can be effectively added through an overlay leveraging a next-generation Internet architecture's basic communication infrastructure [86].

We additionally consider several other problems to be out of scope for a network architecture. A major category of current security problems is software vulnerabilities. While software vulnerabilities of end hosts are clearly out of scope, software vulnerabilities of network components can affect network operation. It is thus important to address these network vulnerabilities through a robust network architecture that can restrain malicious components. Malicious Internet content (e.g., spam or phishing emails, malicious web pages) is preferably addressed by a layer above the communication infrastructure. The architecture, however, should offer mechanisms that assist in defending against these threats.

1.3 Future Internet Architectures

Several efforts at redesigning the Internet have been made over the past two decades to satisfy the new requirements of emerging Internet-based applications. Such requirements include naming, routing, mobility, network efficiency, availability, manageability, and evolvability of the Internet. We discuss several projects in this space based on a loosely temporal order clustered by topics.

The idea of partitioning the network into smaller parts has previously been considered for making network routing more scalable, for instance in hierarchical routing [127, 134], the Landmark hierarchy [242], hierarchies of nodes in Nimrod [47,223], regions in NewArch [57], clusters of computers in FARA [56], isolated regions with independent routing protocols in HLP [232], realms and trust boundaries in the Postmodern Internet Architecture (POMO) [34, 46], and regions in NIRA [259].

The NewArch project [57] describes comprehensive requirements for a new Internet, such as separation of identity from location, late binding using association, identity authenticity, and evolvability. However, it mostly emphasizes a new direction for end-point entities while the packet delivery in the current IP network is left intact. NewArch uses the New Internet Routing Architecture (NIRA) [259] for inter-domain routing, which aims to introduce competition among ISPs in the core by providing route control to the end users, who can choose domain-level paths.

Information-centric networking (ICN) or content-centric networking (CCN) architectures optimize content access through in-network content caches. Since content access across a user population frequently exhibits strong temporal and spatial locality, in-network content caches can serve the same requests made by nearby users. For instance, the Named Data Networking (NDN) [123, 184] architecture decouples location from identity and uses identity for locating the corresponding content. NDN relies on in-network caching of data and is useful for accessing popular static content. The CCNx project proposes a related implementation of content-centric networking, developing detailed specifications and prototype systems [192]. The Publish-Subscribe Internet Routing Paradigm (PSIRP) supports information-centric networking based on a publish-subscribe pattern [237]. It proposes an elegant approach to reduce the state on routers by having packets carry Bloom filters to encode the next hops of a multicast packet [125]. These architectures, however, have a high overhead for point-to-point communication, for ephemeral content (e.g., voice or video calls), or for per-user customized content. Our energy analysis presented in Chapter 14 suggests that content-centric approaches have higher energy utilization than fetching content directly from the origin server, due to the increased power consumption of routers with this architecture.

MobilityFirst [208] is an architecture with the main goal of providing connectivity to billions of mobile devices. At its core is the Auspice system, which provides a highly efficient global name resolution service that can quickly map billions of identities to their locations [220]. NEBULA [9] addresses security problems in the current Internet. NEBULA takes a so-called default-off approach to reach a specific service, where a sender can send packets only if an approved path to a service is available. The network architecture helps the service to verify whether the packet followed the approved path (i.e., supporting path verification). However, NEBULA achieves this property at a high cost. All routers on the path need to perform computationally expensive path verification for each packet and need to keep per-flow state. Serval [186] provides a service abstraction layer for service-ID-based resolution in NEBULA. Serval introduces a service-access layer that enables late binding of a service to its location, which provides flexibility in migrating and distributing services.

XIA [106] proposes an evolvable network architecture that can easily adapt to the evolution of networks by supporting various principal types (where the principal includes but is not limited to service, content, host, domain, and path). Thanks to its flexibility, XIA can use SCION for secure and highly available data forwarding.

The Framework for Internet Innovation (FII) [135] also proposes a new architecture to enable evolution, diversity, and continuous innovation, such that the Internet can be composed of a heterogeneous conglomerate of architectures. The ChoiceNet [253] architecture proposes an “economy plane” to enable network providers to offer new network-based services to customers, providing a network environment for improving innovation and competition.

Several architecture proposals suggest the approach of better path control for senders and receivers, for example i3 [229], Platypus [204, 205], NIRA [259], SNAPP [195], Pathlets [98], and Segment Routing [88]. These proposals enable the source to embed a forwarding path into the packet header, a concept that we refer to as packet-carried forwarding state (PCFS). PCFS provides many beneficial properties, such as enabling multipath communication and protecting packets from unanticipated re-routing.

Forward [81] and SysSec [82] are proposing to build secure and trusted Information and Communication Technology (ICT) systems by engaging academia and industry. Forward is an initiative by the European Commission to promote the collaboration and partnership between industry and academia in their common goal of protecting ICT infrastructures. The Forward project categorizes security threats to various ICT systems including individual devices, social networks, critical infrastructures (such as smart electric grids), and the Internet infrastructure, and it aims at coordinating multiple research efforts to build secure and trusted ICT systems and infrastructures. SysSec aims to bring together the systems security research community in Europe, promoting cybersecurity education, engaging a think tank in discovering the threats and vulnerabilities of the current and future Internet, creating an active research road map in the area, and developing a joint working plan to conduct state-of-the-art collaborative research. Since Forward and SysSec currently focus on identifying and handling threats, we believe our proposed tasks to be a good addition to the projects in that they provide an architecture that would significantly reduce the attack surface. RINA [249] is a recursive inter-network architecture that provides unified APIs across all protocol layers. In RINA, all layers have the same functions with different scope and range, where a layer is a distributed application that performs and manages inter-process communication. We endeavored to design our prototype to fit into this paradigm so that our architecture can support seamless integration with other higher-layer security protocols/mechanisms.

Many researchers are currently studying software-defined networking (SDN), for example in the OpenFlow [171, 189] project. These efforts mainly consider intra-domain communication, which SCION can leverage to communicate within a domain.

Several future Internet efforts provide testbeds for running and testing a new architecture, such as GENI [28], FIWARE [80], and FIRE [79].

We have developed SCION with a focus on security and high availability for point-to-point communication, which is a unique perspective and can contribute to other future Internet efforts. For instance, content-centric networking also needs a routing mechanism to reach the data source. SCION can offer the routing protocol to support that functionality. Once a server is found in a service-based infrastructure or a nearby content cache is found in a content-centric architecture, point-to-point communication between the end host and the server will offer high communication efficiency, as pure forwarding is faster than server-based or content-based lookups. Similarly, SCION can provide the point-to-point communication fabric in a mobility-centric architecture. Consequently, SCION offers mechanisms that complement many previously proposed future Internet architectures.

2 The SCION Architecture

David Barrera, Laurent Chuat, Adrian Perrig,
Raphael M. Reischuk, Pawel Szalachowski

This chapter provides an overview of SCION. The goals to be met by a secure Internet architecture were described in the previous chapter, but to recapitulate briefly, our main aim is to design a network architecture that offers highly available and efficient point-to-point packet delivery, even if some of the network operators and devices are actively malicious. The following chapters describe the SCION architecture in increasing detail.

SCION introduces the concept of an **isolation domain (ISD)**⋆, which is a fundamental building block for achieving the properties of high availability, transparency, scalability, and support for heterogeneous trust. An ISD constitutes a logical grouping of *autonomous systems* (ASes), as depicted in Figure 2.1. An ISD is administered by multiple ASes, which form the **ISD core**⋆. We refer to these as **core ASes**⋆. An ISD usually also contains multiple regular ASes. The ISD is governed by a policy, called the **trust root configuration (TRC)**⋆, which is negotiated by the ISD core. The TRC defines the roots of trust that are used to validate bindings between names and public keys or addresses.

An AS joins an ISD by purchasing connectivity from another AS in the ISD. Joining an ISD indicates an acceptance of the ISD's TRC. Typically, 3–10 ISPs constitute an ISD core, and their associated customers participate in the ISD. We envision that ISDs will span areas with uniform legal environments that provide enforceable contracts. If two ISPs have a contract dispute they cannot resolve by themselves, such a legal environment can provide an external authority to resolve the dispute. All ASes within an ISD also agree on the TRC, i.e., the entities that operate the trust roots and set the ISD policies. One possible model is thus for ISDs to be formed along national boundaries or federations of nations, as entities within a legal jurisdiction can enforce contracts and agree on a TRC. ISDs can also overlap, so an AS may be part of several ISDs. Although an ISD ensures isolation from other networks, the central purpose of an ISD is to provide transparency and to support heterogeneous trust environments. While ISDs may seem to lead to "Balkanization" and prevent an open Internet, they counter-intuitively provide openness and transparency, as we hope to elucidate

A. Perrig et al., *SCION: A Secure Internet Architecture*, Information Security and Cryptography, https://doi.org/10.1007/978-3-319-67080-5_2

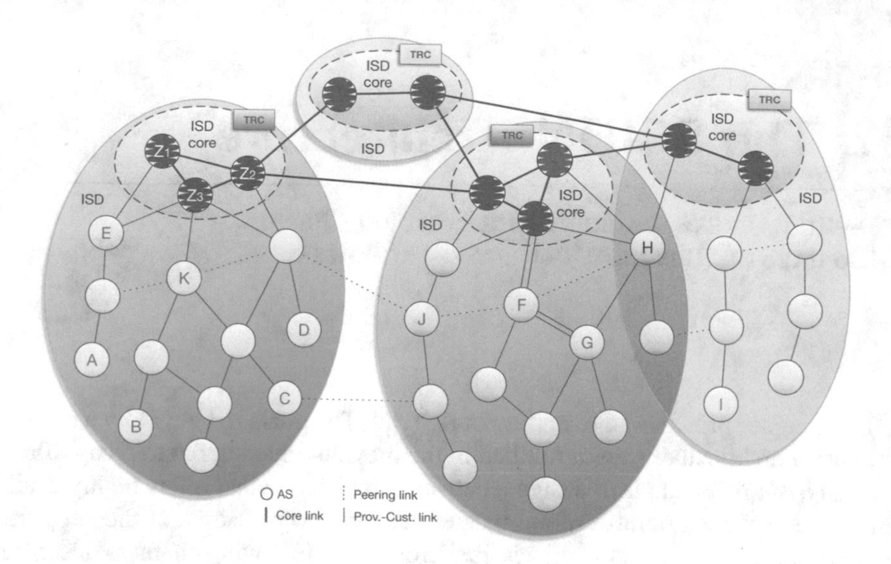

Figure 2.1: Autonomous systems (ASes) grouped into four ISDs. The core ASes are connected via core links. Non-core ASes are connected via customer-to-provider or peering links. AS *H* participates in two ISDs.

in this book (for more information on this point, please refer to the FAQ on Page 409).

SCION uses two levels of routing, intra-ISD and inter-ISD. Both levels utilize **path-segment construction beacons (PCBs)**⋆ to explore routing paths (see Figure 2.2a). An ISD core AS announces a PCB and disseminates it as a policy-constrained multipath flood either *within* an ISD (to explore intra-ISD paths) or *amongst* core ASes (to explore inter-ISD paths). We refer to this process as *beaconing*. PCBs accumulate cryptographically protected AS-level path information as they traverse the network. This information (which we call hop fields (HF)) within received PCBs is chained together by sources to create a data transmission path segment that traverses a sequence of ASes. Packets thus contain AS-level path information, which avoid the need for border routers to maintain inter-domain forwarding tables. We refer to this concept as **packet-carried forwarding state (PCFS)**⋆.

Figure 2.3 illustrates the chronological sequence of operations required to obtain a forwarding (i.e., end-to-end) path. During the *path exploration* or *beaconing* phase, ASes discover paths to core ASes. *Path registration* allows ASes to transform a few selected PCBs into path segments, and register these path segments with a path infrastructure (making them available for other ASes). The *name resolution* process translates a domain name into its associated **SCION address(es)**⋆. The *path resolution* process allows end hosts to create an end-to-end forwarding path to a destination; it consists of (a) *path lookup*,

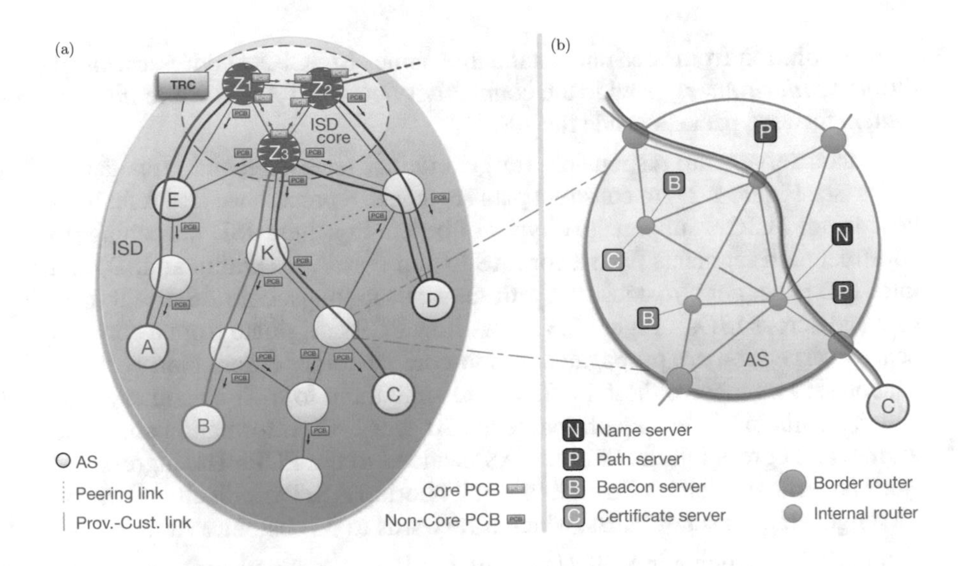

Figure 2.2: (a) ISD with path-segment construction beacons (PCBs) that are propagated from the ISD core to customer ASes, and path segments for ASes *A*, *B*, *C*, *D*, and *E* to the ISD core. (b) Magnified view of a SCION AS with its routers and servers. The path from AS *C* to the ISD core traverses two internal routers.

where the end host obtains path segments, and (b) *path combination*, where an actual forwarding path is created from the path segments. We discuss these phases in this chapter and describe them in more detail in the sections referred to in Figure 2.3.

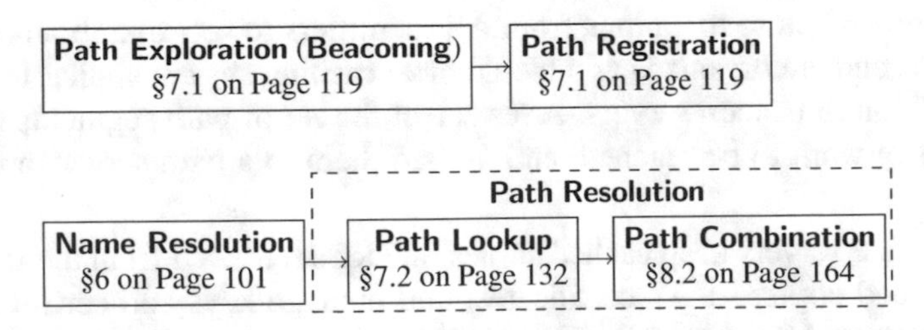

Figure 2.3: Process leading to the creation of a forwarding path.

Servers and Routers

Figure 2.2b shows the main AS components in SCION: **beacon servers**⋆ discover path information, **path servers**⋆ disseminate path information, **certificate servers**⋆ assist with validating path information, and **name servers**⋆ provide name resolution from user-understandable names to SCION addresses. In addition, *border routers* provide the connectivity between ASes, while *internal routers* forward packets inside the AS.

Beacon servers are responsible for generating, receiving, and propagating PCBs (see Figure 2.2a) to construct path segments, a process we also refer to as beaconing. SCION supports two types of beaconing: intra-ISD beaconing (to construct path segments from a core AS to non-core ASes within an ISD) and inter-ISD beaconing (to construct path segments amongst core ASes within an ISD and across ISDs). Figure 2.4 shows how PCBs originate from a core AS beacon server and are propagated to non-core customer ASes. Non-core AS beacon servers receive these PCBs and re-send them to their customer ASes, which results in AS-level path segments. At every AS, information about the ingress and egress interfaces of the AS is added to the PCB. The ingress and egress interfaces identify the link to a neighboring AS. Periodically, a beacon server generates a set of PCBs, which it forwards to its customer ASes.

Inter-ISD beaconing in SCION is similar to BGP's route-advertising process, although in SCION the process is periodic and PCBs are flooded over policy-compliant paths to discover multiple paths between any pair of core ASes. SCION's beacon servers can be configured to implement all BGP route selection policies, as well as additional properties (e.g., control of upstream ASes) that BGP cannot express (see Section 10.9).

Name servers in SCION perform a similar task to DNS servers in today's Internet: translate a human-understandable name into a SCION address. SCION proposes the RAINS system for this purpose Chapter 6. Based on the (ISD, AS) tuple, end-to-end paths can be looked up and constructed. The end-host address and end-to-end path are then placed in the SCION packet header to enable delivery to a given destination.

Path servers store mappings from AS identifiers to sets of announced path segments, and are organized as a hierarchical caching system similar to today's DNS. Through beacon servers, ASes select the set of path segments through which they want to be reached, and upload them to a path server in the ISD core.

Certificate servers keep cached copies of TRCs retrieved from the ISD core, keep cached copies of AS certificates, and manage keys and certificates for securing inter-AS communication. Certificate servers are queried by beacon servers when validating the authenticity of PCBs (i.e., when a beacon server lacks a certificate).

Border routers connect different ASes supporting SCION. The main task of border routers is to forward packets. In the case of a packet containing a service address, the border router forwards it to the appropriate server, and in the case of a data packet the border router forwards it either to a host inside the AS or towards the next border router. Since SCION can operate using any communication fabric inside an AS (e.g., OSPF, SDN, MPLS), the internal routers do not need to be changed.

2.1 Control Plane

We will now discuss the control plane components and mechanisms in more detail. The control plane is responsible for discovering paths and making those paths available to end hosts.

2.1.1 Path Exploration and Registration

Inter-domain beaconing enables core ASes to learn paths to other core ASes. Through intra-domain beaconing, non-core ASes learn path segments leading to core ASes, which enable an AS to communicate with the ISD core. Figure 2.2a shows path segments from ASes *A*, *B*, *C*, and *D* to the core. The beaconing process is asynchronous, i.e., the PCB generation is local, based on a per-AS timer, and PCBs are not propagated immediately upon arrival.

Paths are represented at AS-level granularity, which by itself is insufficient for diversity; ASes often have several connection points, and thus a disjoint path is possible despite the AS sequence being identical. For this reason, SCION encodes AS ingress and egress interfaces as part of the path, exposing a finer level of path diversity. Figure 2.4 demonstrates this feature: AS *F* receives two different PCBs via two different links from a core AS. Moreover, AS *F* uses two different links to send two different PCBs to AS *G*, each containing the respective egress interfaces. AS *G* extends the two PCBs and forwards both of them over a single link to its customer.

An AS typically receives several PCBs representing several path segments to various core ASes. Figure 2.2a shows two path segments for AS *D*, for example. There are three types of path segments:

- A path segment from a non-core AS to the core is an *up-segment*.
- A path segment from the core to a non-core AS is a *down-segment*.
- A path segment between core ASes is a *core-segment*.

However, path segments are typically bidirectional and thus support packet forwarding in both directions. In other words, up-segments and down-segments are invertible: by flipping the order, an up-segment is converted to a down-segment and vice versa. Path servers learn up-segments by extracting them from PCBs they obtain from the local beacon servers. Path servers in core ASes also store core-segments to reach other core ASes.

The beacon servers in an AS select the down-segments through which the AS desires to be reached, and register these path segments at the core path

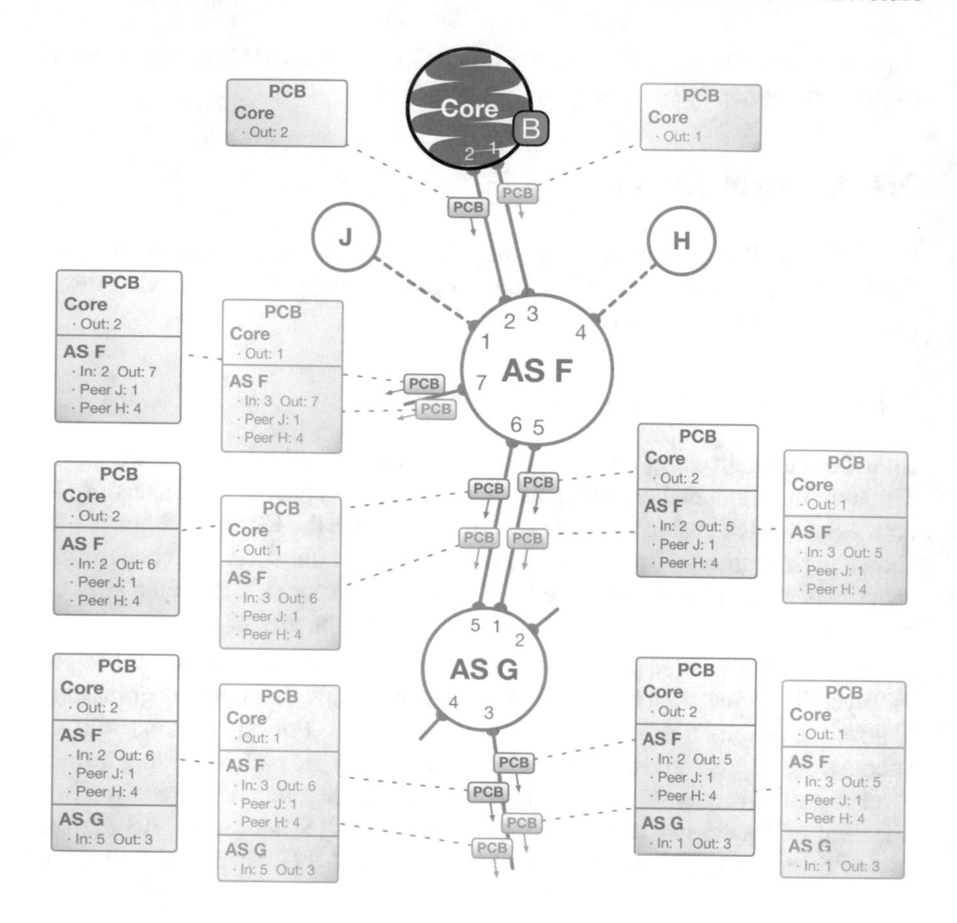

Figure 2.4: Intra-ISD PCB propagation from the ISD core down to customer ASes. For the sake of illustration, the interfaces of each AS are numbered with consecutive integer values. In practice, each AS can choose any encoding for its interfaces. In fact, only the AS itself needs to understand its encoding.

servers. When links fail, segments expire, or better segments become available, the beacon servers keep updating the down-segments registered for their AS.

An important requirement is that SCION also supports peering links between ASes. Consistent with AS policies in the current Internet, PCBs typically do not traverse peering links. However, peering links are announced along with a regular path in a PCB. Figure 2.4 shows how AS F includes its two peering links in the PCB. If the same peering link is announced in the path segments by both ASes adjacent to the peering link, then the peering link can be used to shortcut the end-to-end path (i.e., without going through the core). SCION also supports peering links that cross ISD boundaries, which highlights the importance of SCION's path transparency property; a source knows the exact set of ASes and ISDs traversed during the delivery of a packet.

2.1.2 Path Lookup

To reach its ISD core, a host performs a path lookup at its local path server, fetching up-segments. To reach a remote destination, a host first queries a name server to obtain the ISD-AS-address triplet of the destination. The host then queries its path server for the down-segment of the destination AS. If the local path server has no cached entry for the down-segment, it will query the destination AS's core path server.

Example. Consider a source host in ISD 1 sending a path lookup request to its local path server, which forwards the request to a core path server. If the requested path's destination AS is within ISD 1, the core path server responds by immediately sending up to k down-segments to the local path server. If the requested path's destination AS is in ISD 2, then the core path server first requests the corresponding down-segments from the core path server in destination ISD 2 before responding to the local path server. In both cases, the local path server returns up to k up- and down-segments to the requesting source (where k is a small integer set to 5 in the current implementation). If the up- and down-segments end at different core ASes, then core segments connecting the core ASes are returned as well.

2.1.3 PCB and Path-Segment Selection

Among the received PCBs, ASes must choose a set of PCBs to propagate further, and a set of path segments to register. These PCBs and path segments are selected based on a path quality metric with the goal of identifying consistent, diverse, efficient, and policy-compliant paths. *Consistency* refers to the requirement that there exists at least one property along which the path is uniform, such as an AS capability (e.g., anonymous forwarding) or link property (e.g., low latency). *Diversity* refers to the set of paths that are announced over time being as path-disjoint as possible to provide high-quality multipath options. *Efficiency* refers to the length, bandwidth, latency, utilization, and availability of a path, where more efficient paths are naturally preferred. *Policy compliance* refers to the requirement that the path adheres to the AS's routing policy. Based on past PCBs that were sent, a beacon server scores the current set of candidate path segments and sends the k best segments as the next PCB. To provide some concreteness to this description, we currently use $k = 5$, and send PCBs every 5 seconds to each neighbor over each provider-to-customer link. SCION intra-ISD beaconing can scale to networks of arbitrary size, because each inter-AS link carries the same number of PCBs regardless of the number of PCBs received by the AS.

Inter-ISD beaconing operates similarly to intra-ISD beaconing, except that inter-ISD PCBs only traverse core ASes. The same path selection metrics apply, where an AS attempts to forward the set of most desirable paths to its neighbors. A difference, however, is that an AS forwards k PCBs *per* source AS, with $k = 3$. The periodicity is also reduced; we forward PCBs once a minute or upon path changes. Similarly to BGP, this process is inherently not scalable (as the overhead grows linearly with the number of core ASes); however, as the number of ISDs and the corresponding number of core ASes is small, this approach is viable.

2.1.4 Link Failures

Unlike in the current Internet, link failures are not automatically resolved by the network, but require active handling by end hosts. Since SCION forwarding paths are static, they break when one of the links fails. Link failures are handled by a three-pronged approach that typically masks link failures without any outage to the application and rapidly re-establishes fresh working paths:

- Beaconing occurs every few seconds, constantly establishing new working paths.
- The SCION control message protocol (SCMP) (SCION-equivalent of ICMP) is used for path-segment revocation. As described in detail in Section 7.3, failed links result in rapid erasure of affected path segments from path servers.
- SCION end hosts use multipath communication by default, thus masking link failures to an application with another working path. As multipath communication can increase availability (even in environments with very limited path choice [8]), SCION beacon servers actively attempt to create disjoint paths, SCION path servers make an effort to select and announce disjoint paths, and end hosts compose path segments to achieve maximum resilience to path failure. Consequently, we expect that most link failures in SCION will be unnoticed by the application, unlike the frequent (although mostly brief) outages in the current Internet [131, 144].

2.1.5 Intra-AS Communication

Communication within an AS is handled by existing intra-domain communication technologies and protocols such as IP with Software-Defined Networking (SDN), or Multi-Protocol Label Switching (MPLS). Figure 2.2b on Page 19 shows one possible intra-domain path through the magnified AS.

2.2 Data Plane

While the control plane is responsible for providing end-to-end paths, the data plane ensures packet forwarding using the provided paths. A SCION packet minimally contains a path; source and destination addresses are optional in case the packet's context is unambiguous without addresses. Consequently, SCION border routers forward packets to the next AS based on the AS-level path in the packet header (which is augmented with ingress and egress interface identifiers for each AS), without inspecting the destination address and also without consulting an inter-domain routing table. Only the border router at the destination AS needs to inspect the destination address or packet purpose to forward it to the appropriate local host.

An interesting aspect of this forwarding is enabled by the split of locator (the path towards the destination AS) and identifier (the destination address) [83]: the identifier can have any format that the destination AS can interpret, since only the destination needs to consider that local identifier. In other words, an AS can select an arbitrary addressing format for its hosts, e.g., a 4-byte IPv4, 6-byte medium access control, 16-byte IPv6, or 20-byte accountable IP (AIP [7]) address. A nice consequence is that an IPv4 host can directly communicate with an IPv6 host over SCION.

In the next two sections, we describe how an end host combines path segments into an end-to-end forwarding path, and how border routers forward packets efficiently.

2.2.1 Path Combination

After name resolution and path lookup, the end host obtains path segments that need to be combined into an end-to-end path. A valid SCION **forwarding path*** can be created by combining up to three path segments, in the following ways (all combinations are illustrated with sample paths depicted in Figure 2.5):

- **Immediate combination of path segments** (e.g., $B \rightarrow D$): the last AS on the up-segment (core AS Z_3) is also the first AS on the down-segment. In this case, the simple combination of an up-segment and a down-segment creates a valid forwarding path.
- **AS shortcut** (e.g., $B \rightarrow C$): the up-segment and down-segment intersect at a non-core AS (e.g., K). In this case, a shorter forwarding path can be created by removing the extraneous part of the path.
- **Peering shortcut** (e.g., $A \rightarrow B$): a peering link (e.g., $L \rightarrow K$) exists between the two segments, so a shortcut via the peering link is possible. As in the *AS shortcut* case, the extraneous path segment is cut off. The peering link could be traversing to a different ISD.

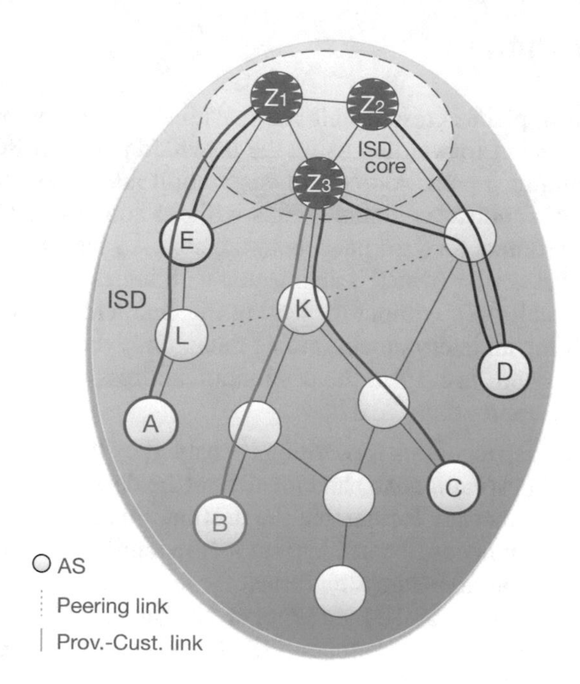

Figure 2.5: ISD with path segments for ASes A, B, C, D, and E.

- **Combination with a core-segment** (e.g., $A \rightarrow D$): the last AS on the up-segment is different from the first AS on the down-segment. This case requires an additional core-segment (e.g., $Z_1 \rightarrow Z_2$) to connect the up- and down-segment. If the communication remains within the same ISD ($A \rightarrow D$), a local ISD core-segment is needed; otherwise (e.g., $A \rightarrow I$ in Figure 2.1), an inter-ISD core-segment is required.
- **On-path** (e.g., $A \rightarrow E$): the destination AS is directly on the path to the ISD core, so a single up-segment is sufficient to create a forwarding path.

Once a forwarding path is chosen, it is encoded in the SCION packet header, which makes inter-domain routing tables unnecessary for border routers: both the egress and the ingress interface of each AS on the path are encoded as packet-carried forwarding state (PCFS) in the packet header. The destination can respond to the source by inverting the end-to-end path from the packet header, or it can perform its own path lookup and combination.

2.2.2 Forwarding

Routers can efficiently forward packets in the SCION architecture. In particular, the absence of inter-domain routing tables and the absence of complex longest IP prefix matching performed by current routers enables construction of faster and more energy-efficient routers, which we discuss in more detail in Chapter 14.

The SCION packet header contains a sequence of hop fields (HF), one for each AS that is traversed on the end-to-end path. During forwarding, each AS inspects its respective HF in the packet header. The HF contains interface numbers of the ingress and egress links, which are essentially descriptors of the links across which the packet is entering and exiting the AS. Figure 2.4 on Page 22 depicts how the HF information is assembled in the PCB as part of the beaconing process.

During packet forwarding, a SCION border router at the ingress point of the AS first verifies that the packet entered through the correct ingress interface corresponding to the information in the HF. If the packet has not yet reached the destination AS, the egress interface defines the egress SCION border router — in which case native intra-domain routing (e.g., OSPF, MPLS) is used to send the packet from the ingress SCION border router to the egress SCION border router.

2.3 Security Aspects

For protection against malicious entities and to provide secure control and data planes, SCION is equipped with an arsenal of security mechanisms.

Similarly to BGPsec [158], each AS signs the PCBs it forwards. This signature enables PCB validation by all entities. To ensure path correctness, the forwarding information within each packet-carried forwarding state (PCFS) also needs to be cryptographically protected, but signature verification would hamper efficient forwarding. Thus, each AS uses a secret symmetric key that is shared among beacon servers and border routers and is used to efficiently compute a message authentication code (MAC) over the forwarding information. The per-AS information includes the ingress and egress interfaces, an expiration time, and the MAC computed over these fields, which is (by default) all encoded within an 8-byte field that we refer to as the hop field (HF). Excluding a few flag bits, the structure of the HF is at the discretion of each AS and requires no coordination with any other AS — as long as the AS itself can extract how to forward the packet on to the next AS.

The specified ingress and egress interfaces uniquely identify the links to the previous and following ASes. If a router is connected via the same outgoing interface to three different neighboring ASes, three different egress interface identifiers would be assigned. The HF's expiration time can be set on the granularity of seconds or hours, depending on the path type.

2.3.1 Algorithm Agility

In terms of cryptographic mechanisms, SCION provides algorithm agility, so that cryptographic methods can be easily updated and exchanged. The

MAC validation of hop fields is per-AS, so an AS can independently (without interaction with any other entity) update its keys or cryptographic mechanisms. We support multiple signatures by an AS, thus, an AS can readily deploy a new signature algorithm and start adding those signatures as well. A component of the path-segment and PCB selection metric will favor creating paths where each AS on the path supports the new algorithm.

2.3.2 Authentication

Authentication in SCION is based on certificates, which bind identifiers to public keys and carry digital signatures that are verified by roots of trust, i.e., public keys that are axiomatically trusted.[1] One challenge is how to achieve trust agility to enable flexible selection of trust roots, resilience to private key compromise, and efficient key revocation.

SCION allows each ISD to define its own set of trust roots, along with the policy governing their use. Such scoping of trust roots within an ISD greatly improves security, as compromise of a private key associated with a trust root cannot be used to forge a certificate outside the ISD. An ISD's trust roots and policy are encoded in the trust root configuration (TRC). The TRC has a version number, a list of public keys that serve as roots of trust for various purposes, and policies governing how many signatures are required for performing different types of actions. The TRC serves as a way to bootstrap all authentications.

We now briefly discuss two properties offered by the TRC. *Trust agility* enables the selection of trust roots used to initiate the validation of certificates. A user can thus select an ISD that she believes maintains a non-compromised set of trust roots. A challenge with trust agility is to maintain global verifiability of all entities, regardless of the user's selection. SCION offers this property by requiring all ISDs with a link between them to sign each other's TRCs — thus, as long as a network path exists, a validation path exists along that network path. *Efficient revocation of trust roots* is the second important property. In today's Internet, trust roots are revoked manually, or through OS or browser updates, often requiring a week or longer until a large fraction of the Internet population has observed such revocations. There is also a long tail of devices and installations that apply revocations very late or never. In SCION, PCBs carry the version number of the current TRC, and the updated TRC is required to validate that PCB. An AS that realizes that it needs a newer TRC can contact the AS from whom it has received the PCB. Following the distribution of PCBs, an entire ISD updates the TRC within tens of seconds.

The authentication of control-plane messages has availability as the main requirement, since the control plane provides communication paths upon which

[1]Our reason for not using self-certifying identifiers [7, 180] for long-term identities is their inherent inability to be revoked and the complexities involved with key updates. For short-term identities, however, we do appreciate their features.

other mechanisms rely. Once end-to-end communication is established, additional entities can be contacted to achieve a more secure authentication of end entities (e.g., web servers). The Attack-Resilient PKI (ARPKI) [23] is a highly secure PKI system based on **log servers*** that keep a public log of all certificates to monitor CAs' operations. In turn, CAs and validators verify the content of log servers. By requiring multiple signatures on certificates, and by adding signatures on all operations, we create a situation where multiple malicious trusted entities within the same ISD are needed to perform a man-in-the-middle attack on a single domain. To further increase security, we combine ARPKI with PoliCert, which enables domains to specify their detailed security policy [235]. By storing the domain policies in an ARPKI log, policy consistency and integrity are ensured. In concert, ARPKI and PoliCert achieve a high level of security, as all PKI attacks we have witnessed in the past decade would have been avoided in this framework.

The ISDs and the ARPKI system used in SCION address the problem of CA compromise, as a CA's authority is scoped to the ISDs in which the CA is active, and as multiple trusted entities need to be compromised to perform a successful man-in-the-middle attack. Moreover, the SCION trust roots update mechanism enables revocation within tens of seconds, enabling quick recovery from compromise.

More details on SCION's authentication infrastructure are provided in Chapter 4.

2.3.3 SCION Control Message Protocol (SCMP)

The control plane includes the SCION Control Message Protocol (SCMP), which is similar to the current Internet control message protocol (ICMP), but authenticated and adapted to SCION. One challenge in the design of SCMP was how to enable efficient authentication of SCMP messages, as the naive approach of adding a digital signature to SCMP messages could create a processing bottleneck at routers when many SCMP messages would be created in response to a link failure. We thus make use of an efficient symmetric-key derivation mechanism called *Dynamically Re-creatable Key* (DRKey, see Section 12.5). In DRKey, each AS uses a local secret key known to its SCION border routers to derive on the fly a per-AS secret key using an efficient pseudorandom function (PRF). Hardware implementations of modern block ciphers enable faster computation than a memory lookup from DRAM, and therefore such dynamic key derivation can even result in a speedup over fetching the key from memory. For verification of SCMP messages, the destination AS can fetch the derived key through an additional request message from the originating AS, which is protected by a relatively slow asymmetric operation. However, local caching ensures that this key only needs to be fetched infrequently. As a consequence, SCION provides fully secured control messages with minimal overhead.

2.3.4 DDoS Defenses

SCION offers several complementary defenses against link-flooding DDoS attacks, which frequently disrupt daily-life communication (e.g., by exploiting vulnerabilities of IoT devices [138] and launching attacks against IT-security blogger Brian Krebs in September 2016 [140], or against the DNS infrastructure causing outages for Twitter, Spotify, and Reddit in October 2016 [137]).

The SCION architecture comes by default with three mechanisms that provide a strong defense against DDoS attacks:

- **Non-registered (or hidden) path segments:** An AS can prevent an adversary from sending traffic to it by not publicly announcing its down-segment on the path servers. A destination thus cannot be reached, unless it explicitly permits a sender to send traffic. This approach, referred to as off-by-default [19], is explained in more detail in Section 7.2.5.
- **Short-lived paths:** Each SCION path segment has an expiration time, which is set in a PCB to provide several hours of validity. A careful administrator of an AS can let a path segment age and only announce it briefly before the expiration time. For instance when a path segment p that expires within 5 minutes is publicly announced at a path server, then p can only be used to attack the destination AS for at most 5 minutes. The approach here is to publicly announce only short-lived path segments, and to provide longer-lived path segments only to trusted and verified senders.
- **Multipath communication:** Because SCION uses multipath communication by default, an adversary has to congest *all* paths instead of only the single path that is currently used. This approach will prevent attacks that are unable to congest all network paths simultaneously: for example consider a multi-homed domain with two providers, with a 1 Gbps link to each provider. In the current Internet, usually only one of the links will be the active link that carries all incoming and outgoing traffic. If the attacker has a capacity of 1.5 Gbps for example, it can congest that link. Once the victim attempts to change to the other link, the attack traffic will simply follow and congest the alternative link. With multipath communication, however, whichever link the adversary clogs, the other link will still be available and thus communication is always ensured. In summary, multipath communication forces the adversary to *simultaneously* clog *all* paths that are available to the victim, which requires a larger attack capacity and access to *all* paths.

Furthermore, SCION offers two extensions to improve availability and defend against DDoS attacks:

- The SIBRA extension (Chapter 11) enables fine-grained inter-domain bandwidth allocations to guarantee availability even during large-scale DDoS attacks. SIBRA enables fine-grained temporal access, in which so-called ephemeral paths expire within tens of seconds, putting a rapid stop to a misbehaving sender.

- The OPT extension (Chapter 12) provides source authentication to prevent attacks with spoofed source addresses. Spoofed victim source addresses are used in reflection-based amplification attacks to disguise the attacker's identity and to redirect the response traffic to the actual victim [214].

2.4 Use Cases

SCION improves many aspects of the current Internet. This section highlights some of the applications and use cases that demonstrate unique properties and benefits offered by the new architecture.

2.4.1 High-Availability Communication

Highly available communication is important in many contexts, in particular for critical infrastructures such as financial networks and industrial control systems used for power distribution. Internet outages have been known to wreak havoc on day-to-day operations, for example preventing ATM withdrawals or payment terminal operations [238]. SCION's control-plane isolation through ISDs, its stable data plane, and its multipath operation all contribute to higher availability.

Business continuity refers to the uninterrupted operation of an organization. Business continuity is currently highly dependent on communication. We can witness the increasing inter-connectedness required for business operations when network outages cause a disruption of a surprising number of operations. For instance, when Telecom Malaysia wrongly announced 179,000 IP prefixes to Level3, it caused global outages for 2 hours, even affecting ATM operations in Sweden [238].

Here are a few examples of sectors where availability is crucial:

- Financial services require highly available communication networks, for instance for the distribution of stock market data, real-time market trading, or transaction processing. While critical communication is often sent over leased lines, it is not economical to pervasively use leased lines between all communicating parties. In this setting, SCION can offer higher availability than the legacy Internet at a lower price than a leased line.

  High availability for communication is also important in blockchain applications such as bitcoin mining, where a disconnected mining pool does not learn of newly mined coins and is wasting processing on finding irrelevant coins. Similarly, a disconnected mining pool cannot post its found coins, which will likely be ignored once connectivity is re-established. Both of these cases occur with high probability if the mining pool's computation capacity is less than half the total mining capacity, which is the case for individual mining pools [12].

- Critical command-and-control infrastructures — such as air-traffic, power-grid, or power-plant control systems, or public safety emergency communication — require very high communication availability. Communication disruptions can lead to outages with significant cost for industry and danger for society.
- Governments require high communication availability especially during crisis situations. Examples of critical communication include cables to foreign embassies, law enforcement communication, or access to databases for verifying documents at a country's border.

With SCION's resilience against network-layer DDoS attacks, prevention of prefix hijacking, and data-plane isolation, communication over the regular SCION network can achieve a level of availability that approximates the availability of leased lines. In addition, the SIBRA extension, as described in Chapter 11, offers an extended level of availability through a concept we call DILL, which stands for dynamic inter-domain leased line. DILLs provide a lower bound on the guaranteed bandwidth at inter-domain scale, regardless of the bandwidth requirements of other ASes.

2.4.2 Path Transparency, Path Control, and Compliance with Traffic Flow Regulations

Packets do not always directly reach their destination via the shortest path. Instead, in current practice, many Internet paths take detours. While some extreme cases of detours are due to prefix hijacking [63, 160, 162], most detours are taken for economic reasons or are simply due to the preferred connectivity of ISPs. As a consequence, traffic that would be expected to stay within a geographic area is often routed through nearby countries. For example, paths connecting sources and destinations within Switzerland are sometimes routed through Frankfurt or London, or traffic that would be expected to stay within continental Europe is routed through London.

Path control and transparency are important properties when a sender wants to influence and learn about the ASes that sensitive data will traverse (for legal, secrecy, or safety reasons). For instance, banking or medical data, which is typically bound by strict data privacy regulations, can be constrained in SCION to traverse only selected authorized ASes: a source knows the AS-level path that a packet will follow based on the hop fields in the packet header. Such packet-carried forwarding state in the packet header provides not only transparency, but also path control by letting the source node select the paths amongst a set of paths provided by path servers. Path transparency and control enable an organization to achieve compliance with laws or regulations that require traffic to be constrained within a jurisdiction. These properties can be further strengthened by SCION's OPT extension (Chapter 12). In a nutshell, OPT provides the receiver with a cryptographically verifiable guarantee that a sequence of ASes were all traversed in the correct order.

2.4.3 Inter-domain Traffic Engineering

In the legacy Internet, only rudimentary forms of inter-domain path control and traffic engineering are possible. For outgoing traffic, one can at best control the next ISP, but only if an AS is multi-homed. A little more path control is available to direct incoming traffic, as an AS can decide to which upstream ISP to send a BGP update. However, to achieve high availability for outages, an IP prefix should be announced to each upstream ISP. AS path pre-pending is a technique that enables a very limited form of path control for incoming traffic; but this technique will not be available in a secure version of BGP, for instance in BGPsec [157, 158].

In intra-domain networks, software-defined networking (SDN) has revolutionized path control; for example, Google has achieved higher network utilization with their B4 system [124]. Analogous to B4's intra-domain path control, SCION makes inter-domain path control available through path registration. An AS can select the down-segments that are announced to the path servers. Hidden paths can be used, which are only communicated to senders who are selected to use them (as discussed in Section 2.3.4). Much path control is available to the sender, who can select which end-to-end path the packet will follow. We anticipate that this level of path control creates a strong reason for adopting SCION.

2.4.4 High-Speed Web Browsing

Current congestion control hinders high-speed communication because the sender and receiver require time to determine their sending rate and to continuously perform congestion control. Consequently, the sending rate is usually below the maximum possible rate. In SCION, through the SIBRA extension (Chapter 11), the sender performs a resource reservation with its initial packet, and the receiver will likely obtain a reservation with a high sending rate, which it can immediately start to use on the reverse path. With such a reservation, a given bandwidth is provided, so no congestion control is needed; consequently, the web server can immediately start sending data at a high rate to the browser.

2.4.5 Mobility Support

With the proliferation of mobile devices, supporting reliable communication can be challenging since these devices frequently connect to and disconnect from (sometimes several) networks. SCION supports high availability and mobility through multipath communication. Moreover, SCION provides a header extension to inform the other party of new down-segments, such that a mobile device that obtains a new address as it connects to a new network can inform the other party about its new down-segment. Failing paths are discarded and new paths are dynamically discovered transparently to users and applications. One challenge, however, is that both sender and receiver might simultaneously move to a new network, and all the previously established paths might fail at the same time. In this unlikely scenario, a name resolution server and a path server need to be contacted to fetch fresh down-segments for the other party [220].

2.5 Incentives for Stakeholders

While SCION offers a wide assortment of security, availability, and performance benefits over current-generation networks, its lack of direct compatibility with BGP may lead to adoption resistance. This resistance can stem from the notion that the cost of changing to the new architecture will be higher than the benefits obtained, or that it is risky to take on a new architecture that may not find widespread adoption. In this section, we discuss deployment incentives to dissipate such reservations.

2.5.1 End Users

End users in SCION benefit primarily from *higher throughput* afforded by the use of native multipath communication, and from *lower latency* due to path control and packet-carried forwarding state. SCION paths are selected based on performance metrics, which translate to better quality of service (such as audio, video, and file transfers) and generally shorter transfer delays. Although the increased size of SCION packets sacrifices goodput, we anticipate that the continuous path optimization of SCION's multipath system will compensate for the higher overhead.

End users also benefit from *higher availability* (i.e., fewer Internet outages) again due to the multipath communication that is used by default. Even if the user's local ISP does not employ SCION, it is possible to provide the benefits of multipath communication via access tunnels as described in Section 10.1.2. Moreover, the SCION-IP gateway (Section 10.3) provides an incremental deployment approach, which enables users to use SCION without requiring changes to software on their devices.

Path control gives users *higher assurance* when performing security-critical tasks such as online banking or shopping. Using SCION, users gain transparency over the communication path to the destination server, allowing them to include or exclude specific paths traversing ASes that are not trusted.

The SCION end-to-end public-key infrastructure offers strong assurance that a contacted web site is the correct entity — fending off man-in-the-middle attacks that could eavesdrop on or alter information sent on a TLS connection. As a consequence, users can perform secure transactions over the Internet with higher confidence.

Finally, SCION extensions (such as Hornet [49] and SIBRA) provide users with a range of additional benefits, such as *high-speed anonymous communication* and *guaranteed bandwidth*.

2.5.2 ISPs

ISPs can create new revenue streams by offering services based on SCION. ISPs that enable SCION can create services for customers who demand higher availability than BGP can provide, but who cannot afford dedicated leased lines. In addition to lower operating cost, SCION gives early adopters increased resilience to network attacks, higher availability, and better path control. ISPs may even offer SCION services to customers of other ISPs through access tunnels. SIBRA, for example, enables inter-domain traffic guarantees, which ISPs cannot offer today unless they operate a global network.

Since SCION PCB propagation policies are more expressive than is possible in BGP, ISPs benefit from finer control of traffic traversing their domain (see Section 10.9), which can help with traffic engineering.

SCION's path transparency properties can provide evidence to regulators and customers that ISPs are not violating network neutrality [194].

Finally, SCION's ISDs and secure operation help to minimize the impact of an ISP's configuration errors, which can simplify ISPs' operations.

2.5.3 Businesses

Businesses or corporations using SCION benefit from path management for incoming and outgoing traffic, path transparency and control, attack resilience, and highly available communication. One particular advantage is that through SCION, a business can ensure that traffic does not leave an ISD. This is important for complying with data privacy laws, which vary from country to country. For example, a recent European Union (EU) ruling declared that companies with an EU presence had to comply with EU data privacy laws, and could no longer make use of "safe harbor" when storing data on servers in approved countries [62]. It is unclear whether forwarding and caching data also falls under this ruling, but SCION allows businesses to specify their traffic policies.

While control over outgoing traffic has so far proven to be an attractive incentive for businesses, control over inbound traffic should also provide an attractive feature. Corporations offering network services to a restricted set of clients (e.g., banks) may want to allow incoming traffic only from those authorized clients or through authorized ISPs. SCION paths are flexible enough to allow this by distributing certain paths to specific authorized entities, rather than announcing them globally.

2.5.4 Governments

Governments using SCION can benefit from the same advantages as businesses, but additionally benefit from avoiding the use of a global trust root. As shown in Section 13.8, a global trust root provides a kill switch that can cause entire networks to be taken offline, which could be particularly damaging in the case of government networks. Like businesses, governments will also value the path control facility that will ensure their traffic traverses ISPs they trust.

The open-source nature of the SCION codebase allows governments to build their own hardware to reduce their reliance on untrusted foreign manufacturers. The codebase can also be inspected and maintained by trusted developers.

2.6 Deployment

Deployability plays a key role in the success of any network architecture. To this end, we have designed SCION to be deployable (by both ISPs and end users) without requiring substantial changes to the existing infrastructure.

2.6.1 Incremental Deployment

As a minimum, an ISP needs to deploy only a single border router capable of encapsulating and decapsulating SCION traffic as it leaves, enters, or traverses its network. SCION ASes must also deploy certificate, beacon, name, and path servers. These servers can run on commodity hardware and can optionally be replicated for increased availability. The current version of the SCION codebase uses IP for internal AS communication, which allows the use of existing intra-domain networking infrastructure and configuration.

We envision that ISDs will grow organically within an area with homogeneous trust. Tier-1 ISPs within those ISDs would become core ASes. SCION facilitates the evolution of ISD and AS structure through efficient updates to the TRC.

Deployment of SCION to end-user sites (e.g., homes or businesses) is designed to require little effort as well, initially needing no changes to hosts or internal network communication equipment. For initial deployment, we achieve customer-friendly conditions through a gateway device that can be installed in a network to enable both SCION and standard Internet communication. The SCION-IP gateway replaces a home access router and transparently enables any type of communication (legacy IPv4/IPv6 or SCION), as described in Section 10.3.

2.6.2 Deployment Caveats and SCION Disadvantages

The deployment and structure of ISDs is hard to predict, as is which ASes within an ISD will or should become core ASes. We envision that among a group of ASes that deploy a top-level ISD, the AS or ASes that can form peering agreements with core ASes in other ISDs should become core ASes in their own ISD. However, SCION itself does not require or impose strict rules regarding the allocation of ISDs; ISDs can overlap, which means an AS can belong to several ISDs. Sub-ISDs are possible as well, offering the flexibility to start an ISD without needing to peer with core ASes of other ISDs and enabling finer-grained control over routing isolation and authentication. In this context, the important properties SCION offers are path control and transparency: as long as communicating hosts can select and inspect the paths of their packets, the question of ISD partitioning is of secondary nature.

A challenge that could arise is that each AS will attempt to be its own ISD or will want to be part of the ISD core. While too many top-level ISDs will pose a problem for SCION scalability, we observe that economically sound decisions will lead to larger ISDs due to economies of scale — because the startup costs of a core AS are higher than those of a non-core AS, the operation of a large ISD will amortize the cost over more non-core ASes. Moreover, ASes preferentially associate with larger ISDs, which can offer better connectivity to other ISDs as well as to other ASes within the ISD. On the other hand, ISD growth is limited to the extent that entities can agree on the ISD's TRC (i.e., roots of trust). Finally, ASes desiring to be part of the ISD core are assessed in the same way in which current ASes assess peering: an AS is permitted into the core if the current core ASes deem it to be large enough to fulfill core AS duties (which include, for example, participating in beacon and path server replication).

SCION ASes need to manage cryptographic keys, which requires additional effort to securely administer. As a security architecture, every AS has to have a public-private key pair, and obtain a signature on the public key. Although managing cryptographic keys can be a challenge for some ASes, it is a necessity for any secure network architecture. In our development, we are building systems to simplify the overhead of managing cryptographic keys, for instance through our CASTLE system [169], which offers a local low-rate CA environment built from off-the-shelf components. To further mitigate the risks associated with the management of cryptographic keys, SCION reduces the effect of key loss and compromise by offering approaches for resilience and quick recovery.

As expected in architectures with PCFS, packet headers are necessarily larger. Larger headers place a limit on goodput, since payload space is traded for header space. The current SCION codebase implements the HF as an 8-byte field. Since every AS on an end-to-end path has to be represented through a corresponding HF, the overhead increases linearly with the number of ASes on the path. However, given that the average AS path in today's Internet is four hops long (and decreasing) [66, 141], the overhead introduced by SCION should not exceed around 50 bytes per packet on average. The performance penalty of transmitting more packets appears reasonable since per-packet forwarding performance can be faster than for forwarding-table-based architectures. While the default header size has not turned out to be a performance disadvantage in our testing environment, many of the proposed SCION extensions further increase the header size.

Due to path dissemination and registration dynamics, SCION beacon and path servers can incur a high overhead under specific circumstances. For example, if a given link's state were to fluctuate frequently between available and unavailable (due to error, hardware fault, or an adversary), the beacon server would need to constantly update the set of paths that include that link, and serve new paths excluding that link. We expect that this case will be rare, but also easily detectable. Additionally, higher quality (uptime, availability) links will have a higher probability of selection, minimizing the impact of rapid path fluctuations.

We believe that the basic building blocks of SCION are relatively straightforward to understand and provide many beneficial properties for applications. However, as more extensions and alternative PKIs are added to the architecture, the operational complexity of the architecture increases correspondingly. We believe that this additional complexity is worth the security, efficiency, and availability guarantees provided by the extensions. It is ultimately up to the networking and research community to decide which of these extensions will be deemed worthwhile for pervasive deployment.

2.6.3 SCION Network Deployment

We have deployed a global SCION network, which we are actively using to vet SCION's functionality and security. The current network has about 50 border routers and servers deployed in ASes around the world, with new nodes joining the network on a weekly basis. The deployment status as of December 2016 is described in more detail in Section 10.1.4. Details and requirements for sponsoring a SCION node can be found on our website. The SCION testbed, enabling any researcher to use the SCION network, is described in more detail in Section 10.7.

2.7 Extensions

SCION's extensible architecture enables new systems that can take advantage of the novel properties and mechanisms provided. As compared to the current Internet, most of the benefits can be afforded through the use of PCFS, path transparency, and control. We briefly describe three systems that have been built as extensions to SCION.

Path validation. SCION, through its use of PCFS, paves the way for the Origin and Path Trace (OPT) mechanism (Chapter 12). OPT enables the sender, receiver, and routers to cryptographically verify the path that the packet traversed. By leveraging the DRKey mechanism (Section 12.5), routers can efficiently derive their key, verify the path, and update the path validation fields.

Anonymity and privacy. PCFS also provides advantages for privacy. With PCFS and path transparency, the source is able to select paths that appear more trustworthy (e.g., those that do not traverse certain ASes). In addition, the packet header can be further obfuscated such that ASes on the path cannot learn identifying details about the source or the destination, unless they are immediately connected to one of them. Proposals such as LAP [113] and HORNET [49] leverage SCION's infrastructure to offer high-bandwidth and low-latency anonymous communication.

DDoS defense. The hierarchical organization of ASes into a manageable number of ISDs enables neighbor-based contracts between pairs of core ASes, which in conjunction with path segments inside the ISDs allows for establishing efficient bandwidth guarantees between any two end hosts (more details are presented in Chapter 11 and Section 13.7.1). Such bandwidth guarantees are provided by the SIBRA extension to prevent DDoS attacks at the architectural level: independent of the number of distributed bots, end hosts obtain protection against Internet-wide link-flooding attacks, one of the major threats in today's Internet. The SIBRA extension offers powerful mechanisms for DDoS defense, as it guarantees a lower bound on the bandwidth between any pair of ASes [22], which cannot be lowered even by a large-scale botnet using new types of DDoS attacks such as Crossfire [129] and Coremelt [231].

2.8 Main Contributions

The SCION architecture introduces many new concepts and contributions. Although prior work has proposed related concepts and methods, many of which we build upon, we believe that SCION has advanced the state of the art by creating a coherent architecture that can be deployed and used in practical environments.

Throughout the book, we highlight some chapters or sections with a diamond symbol in the title to indicate research, engineering, and deployment contributions that we believe are particularly important and interesting. In the remainder of this section, we briefly describe these contributions.

2.8.1 Isolation Domains

The concept of network partitioning and hierarchical domains has been considered since the early days of the Internet [34,46,47,56,57,127,134,232,242,259]. In addition to the scalability sought by previous approaches, SCION's concept of 💎 *isolation domains (ISDs)* (Chapter 3) provides strong security guarantees including meaningful trust roots and the absence of global kill switches. Isolation domains provide control-plane isolation, trust root scoping, and data-plane transparency. Most SCION protocols and extensions rely and build on these properties.

As a design principle, SCION does not require any globally trusted party, and ISDs can operate independently and autonomously. However, there must be a way for them to join the network and be discovered. To this end, in Chapter 5, we present the 💎 *ISD coordination mechanism*, which operates in a distributed fashion without any globally trusted entities. With our mechanism, individual trust decisions made solely by ISDs enable global trust verification, similarly to the PGP web of trust [267], although operating in the constrained environment of large-scale ISPs. The mechanism is based on the rule that trust validations follow routing paths (i.e., commercial relationships). To balance the design tradeoffs, our system allows inconsistencies but makes them visible. It enables determination of network topology and connectivity, from any point of the network, without any central global entity.

2.8.2 Authentication

Another main contribution is SCION's 💎 *authentication infrastructure*, which leverages the properties offered by isolation domains (Chapter 4). 💎 *TRCs* contain the roots of trust of the SCION authentication infrastructure (Section 4.2.1), providing scoped trust, fast and flexible trust root updates, and transparent trust relationships. 💎 *The control-plane PKI* (Section 4.2.3) is a high-availability PKI and is designed to secure SCION's control plane. It ties TRC and certificate distribution to the dissemination of PCBs, thus *removing any circular dependencies* between routing and control-plane PKI operations, which results in efficiency and high availability. On the other hand, 💎 *the end-entity PKI* focuses on achieving high security (see details in Section 4.4). It leverages two recent proposals (i.e., ARPKI [23,24] and PoliCert [235]). First, it provides resilience against a selectable number of compromised trusted parties. Second, it allows domain owners to express flexible policies on their TLS certificates and connections.

The control-plane PKI provides network-level authentication, enabling in-network and end host source authentication, which in turn facilitates construction of a variety of secure network protocols. 💎 *The OPT protocol* (Chapter 12) is a source authentication and path validation scheme. It enables end hosts to enforce path compliance according to their path selection, and moreover, it achieves high-speed and stateless operation on routers. OPT relies on 💎 *the DRKey scheme* (Section 12.5), an efficient key derivation mechanism. DRKey allows network entities (e.g., border routers) to derive symmetric keys (shared with destinations) with a negligible computation overhead and without keeping per-destination state. Due to these properties, we use DRKey for 💎 *the authentication of SCMP messages* (SCMP being SCION's equivalent of ICMP — see Section 4.2.5 and Section 7.6). To the best of our knowledge, it is the first Internet-scale control message protocol with authenticated messages.

As a consequence of scoped trust and isolated control plane, SCION ensures an 💎 *absence of global kill switches* (Section 13.8). No entity can cause an outage of an ISD by performing an operation outside the ISD (such as the revocation of an important key).

2.8.3 Novel Mechanisms and Protocols

Due to its architecture, SCION can intrinsically support multiple novel mechanisms and protocols. For instance, 💎 *RAINS* provides a next-generation name resolution system (Chapter 6). The control plane allows the definition of 💎 *flexible path policies*, enabling implementation of BGP route policies and definition of policies that cannot be expressed in BGP (Section 10.9). Furthermore, SCION's 💎 *data plane* (Chapter 8) provides *highly efficient and secure packet forwarding*. The forwarding path is encoded within each packet and is cryptographically protected. To make a forwarding decision, the border router checks whether the relevant information is fresh and was authorized by its AS. To this end, efficient symmetric cryptography is used. Moreover, the cryptographic mechanisms required are widely supported by modern hardware; thus, a high-speed SCION border router can be built on commodity hardware.

Another example is the 💎 *AS-level anycast infrastructure* (Section 7.5), which provides a service-oriented infrastructure enabling a packet to be delivered to the nearest server of a given service. This infrastructure is an especially powerful mechanism when used for building services that can take advantage of hierarchical caching.

Although path infrastructures have also been explored in other Internet architectures, SCION introduces a novel 💎 *secure path revocation system* (Section 7.3). Our path revocation system works on the link level. Its main novelty is a traffic-driven fault detection and failed-link revocation mechanism. The revocations are disseminated as responses to data packets that encounter a failed link. In this design, the system quickly disseminates revocations only to entities that have used failed paths, thus avoiding the overhead of informing entities that do not use those paths. To the best of our knowledge, it is the first secure and practical inter-domain link revocation scheme. The scheme also provides *authenticated failed-link localization*.

2.8.4 Resource Allocation

Another main contribution is 💎 *SIBRA* (Chapter 11), a SCION extension that implements *global bandwidth resource allocation*. SIBRA's main objective is to provide DDoS attack defense, and it is realized through end-to-end bandwidth allocation. The system provides *botnet-size independence*, a property that no prior DDoS defense system could achieve. A main feature of SIBRA is its per-flow stateless **fastpath*** packet forwarding.

2.8.5 Deployment and Evolvability

Finally, SCION makes the following deployment contributions. 💎 *The SCION-IP gateway* (Section 10.3) provides an easy and flexible way of *interconnecting SCION with the current Internet*. It supports a variety of connection and deployment variants. The gateway can be used by ISPs, organizations, or individual users to bootstrap and benefit from the deployment of SCION even for their legacy clients and legacy IP communication.

Taking into consideration the lessons learned from Internet deployment, SCION is designed to support and deploy new mechanisms. Flexible *extension mechanisms* are built into both the data and control planes (Section 15.1.4 and Section 15.3.4), which enables the architecture to evolve. Furthermore, in the spirit of evolvability and maintenance, SCION supports 💎 *algorithm agility* (Section 17.1), which is crucial in the context of cryptographic algorithms (as over time they become weaker or become vulnerable to a newly discovered attack).

3 Isolation Domains (ISDs) 

LAURENT CHUAT, ADRIAN PERRIG,
RAPHAEL M. REISCHUK, BRIAN TRAMMELL

This chapter discusses SCION isolation domains in more detail. As briefly sketched in Chapter 2, an isolation domain (abbreviated as ISD to distinguish it from the common abbreviation ID) constitutes a logical clustering of the Internet's most coarse-grained organizational unit, namely that of an autonomous system, or AS for short. An AS is a self-contained network administrated by a single entity (e.g., by an Internet service provider (ISP) or a university) and communicates with other ASes through well-defined interfaces based on contractual business relations. Figure 2.1 on Page 18 sketches how ASes are grouped into ISDs.

To join an ISD (i.e., become a member of an ISD), an AS needs to be connected to it, and needs to accept its regulations and policies. An ISD specifies accepted authorities, which are commissioned and authorized to provide digital identities and cryptographic keys for the entities inside the ISD.

As we will see in more detail, the term *isolation* refers to a property of ISDs that applies to the network's control plane only. Regarding the network's data plane, the important properties of ISDs are *transparency* and *control*. In other words, SCION does not isolate end hosts, nor does it limit communication or facilitate censorship, as we explain in more detail in the FAQ on Page 409. SCION rather provides members of ISDs with communication guarantees, with control over packet routes, and with transparency over forwarding paths.

The natural questions to ask are thus: How can isolation in the control plane achieve transparency in the data plane? Why is isolation in the control plane necessary at all? How can the current Internet be structured to best achieve a desirable level of isolation? We attempt to answer these and related questions in this chapter.

3.1 Why Isolation?

Before considering the details of how ISDs are implemented in SCION, we are going to step back and take a look at the rationale behind structuring ASes into

A. Perrig et al., *SCION: A Secure Internet Architecture*, Information Security and Cryptography, https://doi.org/10.1007/978-3-319-67080-5_3

ISDs. We first note that the concept of letting each ISD agree on individual policies, keys, and authorities naturally provides an isolation property among groups of ASes. To indicate the benefits of such a property, we observe that isolation between ASes is lacking for most features of today's Internet. Using the two examples of authentication and routing, we then illustrate why the lack of a suitably granular isolation property is the main reason for the security and availability issues that plague the current Internet.

3.1.1 Isolation for Authentication

To understand the problems related to the current Internet's lack of isolation, one may consider its authentication infrastructures. At a high level, authentication infrastructures enable users to verify digitally signed information (such as names, addresses, routes), assuming that the cryptographic keys necessary for the verification of such information are correctly distributed. Distributing and authenticating cryptographic keys in environments with heterogeneous trust, however, poses a major challenge. This holds for the two most prominent models of existing authentication infrastructure: *monopoly* and *oligopoly* (also referred to as *oligarchy* in the literature).

Monopolistic Infrastructures

Infrastructures based on a *single* root of trust (or a small number of keys held by a few entities), such as DNSSEC, suffer from the innate problem that *all* involved entities must agree on a common root of trust and on the entity that should manage the root of trust. In the case of DNSSEC, no less than the entire world has to agree on a common root. The fear of global surveillance paired with an increase in power of individual nation-states has led ICANN, the organization responsible for allocating and assigning names in the root zone of DNS, to issue a statement recommending globalization of Internet governance [121]. On 1 October 2016, ICANN officially entered the private sector and transitioned to a "multi-stakeholder model" as its contract with the U.S. government expired [122].

Besides the administration problem, there is also a serious security hitch with monopolistic infrastructures: a single root of trust evidently constitutes a single point of failure. In August 2016, Microsoft inadvertently leaked a highly permissive signed policy, which was then referred to as a "golden key". This policy could not only be used to unlock tablets and phones sealed by Windows Secure Boot (e.g., to install an alternative operating system), but also to enable backdoors for mass surveillance purposes [69, 173, 191]. It is even believed that Microsoft will be unable to fully revoke the policy [252]. In the case of DNSSEC, a compromise of the global trust root can cause severe damage to the entire world, essentially to each host worldwide that relies on

DNSSEC, directly or indirectly. We observe a significant *kill switch* here: the revocation of a DNSSEC certificate for a top-level domain name (such as `.com`) would remove that entire top-level domain since its validity would no longer be verifiable and thus the name resolution would fail. Interestingly, this kill switch becomes more severe when more cryptographic protection is added to today's domain name system. More details on Internet kill switches are provided in Section 13.8 on Page 325.

Oligopolistic Infrastructures

Infrastructures based on *multiple* roots of trust, such as the TLS infrastructure, suffer from weakest-link security — that is, *any* of the multiple roots, when compromised, may cause severe damage to *any* of the entities in the infrastructure. In other words, each member of the oligopoly has global authority. For the case of TLS, any certification authority (CA), possibly run by a national intelligence agency or by a malicious organization, may issue rogue certificates for any TLS domain. These rogue certificates will be recognized as valid by today's standard browsers.

Both the monopoly and the oligopoly model have in common that the scope of keys is unrestricted. The compromise of any cryptographic signing key enables man-in-the-middle attacks against billions of hosts around the world. The attack vector of these large-scale attacks can meaningfully be diminished through the concept of isolation by structuring the large number of existing entities into isolated domains, each with its authorities and individually managed keys, and by limiting the scope of the keys to the respective domains.

SCION resolves these issues by restricting the scope of root keys to ISDs, and enabling clients to select the TRC(s) they want to use.

3.1.2 Isolation for the Propagation of Routing Information

The process referred to as *inter-domain routing* is carried out in today's Internet by the Border Gateway Protocol (BGP). At a high level, every AS advertises to other ASes the IP address space for which it is responsible. The information is propagated to other ASes such that, after some convergence time, every AS should have learned how to reach any other address in the Internet.

This design works well in most cases, but is vulnerable to misconfigurations and attacks: a misconfigured AS can unintentionally attract traffic by advertising wrong addresses to its direct neighbors. More severely, a malicious AS can launch IP prefix hijacking attacks by deliberately advertising addresses that the AS does not control. The lack of isolation can lead to problems such as unavailability and espionage and affect virtually every host in today's Internet (see Section 1.1.4 on Page 6 for examples of concrete incidents).

By leveraging the isolation principle, SCION separates the routing infrastructure in one ISD from those of other ISDs and thus removes a cause of many instances of unavailability in today's Internet. More precisely, SCION addresses and routes to entities are valid only within the respective ISD. This means, in particular, that entities outside an ISD cannot affect communication within that ISD.

An interesting question is whether the isolation of failures and misconfigurations may result in undesirable confinements, such as the unreachability of destinations outside the source ISD. Fortunately, the opposite is true. Not only is the availability of communication increased due to the impossibility of external attackers intruding into isolated routing planes, but also because a well-manageable number of ISDs permits the scalable execution of a secure inter-ISD routing protocol with cryptographically protected route advertisements. Thanks to the limited number of ISDs, cryptographic keys are easily disseminated across all ISDs. These keys are used to validate the authenticity of routing updates across ISDs.

Definition: Isolation Principle

Intuitively, the isolation principle separates the control plane of a domain (e.g., an ISD) from outside influences. More formally, let $\mathcal{V}_D^E$ be the *view* of the control plane for a given domain D residing in an environment E, i.e., $\mathcal{V}_D^E$ is the set of all messages exchanged *inside* D's control plane in environment E. By environment, we denote the set of outside entities with which D can communicate. We say that the isolation principle holds for domain D if for all environments E, we have

$$\mathcal{V}_D^E \approx \mathcal{V}_D^{\varnothing}$$

where $\varnothing$ is the empty environment and $\approx$ denotes indistinguishability between two views with respect to intra-domain routing messages.

Definition: Isolation Domain (ISD)

An ISD is a set of connected ASes (i.e., forming a connected graph) that satisfies the following conditions:

- All member ASes accept the trust roots and policies described in the *trust root configuration* (see Page 63) as managed by the ISD core.
- The ISD satisfies the isolation principle, i.e., its control plane is protected against outside influences.

3.2 The ISD Core

Each ISD is administered by the ISD core, a consortium of one or multiple autonomous systems referred to as core ASes.

Definition: ISD Core

The ISD core is formed by a set of directly connected ASes, the *core ASes*. All members in that set agree that they form the ISD core. They also agree to perform the following functions:

- manage and distribute the ISD's TRC;
- sign the TRCs of neighboring ISDs and endorse other ISDs;
- issue certificates to all ASes in the ISD;
- provide connectivity to neighboring ISDs;
- generate and disseminate inter-ISD path-segment construction beacons (PCBs), also called core PCBs;
- generate and disseminate intra-ISD PCBs;
- provide highly available services (beacon, name (RAINS), path, certificate, SIBRA, and time servers); and
- maintain a list of all recognized ISDs.

The tasks of the ISD core are broadly divided into two categories: manage the control-plane public-key infrastructure (PKI), and provide global (inter-ISD) and local (intra-ISD) connectivity. For the following discussion, we assume familiarity with the basic SCION concepts described in the previous chapter.

As a foundation for the control-plane PKI, the core ASes establish the trust root configuration (TRC). Specifically, a TRC defines the roots of trust that are used to validate bindings between names and public keys or addresses, and defines a policy on how the TRC can be updated. The ISD core manages and distributes the TRC. For TRCs to be accepted by other ISDs, they must be signed by trust roots of neighboring ISDs. Sections 4.1 and 4.2.1 provide more information on TRCs, their creation, and their dissemination. The core ASes issue certificates for other ASes in the ISD — the TRC contains the root of trust public keys to verify these certificates. To enable these operations, core ASes operate the core certificate servers.

The second major task of the ISD core is to provide local and global connectivity. Core ASes connect to core ASes in other ISDs. To discover paths, intra-ISD and inter-ISD path-segment construction beacons (PCBs) are emitted periodically. For path exploration, path registration, and path resolution, core ASes run the core beacon and path servers. Moreover, every core AS runs a time synchronization service.

When a new ISD is created, its core must make an announcement to other ISDs. Since we do not want to rely on any centralized entity to decide on the fate of a new ISD, we use a distributed ISD coordination process, which is presented in the following section. Through the ISD coordination process, the ISD core maintains a list of existing ISDs (and their TRCs).

3.3 Coordination Among ISDs

One of the distinguishing properties of the SCION architecture is that it was designed to operate without any global authority. Developing and deploying a decentralized authentication infrastructure on a global scale has been a long-standing problem [35, 97]. In SCION, each ISD must be able to derive a list of existing ISDs. This would be straightforward if we could assume the existence of a trustworthy authority, or if all participants could agree on who should be able to join the network, but global consensus is hard to achieve in an environment with mutually distrusting entities. If consensus was required, some ISDs could collude and prevent a new ISD from joining the network; a single entity could also create and control multiple ISDs with the only intention of gaining influence, which is referred to as a Sybil attack [72]. Instead of relying on consensus amongst all existing ISDs, we focus on providing transparency and accountability to deter misbehavior.

The mutual discovery of ISDs follows a distributed approach in which every ISD builds its own local view of the global ISD topology. Our approach relies on local consistency and on neighbor-based propagation of authenticated information. More precisely, new ISDs are announced in advance — to avoid identifier collisions — by neighboring ISDs, through beacon extensions. This approach tolerates bogus ISDs (i.e., ISDs with globally unique identifiers but without legitimate purpose). Transparency, however, allows such illegitimate ISDs to be detected and ignored.

Each ISD is identified by a unique integer and a description. If a dispute arises between two or more new ISDs regarding the attribution of an identifier, these ISDs need to pick a new identifier to announce, or the other ISDs need to decide which announcement they want to support for a given identifier (if no agreement is reached between the conflicting ISDs). We present the details of ISD coordination[1] in Chapter 5.

3.4 Name Resolution

The mechanism for ISD coordination that we sketched in the previous section allows each ISD to obtain a list of other ISDs; in this section, we describe how

[1] In distributed systems, consensus can only be achieved by assuming either (a) the existence of a trusted centralized authority, or (b) resource parity and *coordination among entities* [72]; the term “coordination” as used in this context, however, does not imply that all ISDs must reach a complete agreement. ISD coordination only designates the mutual discovery and the announcement of ISDs.

consistency is maintained for name resolution across ISDs without any global authority. In summary, SCION uses a DNSSEC-like protocol called *RAINS* (described in detail in Chapter 6), in which delegation from one zone to another is performed by a signature identifying a zone key (ZK) for the subordinate zone, with unique root zone keys (RZK) per ISD. Each such root zone contains delegations to the authority for each top-level domain (TLD), which in turn handles resolution for second-level names, and so on.

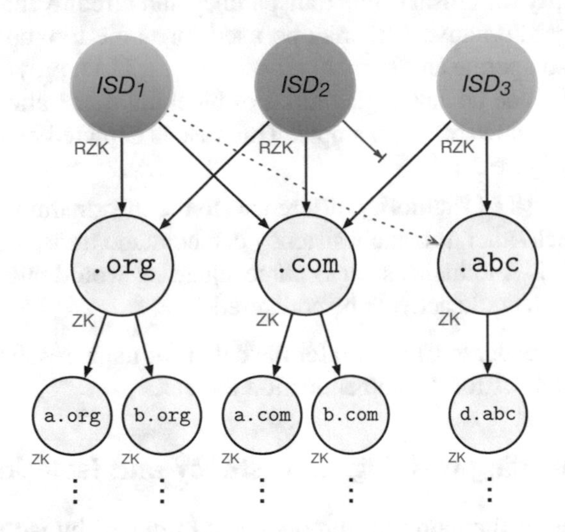

Figure 3.1: ISDs delegate name resolution to the TLD authorities. ISD_2 has refused to delegate to the authority of TLD `.abc`, while ISD_1 operates a shadow authority for TLD `.abc` (see text below). Delegation assertions are discussed in Section 6.3 on Page 106.

Clients connected to the SCION Internet from different ISDs may therefore have different views of the global namespace (for example, ISD_1 and ISD_2 in Figure 3.1 have no direct delegations to the `.abc` TLD). This is an unavoidable consequence of isolation as an architectural principle: it makes little sense to build inviolate isolation into each ISD, then delegate the first step in most communication establishment (name resolution) to a non-isolated global DNS root. On the other hand, a globally consistent namespace is one of the advantages of the Internet as a platform.

This inconsistency is mitigated by three factors, which the governance models presented in the next section are intended to support:

- In the typical case, ISDs simply certify authorities for TLDs as shown by the black arrows in Figure 3.1. In other words, each ISD will delegate to the same authority for a given TLD, and the chain of signatures will be identical beyond the ISD root signature of the TLD authority. This replaces a *single* global root with a *collection* of global roots, which are mostly, but not completely consistent.

- The remaining inconsistency is generally a consequence of applications of isolation: (a) An ISD may *refuse to delegate* to an authority for a given TLD, because that TLD is of no use to clients connected to it. This case is depicted by the interrupted red line in Figure 3.1. The possibility of refusal offers transparency and circumvents censorship in that any alternative ISD may be used for name lookups. (b) An ISD may also operate an *alternate authority* for a TLD, providing additional due diligence on new registrations or blocking registrations intended for abuse by malware, for example. This case is depicted by the dotted gray line in Figure 3.1.
- By adding TLD authority differences to the information that ISDs learn from each other, this inconsistency can be made transparent at the inter-ISD level. Anomalies such as name squatting would thus become transparent and could actively be countered.

We refer the reader to Chapter 6 for the details of name resolution in SCION, and to Section 4.3.1 for the authentication thereof.

3.4.1 Reconciling Naming Consistency and Isolation

The properties of the name resolution system implied by isolation as a first principle of the architecture mean that, while the information associated with a domain name (addresses, authorities, etc.) can be guaranteed to be consistent within an ISD, since each ISD has a global root, naming consistency cannot be guaranteed across ISDs. Some of this inconsistency can serve to implement the policies of each ISD (e.g., filtering malware domains published in TLDs), but other inconsistency is not desirable (e.g., an ISD creating and reserving a large number of TLDs through name squatting). Managing this inconsistency requires inter-ISD coordination. This process is detailed in Section 6.5, which describes the Naming Consistency Observer (NCO), a process run cooperatively by all ISDs to make isolation-based inconsistency transparent to all participants in the SCION Internet, thereby providing a method to deter non-desirable inconsistency.

The NCO also provides a way for the SCION Internet to cleanly inherit the current global naming root. The set of TLDs accessible through SCION will necessarily be inherited from the ICANN global root, and changes to this set of TLDs will continue to be made according to ICANN's policy development process. The set of TLDs that each ISD is presumed to start from comes from this global root. ISDs can, of course, sign additional TLDs not present in the global root, and the visibility of these TLDs through the NCO makes it possible for other ISDs to determine whether they want to sign them as well; it may also make isolated versions of these TLDs available.

Whether this mechanism will eventually replace the ICANN management of the global root, or act as an input to ICANN's process for adding new TLDs to the set in the global root is a future question for the Internet community at large. In any case, the global ISD proposed in Section 3.5.4 provides unmodified access to the ICANN global root.

3.5 ISD Governance Models

Given how central ISDs are to the SCION architecture, the qualities of a SCION-enabled Internet are in part determined by the policies by which ISDs are created and connected to the Internet, and how policy-level conflicts among ISDs are resolved. In this section, we examine several possible sets of policies for inter-ISD governance, and their implications for SCION's operation, incremental deployment, and transition to a SCION-based Internet.

These models are presented primarily to explore the space of possible governance structures for a SCION-based Internet; we do not envision or condone any one model as the way forward. Since they are concerned solely with non-technical conflict resolution among ISDs, elements of different models can and will be combined. We anticipate that ISD creation will occur organically following a combination of these models.

Some ISDs may evolve from existing tier-1 ISPs, indicating they will operate largely as described in Section 3.5.1. On the other hand, jurisdictions may insist on sovereign authority as in Section 3.5.2, as the root of trust for routing is a matter of law or regulation; in these jurisdictions, only the national ISD would be available. In any case, it is likely that the initial governance structures will at least bootstrap off the current multi-stakeholder model as embodied by the Internet Engineering Task Force (IETF), the Internet Assigned Numbers Authority (IANA), regional Internet registries (RIRs), and the Internet Corporation for Assigned Names and Numbers (ICANN), even if new SCION-centric governance organizations also evolve. A SCION Internet, in which ASes are free to be members of multiple ISDs, may evolve both small *isolation service providers* (IsSPs) (Section 3.5.3) and a global default domain. These models will interact with the transition mechanisms described in Chapter 10.

3.5.1 A Bottom-Up Model: Grassroots Deployment

In a bottom-up model, some existing ISPs would begin by creating ISDs and offering SCION services within their isolation domains to their customers. Full connectivity within the SCION Internet is therefore provided by tunnels between SCION islands over a traditional Internet substrate. Eventually, interconnections or mergers between ISDs will lead to organic growth and increased availability of SCION connectivity and decreased reliance on tunneling, as experience with the transition to the IPv6 Internet has shown. ASes would become connected to the SCION Internet through their existing transit relationships with upstream providers, and existing peering relationships at Internet exchange points (IXPs). This incremental deployment model is described and analyzed in Section 10.1.2.

In terms of governance, this model leans on existing structures to bootstrap itself. ISD numbers would be administered by IANA according to procedures established through the IETF standards. The SCION core protocols would therefore need to be published as IETF standards as well. Given the differences in areas of policy expertise and the relative scarcity of ISD numbers, IANA would then delegate this assignment to the RIRs. AS numbers would continue to be assigned by RIRs according to current policies; an Internet-connected ISP's AS numbers could be used in SCION as well. Addresses for SCION-attached networks would be administered as addresses in the legacy Internet, and assigned by the RIRs according to their own policies and subject to their policy development processes. Given exhaustion of IPv4 addresses in each region, growth in the SCION Internet would therefore predominantly happen using IPv6 addressing. The naming root for each ISD would be provided by ICANN, according to its policy process.

The primary advantage of this model is ease of transition and the relatively lightweight coordination required. ISPs and ASes can each decide according to their own requirements to join the SCION Internet. To do so, downstream ASes may either wait for their upstreams to join, or purchase transit from an existing member of an ISD and tunnel SCION traffic to it. Large ISPs may enter into agreements with others to form a consortium to operate an ISD; tier-1 ISPs may even decide to operate ISDs on their own.

These actors would interact with each other in regulatory, governance, and technical forums in which they already participate: IETF, RIR, ICANN, and regional network operator groups. Multilateral conflict resolution would become a matter of each RIR's policy development process, and bilateral conflict resolution a matter of national or international contract law; new governance organizations may eventually emerge to take over parts of these roles, as necessary.

This model would mirror the present Internet, which may be seen as both advantageous and disadvantageous. While the present model does scale well in terms of administrative overhead, an organic transition inherits all the strengths and weaknesses in Internet governance of the present Internet, and current vested interests would retain their advantages. However, given the inertia inherent in the Internet industry, we envision this model as the default one for establishment of a SCION Internet.

3.5.2 A Top-Down Model: Sovereign Authority

The grassroots model may not be acceptable within some jurisdictions, which may insist on sovereign authority over Internet traffic and interconnections.

Since one property of an ISD is a coherent set of policies and regulations for managing the addition and removal of ASes from the ISD, and since one of the widely accepted rights and responsibilities of sovereign entities in international law is the resolution of conflicts within their territory, it is natural to assume that some sovereign states will be willing to form ISDs, i.e., *country-based ISDs*.

A sovereign authority ISD would be created by an internationally recognized sovereign power. Interconnections between sovereign authorities would be governed by bilateral or multilateral treaty. A multilateral SCION Internet treaty could be overseen by an existing international body, for example a United Nations agency such as the International Telecommunication Union (ITU). Connection to a nation's ISD is wholly a matter of national law and regulation, subject to the terms of the treaties governing interconnections. By contrast, interconnections between sovereign authority ISDs and other ISDs would be a matter of international contract law.

This model has several apparent advantages:

- National-level lawmaking and regulatory bodies for telecommunications already exist in most sovereign states, and they already have competence for monitoring the activities of telecommunication service providers (ASes) within their territory. National isolation makes enforcement of Internet law much easier, as the ambiguity about the law in effect at the source of traffic or location of content is removed by the nature of the routing topology.
- Countries with policies on cross-border traffic routing, whether to better regulate the handling of Internet traffic or to defend their citizens against surveillance or other malfeasance not subject to that country's law, will have in-country routing by default, since all communications between two ASes in an ISD stay within the ISD. If a country has laws restricting the circulation of certain types of data (e.g., stating that medical records cannot leave the country), then an ISD following this model can be used to achieve compliance.
- Some national authorities already act as roots of trust for their citizens and registered corporations, and there are advantages in identity management if a citizen's or corporation's identity on the Internet is vouched for by the government, which usually has mechanisms for real-world identity verification.

Unfortunately, a top-down approach also has some severe drawbacks:

- This model would require sovereign entities to manage the networks in the ISD core. However, most countries are not in the business of providing Internet service, and would need to develop the competence for the technical management of the ISD core. In countries with an incompletely privatized or former incumbent national telecom provider, the contract for running the ISD core could naturally be given to that provider. Countries with robust protection of commercial competition would need some other mechanism to select the ISD core operators.

- This model negates the advantages of isolation for transnational entities. An entity with presence in multiple countries, such as a multinational corporation, would need separate ASes in each ISD, and would need to build a private network between its own ASes to prevent internal traffic from crossing the inter-ISD links at the international level.
- It would require a significant transition in both routing topology and governance structure from the present Internet, which would be hard to achieve incrementally. With respect to routing, coordinated migrations from one type of interconnection to another are virtually impossible to implement at Internet scale, so both routing technologies would need to coexist for an indefinite period of time. Any failure in international coordination would lead to widely varying views of the SCION vs. non-SCION Internet depending on which country's ISD one is connected to. With respect to governance, a relatively technically complicated multilateral international treaty would need to be negotiated to set the technical framework for international interconnection; this would take time. Existing governance structures (e.g., ICANN and the RIR system) would either need to evolve to derive their authority from international treaties, or they would need to be dismantled and replaced with new organizations under the new treaty arrangement.

3.5.3 The Isolation Service Provider (IsSP) Model

The models above assume that ISDs must provide core routing services and act as trust anchors. Regarding routing services, however, this is not necessarily the case, especially considering the fact that the current Internet topology is increasingly dominated by peering links as opposed to the textbook model of tiered transit [2]. We therefore consider a SCION Internet with "stripped-down" ISDs that provide primarily trust root and infrastructure services for isolation over networks operated by other entities.

In this model, the links within and between ISD cores primarily handle control traffic. Almost all up-segments and down-segments will be joined either by an AS below the core (which itself might be an existing tier-1 ISP), by a peering link between ASes below the core, or by a peering link across ISD boundaries (see Figure 3.2). The ISD core then provides low-bandwidth, last-chance default routing for address pairs without an existing high-bandwidth peering link between ASes in different ISDs.

There are two reasons to consider this model. First, the amount of trust placed in ISDs by ASes within them makes the ISD a target for compromise. The expense associated with an AS leaving an ISD in a model where the ISD provides that AS's sole connectivity to the Internet is high, as it may be

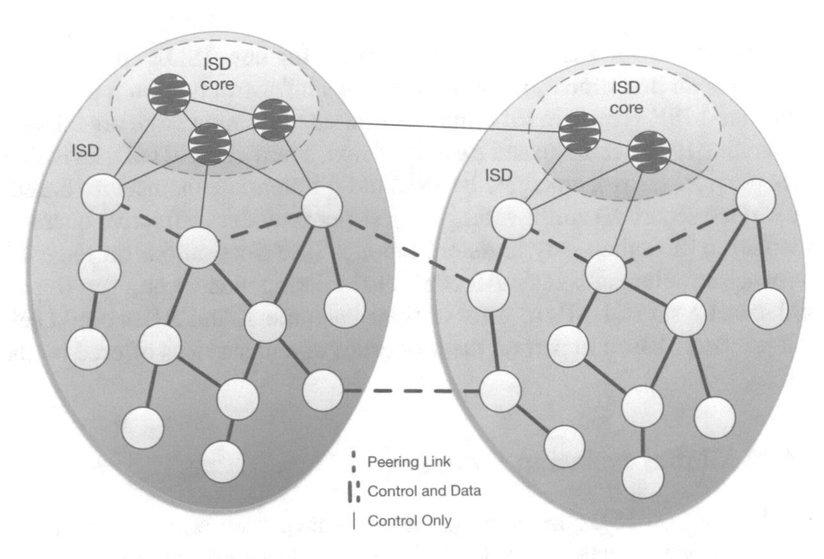

Figure 3.2: Example of the Isolation Service Provider (IsSP) Model.

associated with expensive-to-modify physical infrastructure. The trust the AS places in the ISD is also great. This provides incentives both for malfeasance on the part of the ISD — since it cannot realistically be punished for bad behavior — as well as compromise of the ISD by an external entity. Separating an AS's Internet connectivity from its trust root moves the Internet to a model where most top-tier ASes belong to multiple ISDs, and handle inter-ISD traffic through "internal" peering links. This multi-membership (illustrated in Figure 3.3) allows an AS to react to malfeasance, incompetence, or compromise of an ISD by leaving the ISD without any penalty to its connectivity. This "big red button" is an important tool in holding the ISD core accountable.

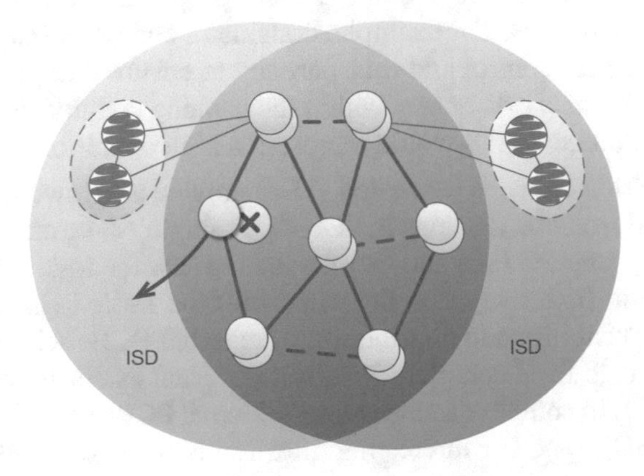

Figure 3.3: Example of multi-membership in the IsSP Model.

This model also lowers the barrier to entry for new ISDs. If creating an ISD requires an investment in communications infrastructure on a par with an existing tier-1 ISP, growth in ISDs may be limited to existing large organizations. By separating traffic carriage and trust provision, a new class of ISD, an isolation service provider (IsSP), emerges. IsSPs could differentiate themselves based on name service security, careful vetting of first-tier providers, effective quarantine and isolation of maliciously registered names (see Section 6.5 on Page 116), governance structures for inter-AS conflict resolution, and so on. Top-tier ISPs would select a set of IsSPs to offer to their customers, and ASes could select transit providers based in part on the properties of and services offered by these IsSPs.

3.5.4 A Global Isolation Domain

Regardless of how ISDs are created, once we have a situation in which every AS is routinely connected to multiple ISDs, it makes sense to create a default, global isolation domain with relatively permissive policies. This ISD would be roughly equivalent to the legacy Internet in a single ISD, but running SCION protocols. This arrangement would allow other ISDs to operate arbitrarily restrictive isolating policies with respect to other ISDs, while allowing ASes that are also connected to the global ISD to maintain default connectivity. Moreover, the global ISD would offer a view of the name hierarchy that is similar to the current DNS namespace.

3.6 Nested Isolation Domains

ISDs provide control-plane isolation, path transparency, and the ability to control paths. It is desirable to achieve these properties at a finer granularity than that of global ISDs. For example, consider a conglomerate of banks that desire stronger path control and transparency to ensure that packets will stay within one bank's network, or stay within the banking conglomerate's networks. Setting up several new ISDs would represent a high operational overhead.

Nested isolation domains (or nested ISDs) provide a lightweight mechanism for hierarchical isolation domains in SCION. A single AS or multiple ASes can decide to set up a nested ISD, and they can define how the routing infrastructure of the enclosing (external) ISD interacts with the nested (internal) ISD. In particular, the visibility and distribution of external PCBs within the internal network can range from complete isolation (external PCBs are not sent inside the internal ISD) to complete transparency (external PCBs are propagated inside the internal ISD). Another interesting question is whether internal PCBs are visible externally or not, allowing the nested ISD to achieve some level of secrecy for its internal network structure. For the visibility of the internal ISD structure, we propose three levels of transparency:

- **Transparent:** All paths of a transparent internal ISD are announced to the external ISD. All communication leaving the transparent internal ISD contains the hop fields of the internal ISD.
- **Translucent:** A translucent internal ISD is visible, but its internal paths are not publicly announced outside the internal ISD. The hop fields contained in the packet header refer to the internal ISD, thus some topological information about the internal ISD is leaked.
- **Hidden:** A hidden internal ISD is invisible to the outside. All communication leaving the hidden internal ISD has the internal hop fields removed (similarly to the source address of devices behind a NAT device). The structure of a hidden internal ISD, including ASes, paths, devices, and certificates is thus not exposed externally.

Because the SCION data plane uses info fields to designate each path-segment transition, each SCION packet header provides ISD-level path transparency. Therefore, nested ISDs can help enforce interesting policies: a sender can ensure that a packet cannot leave a corporation, or a firewall at the border of a corporation can ensure that a packet will not leave a network defined by a conglomerate of ASes. These properties are in stark contrast to today's Internet, where a destination IP address cannot provide strong properties for the scoped propagation of a packet.

The details about nested ISDs will be specified in a future version of SCION.

Part II

SCION in Detail

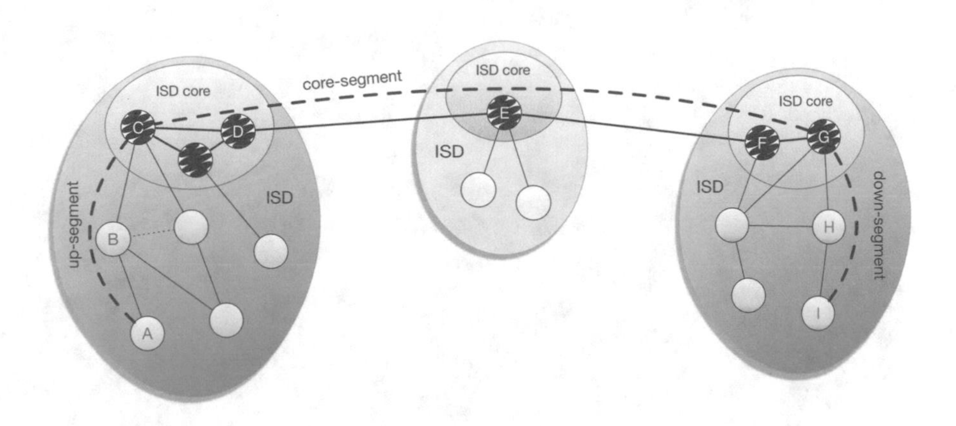

4 Authentication Infrastructure 

LAURENT CHUAT, ADRIAN PERRIG,
RAPHAEL M. REISCHUK, PAWEL SZALACHOWSKI

In this chapter, we discuss the authentication infrastructure of SCION, which enables verification of identities and assertions that data did indeed originate unchanged from the claimed entity. SCION offers built-in support for various types of authentication and various uses, and thus provides several infrastructures to support authentication.

We start this chapter by providing an overview of the SCION authentication infrastructure. We then present a public-key infrastructure (PKI) for the control plane, and we describe the details of managing trust root configurations (TRCs), which includes how TRCs are created, updated, and disseminated. Finally, we explain how control-plane messages, names, and end entities are authenticated.

Chapter Contents

4.1 Overview

As a foundation for authenticating messages, names, and entities, each ISD core has a set of *trust roots*. Neighboring ISDs sign each other's trust roots to guarantee global verifiability of authentication information. To decrease the number of trusted entities on long verification chains and thus increase security, mutually trusting ISDs can additionally sign each other's trust roots even when they are not directly connected. In comparison to today's authentication infrastructures such as BGPsec's RPKI [6] or TLS's PKI, SCION offers the following improvements in terms of security and flexibility:

A. Perrig et al., *SCION: A Secure Internet Architecture*, Information Security and Cryptography, https://doi.org/10.1007/978-3-319-67080-5_4

- **Efficient updating of trust roots:** Even after key loss, disclosure, or compromise, trust roots can be rapidly updated, without software updates.
- **Resistance to compromised entities and keys:** Compromised or malicious trust roots outside an ISD cannot affect operations that stay within that ISD. Moreover, SCION's authentication infrastructure can be configured to withstand any single compromised key for certain critical operations. In particular, in the case of end-entity certificates, higher levels of security can be achieved and the system can be configured in such a way that at least three independent trust roots need to be compromised to forge a certificate.
- **Decentralized trust model:** Authentication relies on local trust roots. This enables limiting the scope of authorities and preventing global kill switches, as we describe in more detail in Section 13.8 on Page 325.
- **Flexible trust policies and trust agility at several levels:** Each ISD can define its own trust policy. ASes need to accept the trust policy of the ISD(s) they are in, but they can decide which ISDs they want to join, and they can also participate in multiple ISDs. End entities can decide which ISD they want to rely upon for resolving names and verifying the association between names and public keys. They can also define the set of trust roots they want to use for signing their entity certificates, irrespective of which ISD they connect to (although if trust roots of a remote ISD are desired, then the name will need to be selected from a namespace for which the remote ISD is authoritative). This flexibility enables fine-grained management of today's heterogeneous trust environments.
- **Highly available authentication infrastructure for the control plane:** Authentication is possible without circular dependencies on the availability of routing to verify a certificate's revocation status, for example.
- **Scalability:** The authentication infrastructure scales to the size of a global Internet and is adapted to the heterogeneity of today's Internet constituents.
- **Transparency:** A verifier always knows the exact set of entities that need to be trusted for a given authentication operation, and knows that any other entity cannot influence the operation.
- **Algorithm agility:** SCION offers algorithm agility by providing support for multiple signatures, so that a new cryptographic algorithm can readily be used in addition to the current algorithm.

To our knowledge, no previous system has achieved such a strong set of properties. Although global authentication services without global trust have been studied for decades, previous work still relied on a globally consistent name hierarchy [35, 97].

4.1.1 Trust Root Configuration (TRC)

The foundation of the SCION authentication infrastructure is the trust root configuration (TRC), which expresses the trust roots of each ISD. The TRC defines the trust roots that are used for all authentication procedures in SCION, thus all ASes, services, and end hosts need a TRC to use SCION. In short, a TRC contains

- a version number, a creation timestamp, and an expiration timestamp;
- trust roots for SCION's control-plane, name-resolution, and end-entity PKIs;
- parameters specifying (a) the quorum of core ASes (from the local ISD) required to sign a new TRC, (b) the quorum of CAs required to change the end-entity PKI's parameters and trust roots;
- signatures of core ASes to certify the authenticity of the TRC; and
- signatures of remote ISDs' trust roots (at least one core AS, one CA, and a name root key). All neighboring ISDs have to cross-sign, so that each routing path has a corresponding chain of trust that can be followed. Non-neighboring ISDs can sign TRCs as well (to create shortcuts in trust paths).

See Section 16.1 for the complete list of items contained in a TRC.

We assume that all entities can initially obtain an authentic TRC, e.g., with an offline mechanism such as a USB flash drive provided by the ISP, or with an online mechanism that relies on a trust-on-first-use (TOFU) approach.

Dissemination of TRCs

Information about a TRC update is disseminated via SCION's beaconing process. Each PCB contains the version number of the currently active TRC, and if the TRC version number of a received PCB is higher than the locally stored TRC, a request is sent to the AS that sent the PCB to obtain the new TRC (see details in Section 7.1). The new TRC is verified on the basis of the current one, and is accepted if it contains at least the required quorum of correct signatures by trust roots defined in the current TRC. This simple dissemination mechanism has two major advantages: it is very efficient (as fresh PCBs rapidly reach all ASes), and it avoids circular dependencies with regard to the verification of PCBs and the beaconing process itself (as no server needs to be contacted over unknown paths in order to fetch the updated TRC). The details of the TRC dissemination process are described on Page 72.

Revocation of Trust Roots

The TRC dissemination mechanism also enables rapid revocation of trust roots. When a trust root is compromised, the other trust roots can remove it from the TRC and disseminate a PCB with a new version number. The size of the quorum needed to sign a new TRC must be larger than one to prevent any single compromised root of trust from creating a new TRC.

TRC Verification

The TRC contains the roots of trust to verify all certificates and statements made by an ISD. We now briefly describe several verification cases and give a detailed list of verifications in the remainder of this chapter. To visualize the "flow of trust" in the sequence of verifications, we draw diagrams as follows. Each circle represents a cryptographic key that is used to certify another cryptographic key, for instance by using a digital signature. The key (or set of keys) inside the *double circle* is the root of trust that is axiomatically trusted to establish trust in other keys. An arrow depicts the flow of certification, where the key corresponding to the first node certifies the key corresponding to the node pointed to by the arrow. Intuitively, the arrow indicates the "flow of trust" so that when the first node is trusted, the second node pointed to by the arrow is also trusted. We use this depiction to convey the intuition; later in the chapter we use a more formal representation.

(a) Update (b) Cross-signing

Figure 4.1: TRC verification mechanisms.

Figure 4.1a represents the creation of a new TRC with version T_{i+1}, which is signed using (at least) the quorum of roots of trust defined in the TRC with version T_i. Figure 4.1b shows the cross-signing of the TRCs of two ISDs. SCION requires that every pair of connected ISDs also cross-sign each other's TRC. This is an important requirement as it guarantees that if a forwarding path exists, then verifying a destination's statement is always possible. For instance, given a sequence of ISDs that need to be traversed from a source to a destination, then the respective sequence of cross-signed TRCs enables the source node to verify any statement made by the destination's ISD.

The TRC is explained in more detail in Section 4.2.1 and the format is specified in Section 16.1 on Page 369.

4.1.2 Public-Key Infrastructures in SCION

SCION offers the following three PKIs, which we briefly describe next:

- a control-plane PKI (details in Section 4.2),
- a name-resolution PKI (details in Section 4.3),
- an end-entity PKI (details in Section 4.4).

Why more than one PKI?

Ideally, SCION would use a single, highly secure public-key infrastructure. Unfortunately, the infrastructure we describe in Section 4.4 would introduce a circular dependency if used in the control plane, and would therefore not provide high availability guarantees.[a] To obtain intuition on this point, consider "rebooting" the Internet: when routes are initially established through routing updates, all information to verify these routing updates must be available locally or obtainable from the entity that sent the routing update.[b] Our end-entity PKI requires end entities, CAs, and logs to be able to communicate; therefore, circular dependencies would arise if such a PKI were used to authenticate the control-plane PKI, as the verification of routing messages would rely on routing and vice versa. For this reason, the control-plane PKI is based on trust roots that include the core ASes so that no additional entities need to be contacted to issue AS certificates or new TRCs.

[a] Bobba et al. [37] previously described such circular dependencies in the context of wireless networks. Their solution is to rely on self-certifying identifiers, which unfortunately are not easily applicable in SCION because of the general difficulty of updating or revoking self-certifying identifiers.

[b] The current RPKI system of BGPsec has a circular dependency since querying the revocation status of a certificate requires the reachability of RPKI servers [59].

The control-plane PKI is a simple infrastructure that creates short-lived certificates for the ASes of an ISD. The purpose of these AS certificates is to enable the validation of signed beacons and path segments and to establish secret keys with other ASes (e.g., through a Diffie-Hellman key exchange). As described above, the root public keys are defined in the TRC of the respective ISD. Each core AS operates one online and one offline key pair. (Potentially, CAs could also participate under the condition that their servers be accessible without introducing a circular dependency with beaconing.) The revocation of root keys is accomplished through a TRC update operation. The revocation of AS certificates, however, would introduce additional complexity because each usage of an AS certificate would require a revocation check. We therefore make use of short-lived certificates for ASes, with a lifetime on the order of a few days. AS certificates are not directly signed with root keys contained in TRCs; instead, root keys sign *core AS certificates*, which, in turn, are used to authenticate the regular AS certificates that both core and non-core ASes use for their daily operations.

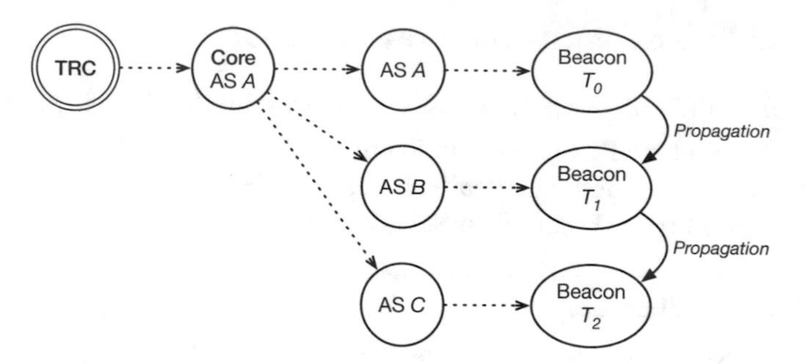

Figure 4.2: Authentication of beacon messages. Core AS *A* has both a core AS certificate for issuing AS certificates, and a regular AS certificate for signing PCBs.

Figure 4.2 depicts the chain of trust used for the verification of beacon messages that are created by core AS *A*, forwarded to non-core AS *B*, and then forwarded further to non-core AS *C*. The AS certificates are verified based on the trust roots in the TRC, then the signatures in the beacon message are verified based on those AS certificates.

AS certificates are also used for issuing certificates for the hosts inside the ASes, for instance, by the OPT protocol (see Section 12.3).

The name-resolution PKI also has its trust roots embedded in the TRC. Root keys are used to sign the root zone files of a DNSSEC-like infrastructure. To achieve a higher level of security than DNSSEC, a domain's name resolution key is also signed with the end-entity certificate, as shown in Figure 4.3. In standard DNSSEC, one has to trust all entities from the root to the leaf of the name resolution tree: if any of those keys is under the control of an adversary, then the final name resolution entry can be fabricated by the adversary. The DNSSEC authentication still provides an initial authentication, but the strong end-entity PKI validation will provide a high level of assurance that the domain key is correct and in turn the final entry is correct.

The end-entity PKI is a high-security infrastructure used in SCION for end-entity certificates, similarly to the TLS PKI used today for HTTPS. For this PKI, we assume that the routing and forwarding infrastructures are operational. Consequently, clients can contact additional services for the verification of a certificate, for instance for the verification of its revocation status. The main goal of this PKI is to achieve high robustness against compromised trust roots and malicious CAs, which we achieve through three approaches: (a) use ISD-scoped trust roots, such that a CA outside an ISD cannot create a fake certificate for an entity inside the ISD, (b) record all certificate-related events in a publicly verifiable append-only log, and (c) require multiple CAs and log servers to sign each certificate. The left box in Figure 4.3 depicts the verification of an

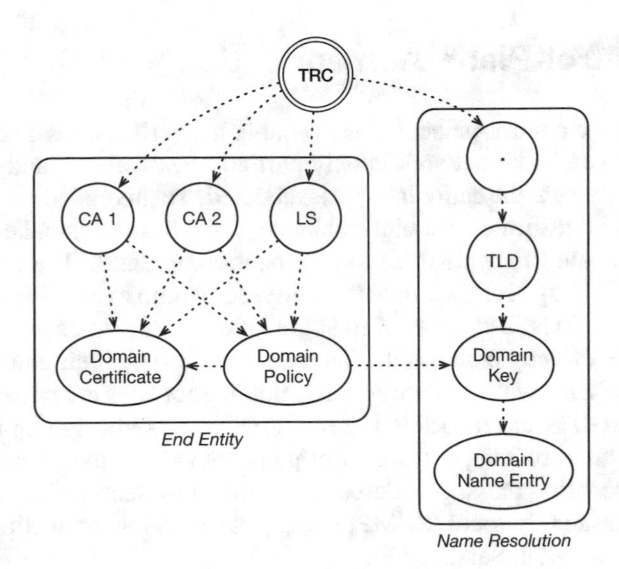

Figure 4.3: Authentication of an end-entity certificate and a name-resolution entry.

end-entity certificate, which in this case relies on signatures by two distinct CAs, and certificate registration by one log server (LS).

Through such a construction our PKI provides the following property for a client-server communication: if the client has a TRC set as its trust root, an adversary can spoof the server's identity only if she is able to compromise a threshold number of trusted entities (set in the TRC). In the case of a name resolution pointing to another ISD, the client has to obtain a proof (asserted by the threshold number of trusted entities) that the server's domain does not have a registered policy in the local ISD.

4.1.3 Catastrophe Prevention and Recovery

In case of catastrophic events, such as several private root keys being disclosed due to a critical vulnerability in a cryptographic library, SCION is equipped with a recovery procedure. The procedure consists in creating a new TRC with fresh trustworthy keys (and potentially new algorithms), and re-sending the TRC to all entities in the ISD and cross-signing entities.

No single malicious entity (e.g., AS, name trust root, CA, log server) can take down the entire SCION network or impersonate an end entity, since critical actions require signatures from multiple parties. Even in the event of several entities forming a coalition to carry out an attack, the effects of that attack would be limited to one or a few ISDs. Moreover, all actions are publicly visible, which will deter participants from misbehaving.

4.2 Control-Plane Authentication

The goal of the control-plane PKI is to enable the verification of control-plane messages, even if the network is only partially available — and even in the extreme case where the entire Internet is rebooted. To this end, the control-plane PKI provides certificates that bind public keys to ASes, and handles all aspects of the certificate life cycle (creation, revocation, update). The control-plane PKI is operated by each ISD independently, so that no external ISD can affect internal operations. The roots of trust are held by core ASes and by selected CAs that are directly connected to a core AS or deploy their equipment with their roots of trust within a core AS. The public root keys are embedded in the TRC. The update and revocation of root keys occurs via signing a new TRC with a quorum of root keys. The control-plane PKI also provides authentication for path revocation messages (described in detail in Section 7.3) and SCION Control Message Protocol (SCMP) messages (overview in Section 4.2.5 and described in detail in Section 7.6).

4.2.1 Trust Root Configuration (TRC) Life Cycle

In this section, we describe the life cycle and management of trust root configurations (TRCs), i.e., we illustrate how TRCs are created, updated, and disseminated, and we show how consistency is enforced.

Creating a TRC

Initially, a TRC is created when an ISD is created and joins the SCION network. Details on how a new ISD can join the network are provided in Chapter 5. In short, the process of creating a new TRC is conducted as follows:

1. A new ISD sets trust anchors for the authentication of
    a) control-plane messages (the root keys of core ASes),
    b) names (the key of the name resolution root zone),
    c) end entities (the keys of root CAs and log servers).
2. The ISD specifies all TRC parameters (see Section 16.1 on Page 369). In particular, the ISD sets the value of the `QuorumTRC` parameter, which defines how many of the current core ASes will need to sign the next TRC. The ISD also sets the quorum required to change end-entity trust roots (i.e., `QuorumCAs`). The version number of the first TRC is set to 0.
3. The TRC is first signed by at least a quorum of core ASes in the ISD (with their online root keys), which is represented by the `QuorumTRC` parameter. At this point, the TRC is operational within the local ISD.
4. To be accepted by external entities, the TRC is signed by trust roots of other ISDs. Specifically, at least one core AS, one root CA, and a name

trust root from a remote ISD sign the TRC. As chains of trust follow networking paths every neighbor ISD has to sign the TRC. Non-neighbor ISDs can sign the TRC as well, to create shortcut trust chains.

5. Similarly, to accept external TRCs, the ISD (i.e., at least one core AS, one root CA, and the name trust root) signs the TRCs of its neighboring (and optionally non-neighboring) ISDs after verifying the identities of the ISDs. This and the previous steps can be combined, and may be part of a *cross-signing ceremony*, where administrators physically meet each other, or may be executed through out-of-band verifications.

The initial TRC should be delivered to all ASes and end hosts (by a trusted software vendor or a local ISP, for example) via an authentic channel.

Updating a TRC

A TRC update can be conducted for recovery or operational reasons, such as changes in an ISD core or key updates. Updating a TRC is similar to creating it. The only difference is that the version number of the new TRC is the version number of the current TRC plus one, and the new TRC must be signed by at least a quorum of core ASes as specified in the current TRC.

TRCs are signed with online root keys. However, any change to the section of the TRC related to core ASes (i.e., the list of AS keys and quorum parameters) must be approved with offline keys. Such changes can happen in case of key rollover, loss, or compromise, or in case of addition/removal of a core AS. To do so, special requests called *update tickets* must be sent to a quorum of core ASes who will sign the ticket (if they approve it) with their offline key. Once a sufficient number of offline keys have signed them, the update ticket(s) must be attached to the new version of the TRC, which will itself be signed with online keys. Update tickets enable two properties to be efficiently achieved: (a) *simultaneous updates* (e.g., of several root keys, to decrease the number of TRC updates and thus to keep the overhead low), and (b) *asynchronous updates* (obtaining all the required signatures might take some time; in the meantime, other updates can be applied to the TRC). During the TRC verification procedure, if any ticket is attached to the TRC, one must check that the ticket is signed by a sufficient number of legitimate entities and that the update is compatible with the previous version of the TRC.

Parameters of naming and end-entity PKIs are governed by a naming trust root and CAs, respectively. To update the parameters, again update tickets are used. If CAs want to change their section in a TRC (e.g., by adding/removing a CA or log server), an update ticket is sent to root CAs, and if `QuorumCAs` many of them sign it, the ticket is passed to the core ASes that check the quorum and sign the new TRC (with the proposed change). In the case of a change in the name resolution section of the TRC, a ticket is created and signed by the name trust root (as there is a single name trust root key, no quorum is required).

If the new TRC is signed according to the quorum parameters of the previous TRC, the new TRC is valid. However, to allow verification of remote messages and entities, the new TRC has to be cross-signed with neighbor TRCs. Trust roots (ASes, CAs, and name roots) use their online keys for cross-signing, and the process of cross-signing updated TRCs can be automated in most cases. In particular, provided trust anchors are unmodified or added, the new TRC can be submitted to neighboring ISDs, which verify whether the TRC is signed in accordance with the value of the current `QuorumTRC` parameter of the current TRC. Note that in this case, trust roots of the new TRC do not have to sign neighbors' TRCs (as their keys remain unchanged). For instance, when an ISD A updates its TRC from TRC_i^A to TRC_{i+1}^A by adding a new AS, all neighbor ISDs of A have to sign TRC_{i+1}^A, but A does not need to sign their TRCs as the previous signatures are still valid (no trust root of A was removed with the TRC update).

However, TRC updates enable an ISD to remove trust anchors that were involved in signing neighbor TRC(s). In such a case, the TRC update must be combined with re-signing all TRCs that were signed by the removed anchor(s). In both cases, the new TRC has to be signed by trust anchors of neighbor ISDs. However, to automate that process, ASes, CAs, and name roots can implement a default policy to sign a remote new TRC if the parameters they are interested in are unchanged (e.g., for instance a CA can immediately sign the TRC if its CAs remain unchanged). In extraordinary cases, remote ISDs can refuse the cross-signing request and negotiate the TRC update process out of band.

After the new TRC is created and cross-signed, it is first loaded onto the core certificate servers, which propagate the TRC among the beacon servers. New TRCs are then disseminated via the beaconing process. The TRC update can remove trust anchors used in routing, name, and end-entity validation; thus it can create potential availability issues within the ISD. To maintain availability, an old TRC can be used for a grace period as specified by the `GracePeriod` parameter. Certificates issued within this period by a removed trust anchor are still valid, but should be re-issued (see details on Page 74). Grace periods also limit the time between TRC updates. Two consecutive TRC updates cannot be conducted until the previous TRC grace period expires. For example, TRC_{i+2} cannot update TRC_{i+1} while TRC_i can be used. This rule restricts the number of TRCs that can simultaneously be used in the validation to two.

Quorum Size for TRC Update

The approach for the TRC update is designed to withstand one malicious core AS. Although it is possible to extend the approach to tolerate multiple compromised core ASes, the system would need more complex operations, such as consensus algorithms, which would impact availability. In the interest of keeping our description simple and short, we defer more complex attacker models to subsequent efforts. We emphasize that for the system we describe here, even in the catastrophic case of a complete compromise of all cryptographic keys, SCION provides insulation of all other ISDs and enables a new TRC to be bootstrapped. As long as only a single core AS is compromised, integrity and consistency of the sequence of TRC updates is not affected. Since we assume that core ASes are going to be a relatively small number of entities maintaining active business relationships with each other, misbehavior against each other is expected to be rare and can be handled through an out-of-band mechanism. In particular, if a core AS misbehaves, another core AS can collect evidence, convince other core ASes of the misbehavior, and exclude the malicious AS from the ISD core. For these reasons, we assume that core ASes extend a certain level of trust toward each other. Despite being competing entities, core ASes benefit from cooperation to keep the network operational. Moreover, the control-plane PKI operations are accountable as all operations are signed, so that misbehavior will lead to a trail of evidence that can be used in remediation. It is reasonable to assume that a certain degree of mutual trust can exist in a small group of entities that are cooperating in an economic environment. However, we stress that SCION does not require strict consensus as required in the current Internet. Due to the introduction of isolation domains with local TRCs, core ASes that disagree on fundamental parts of their TRC (such as root CAs) can form their own ISD.

We now discuss how to pick the quorum size, i.e., the number of entities needed to perform certain operations. Let G represent the number of legitimate "good" core ASes, B the number of malicious "bad" core ASes we want to be able to tolerate, and U the number of unavailable good core ASes we want to be able to tolerate. Then we obtain the following formulas for the quorum size: $QS = G - U$, $QS \geqslant \lceil \frac{G+1}{2} \rceil + B$, and $\lceil \frac{G+1}{2} \rceil > B$. The rationale is as follows. If not all good ASes participate, the quorum size needs to be smaller than the total number of good ASes. In the case of simultaneous quorums [148], to ensure that there is at least one good node that participates in all quorums, we need to have at least $\lceil \frac{G+1}{2} \rceil$ good nodes per quorum. In the worst case, all bad ASes always participate in each quorum, thus the quorum size is at least the number of required good ASes plus the number of bad ASes. The third inequality ensures that, when taking the majority of responders, the bad ASes cannot outnumber the good ones.

Consequently, to tolerate one malicious AS and one unavailable good AS, we would need at least six core ASes with $G = 5$, $B = 1$, $U = 1$, and $QS = 4$, so even if the bad AS refuses to participate, we can ensure that progress can be made. With fewer than six core ASes, we cannot tolerate a malicious core AS and one unavailable good core AS. Assuming that all good core ASes are available, a satisfactory assignment is $G = 3$, $B = 1$, $U = 0$, and $QS = 3$.

Disseminating TRCs

We assume that all entities within an ISD are pre-loaded with a recent TRC. Moreover, on startup, all servers and end hosts obtain all missing TRCs (from the TRC they possess to the most recent TRC) of their own ISD from a local certificate server. However, we impose a restriction on this catching-up operation: it can only start from a TRC that is at most 1 year old. Then, the TRCs are validated (i.e., to verify whether subsequent TRCs respect their update policies), and the most recent TRC is employed for all subsequent operations. There are two main requirements for the dissemination of TRCs: (a) *efficiency*, a new TRC must be rapidly disseminated (to all other parties that need it); and (b) *avoiding circular dependencies*, an unverified path should not be used to fetch a TRC. The first requirement is met by tying TRC dissemination to the beaconing, path registration, and path lookup processes. The second requirement is achieved by applying the following rule: a party that sends a signed object (beacon, path segment, name resolution entry, or end-entity certificate) must have the complete trust information (i.e., TRCs and certificates) required to successfully verify it.

More precisely, TRCs are disseminated via SCION's beaconing process and verified based on the current or the previous TRC. While beaconing, each AS adds to the beacon the version number of the TRC it is currently using. A beacon server receiving a new beacon checks the version number contained in the beacon against its locally stored TRC. If the TRC version number within the received beacon is higher than the locally stored TRC, the beacon server sends a request to the beacon server that sent the beacon to request the new TRC. Similarly, the sender can be asked for the missing TRCs of remote ISDs, as they are required to verify beacons. The sender returns the new TRC, which is then verified by the receiving beacon server. The conditions under which the new TRC is accepted depend on the ISD that updated it. If the local ISD has updated the TRC, then the new TRC is accepted if

1. it contains at least the required quorum number (i.e., `QuorumTRC`) of correct signatures of trust roots defined in the current TRC, and
2. the receiving AS's certificate is valid against the new TRC.

As a beacon server has previous TRCs from its ISD, it is able to verify their consistency. If the last condition is not met, the AS postpones the TRC acceptance until its own certificate is re-issued and valid against the new TRC. In such a case, the AS continues beaconing using its current TRC and certificate, meanwhile contacting its core to obtain a new certificate. Note that other ASes will learn about the new TRC, as the version is in the beacon (at least in the first AS entry, i.e., the AS entry of the core AS that initiated the beaconing).

In the case of a new TRC from a remote ISD, the update is accepted if

1. it contains at least the required quorum number (i.e., `QuorumTRC`) of correct signatures of trust roots defined in the previous TRC; and

2. there exists a chain of trust from (a) the local TRC to the new TRC, and (b) from the new TRC to the local TRC.

Note that TRC cross-signing follows the physical network connections, thus the latter condition is satisfied when all subsequent TRCs in the core beacon are cross-signed.

After the new TRC is accepted, it is submitted by the beacon server to a local certificate server. Finally, the TRC is re-distributed internally. To this end the beacon server replicates the TRC among all beacon servers within its AS, and similarly the certificate server replicates the TRC among all local certificate servers.

Path servers discover new TRCs via path-segment registration messages (see Sections 7.1 and 4.2.3). This occurs when the beacon server registers path segments (authenticated with the new TRC) with its local path server and a core path server. The path servers check the version of the TRC with which the segments were authenticated. If a new TRC is detected, the path servers query the beacon server for this TRC. After the new TRC is returned, the TRC and path segments are validated accordingly, and after a successful validation the TRC and path segments are saved.

Similarly, end hosts learn about new TRCs through the path lookup process (see Sections 7.2 and 4.2.3). End hosts, at the end of the process, obtain a set of path segments that can be combined into a forwarding path to the destination. For a path segment authenticated with an unknown TRC, this TRC is requested from the local path server, which has returned the path segments. The path server is obliged to possess all TRCs needed to verify every path in the set.

An example of TRC dissemination is presented below in Section 4.2.4.

TRC Update Frequency

Revocation of root keys happens through TRC updates. The update frequency of the TRC should be very low, for the following reasons:

- Each update requires a TRC dissemination to all entities that intend to communicate with a host inside that ISD.
- If a host was offline for a while and has an old TRC, it needs to fetch all intermediate TRCs to verify the current one.
- By making the TRC update a rare operation, we can make minimal use of the TRC signing keys and shield them from regular operations, which enhances security.

We thus aim for a TRC update that occurs at most on a weekly basis, ideally on a monthly or longer frequency. However, TRCs have to be updated before they expire (expiration timestamps are set by ISDs themselves).

TRC Lifetime

TRC updates can create inconsistencies in chains of trust. For instance,

- a beacon server receives a beacon where ASes use different TRC versions for authentication,
- an end host fetches up-segments and down-segments authenticated using TRCs with different version numbers,
- an end host possessing a new TRC obtains a name resolution response authenticated with an old TRC.

To achieve high availability, such inconsistencies have to be handled without breaking validation. To this end, an ISD can define a *grace period* (during which a previous TRC can still be used). This period is defined in the TRC with the `GracePeriod` field. Then, TRC_i is valid effectively until

$$TRC_{i+1}.\texttt{CreationTime} + TRC_i.\texttt{GracePeriod}\,. \tag{4.1}$$

4.2.2 AS Keys and Certificates

To achieve a high level of security for the control-plane PKI operation, we propose that core ASes have online and offline asymmetric key pairs. In this design, *offline* keys are used for infrequent safety-critical operations that will require administrator involvement to cross an air gap, and *online* keys are used for frequent automated operations that do not require administrator involvement.

The renewal of AS certificates is an example of a fully automated operation that occurs every few days and only requires online keys. For the addition or removal of a core AS, offline keys are required to ensure human operator involvement and that even a complete compromise of all online keys does not permit sensitive changes to the TRC. So even for an event such as the Heartbleed vulnerability [75], which may simultaneously compromise all online keys, the basic infrastructure (i.e., the control-plane PKI) remains safe and all online keys can efficiently be updated to fresh keys after the vulnerability has been patched.

The revocation of AS certificates is challenging. As AS certificates are involved in the availability-critical operation of beacon dissemination and beacon validation, checking the revocation status of an AS certificate would result in a deadlock. Consider a revocation service operated by the ISD core, which keeps a list of revoked AS certificates. When verifying an upstream AS's signature contained in a beacon, a downstream AS would need to contact the revocation service to ensure that the certificate is still valid. In the case of a network reboot, however, an AS has no paths yet to the core and thus cannot validate the first beacon message it receives. If one used an unverified path to reach the revocation service, then the system would not operate safely in the initial phase. We propose a more elegant approach: *short-lived certificates*, which do not need any revocation if we can tolerate a short period of continued validity when a certificate is compromised [212].

Table 4.1 gives an overview of the different keys and certificates used in the control-plane PKI. The TRC contains the offline and online keys of all core ASes and is signed with a quorum of root keys (online or offline, depending on the context); as such, it can be considered to be a self-signed root certificate, except that multiple parties are involved. Online and offline root keys are included in TRCs while other keys are authenticated via certificates. All ASes (including the core ASes) use short-lived AS certificates to carry out their regular operations (such as signing beacons and path segments). Core ASes hold an additional certificate whose only purpose is to authenticate (other ASes' and their own) AS certificates.

Name	**Notation**	**Auth.**[1]	**Validity**[2]	**Usage**
Offline root key	$K_{i,offline}$	TRC	5 years	Critical TRC update: - Addition/removal of core ASes - AS quorum parameter change - Update of root keys
Online root key	$K_{i,online}$	TRC	1 year	Signing core AS certificates TRC creation, non-critical update Cross-signing
Core AS key	$K_{i,core}$	$C_{i,core}$	6 months	Signing AS certificates
AS key	K_i	C_i	3 months	Beacon authentication Path-segment authentication DRKey (used within SCMP)

(a) Private Keys

Name	**Notation**	**Signed by**	**Associated Key**	**Validity**
Core AS certificate	$C_{i,core}$	$K_{i,online}$	$K_{i,core}$	1 week
AS certificate	C_i	$K_{i,core}$	K_i	3 days

(b) Certificates

Table 4.1: Summary of keys and certificates used in the control-plane PKI.

The TRC and certificate lifetimes are selected in such a way that frequently used online keys that are more exposed to potential compromise are rolled more frequently than infrequently used offline keys. The root key pairs of an AS must be updated regularly through TRC updates. To ensure that key update is a

[1]Location of the corresponding (authenticated) public key.
[2]Recommended usage period before key update (best practice).

periodically practiced function (for good security hygiene), all keys are updated at least every few years. Also, we suggest a validity period of three days for regular AS certificates and one week for core AS certificates. (Please note that the key validity period indicates the key lifetime, whereas the certificate validity period indicates how often the certificate needs to be re-signed.)

When an AS joins an ISD, it obtains from the ISD core:

1. a unique and permanent AS identifier (within the ISD),
2. a certificate for the AS's public key.

The AS certificate is issued by a core AS, which is obliged to check the identity of the AS. AS certificates are short-lived, by default valid for 3 days, although each ISD can set different policies on the certificate lifetime. This design decision is motivated by simplicity and availability requirements, as short-lived certificates can be validated without any additional information (such as certificate revocation status information) and no revocation system is required [212].

On the other hand, a consequence of short-lived certificates is a need for frequent re-issuing of certificates before they expire. This process is automated in SCION. An AS that needs to re-issue its certificate contacts the core AS that issued it before, with a re-issuance request. The re-issuance request proves that the requesting AS still possesses the private key that corresponds to the public key included in the current certificate. The core AS verifies the request, copies the fields of the current certificate, sets a new lifetime for the new certificate, and signs it.

Only the core AS that issued a certificate can re-issue it. With such an approach, every core AS keeps a one-to-one mapping between ASes and their current public keys (that were certified by this core AS). The re-issuance requests are introduced to prove that this mapping is still valid. However, in the case of key loss or key compromise, an affected AS has to contact the corresponding core AS to change its mapping as the old public key should no longer be used. The AS generates a new key pair and contacts the core AS (out of band). The core AS checks the identity of the AS, and subsequently issues a new certificate (with the new public key), and changes the mapping so that old certificates cannot be re-issued anymore. We emphasize that short-lived certificates are irrevocable during their lifetime; thus if a private key is compromised, the adversary can use the corresponding certificate until it expires.

Although core ASes have core AS certificates, these are only used for issuing other certificates. To separate certificate and key usage, control-plane operations (such as beacon and path-segment signing) are performed by core ASes with AS keys and corresponding AS certificates. To this end, a core AS periodically creates its own AS certificate and signs it with its core AS key. The format of AS certificates is presented in Section 16.2 on Page 370.

A TRC update may invalidate an AS certificate, e.g., when a certificate-issuing core AS has been removed from the TRC. In such a case, the affected AS can still use its certificate in combination with an old TRC. However, this is possible only during a grace period as described above, and thereafter, the AS has to obtain a new certificate from one of the current core ASes. In case of large-scale compromise, the core AS can revoke its online root key that was used to sign the core AS key that signed the AS key. Revocation of the online root key through a TRC update would thus invalidate all the underlying AS keys.

4.2.3 Control-Plane Authentication

PCBs and path segments are authenticated with AS certificates. We now describe these operations in more detail.

Beacon Authentication

The authentication of PCBs is especially important as PCBs are used in building path segments. The details of PCB creation and the PCB format are described in Section 7.1.1 on Page 120 and Section 15.3 on Page 356.

The beaconing process is initiated by a core beacon server, which creates a PCB, appends its AS entry, and signs the PCB with a private key that corresponds to an AS certificate that can be validated based on the current TRC. Certificate and TRC versions are included within the AS entry; thus they are signed as well. Then, the PCB is sent to a beacon server of a neighboring AS. The receiving beacon server first checks whether it has the TRC and certificate with the versions announced in the PCB. If the TRC is new, then the beacon server updates it. If the certificate version is unknown to the receiving beacon server, the sending beacon server is queried, and the correct certificate is returned. (New certificates are replicated across the AS's certificate servers.) At this point, the receiving beacon server has all the information necessary to verify the PCB. It verifies the certificate based on the TRC, and finally verifies the PCB's signature based on the certificate.

The receiving beacon server continues the beaconing process by appending its own AS entry. As before, the beacon server states versions of the used TRC and certificate. Finally, the beacon is signed and sent to neighboring ASes. The next receiving beacon server checks the versions of all TRCs and certificates involved in the beacon authentication. If any TRC or certificate is missing, the sender is queried. With such a beaconing design, the relevant TRCs and certificates are disseminated step by step with the beacon dissemination. The beacon server continues the validation (and beaconing) after all necessary TRCs and certificates are provided. Namely, it verifies the signature of each AS entry individually. The beacon continues to propagate until it reaches a **leaf AS***.

Path-Segment Authentication

Beacon servers turn PCBs into path segments in every registration period (see details in Section 7.1 on Page 119). Path segments have the same format as PCBs, except they may be without some optional metadata. Each AS entry within a path segment is signed, and it contains information about the TRC and certificate used to protect the entry.

When a beacon server registers path segment(s) with a path server (local and/or core), the path server(s) can query the beacon server for the TRC(s) or the certificate(s) (if any are missing). Path servers accept path segments if their authenticity is verified, i.e., if

1. the required TRC(s) and certificate(s) are provided,
2. if there is a new TRC, it is consistent with the previous one,
3. the certificate(s) are valid with respect to the corresponding TRCs,
4. the path segment's signatures are valid with respect to the certificates.

Note that through such a registration process, path servers learn new TRCs and certificates.

The way in which end hosts verify path segments is similar to the way path servers do so. An end host sends a path request to its local path server, which then resolves the request recursively (see details in Section 7.2 on Page 132). For the path segments obtained from the local path server, the end host conducts the verification. Hence, it first asks for all missing TRCs and certificates used in the authentication of any path segment. Then, it verifies each path segment as described above. Similarly, in this way, the end host learns new TRCs, which finally replace the old ones.

Besides path registration, SCION allows for removal of path segments from the path servers and end hosts that cache them. The path revocation mechanism is described in Section 7.3 on Page 138.

4.2.4 Authentication Examples

Intra-ISD Beaconing

An example of intra-ISD beacon authentication is presented in Figure 4.4. Core ASes update the TRC from TRC_{i-1} to TRC_i, after AS F has been removed from the core. The figure shows how this update is disseminated across the ISD using a single beacon.

First, core AS D initiates the beaconing process, by creating a fresh PCB with D's entry. The entry includes the version number of the new TRC (i.e., i) and the version number of D's certificate. The entry is signed with D's AS private key, and the corresponding public key is included within D's certificate C_D, which in turn can be validated based on TRC_i. When AS C receives the PCB, it learns that there is a new TRC. C asks D for TRC_i (and for D's certificate if

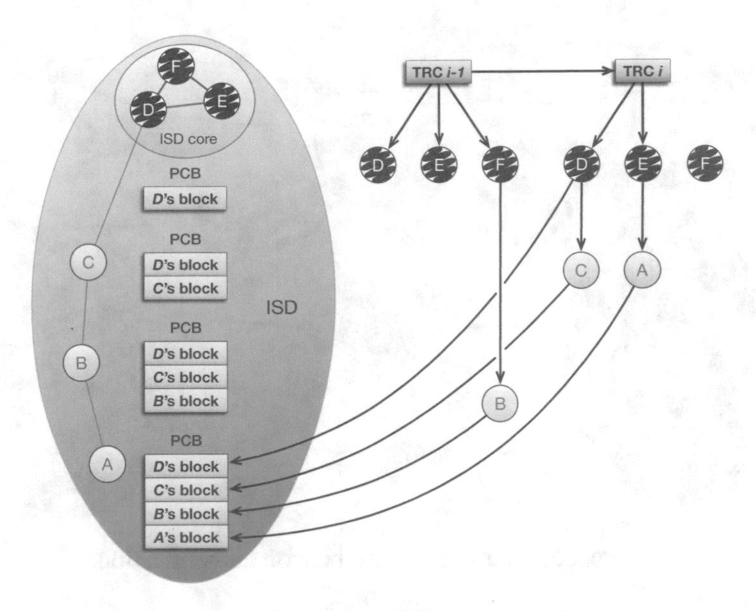

Figure 4.4: Example of intra-ISD beacon authentication.

it is missing), and then verifies whether the TRC update was correct. Further, *C* verifies the signature over the PCB using *D*'s public key contained in *D*'s certificate, and validates this certificate with the new TRC. The PCB is accepted, and the new TRC is propagated among *C*'s PCB and certificate servers. Finally, AS *C* appends its own entry to the PCB, and signs it. *C*'s certificate is issued by *D*, which is in the new TRC, so *C* uses *i* as the TRC version number for its entry.

Then, the PCB is sent to AS *B*, which verifies the new TRC (obtained from *C*), and the first entry in the same way as *C*. Then *B* verifies the next (i.e., *C*'s) entry, which is signed with *C*'s key, and which can be successfully validated based on the new *i*-th TRC. Finally, *B* creates its own entry; however this entry cannot be marked with TRC_i's version number because *B*'s certificate is signed by AS *F*, which was removed from the core by the TRC update. In such a case, *B* is forced to temporarily use TRC_{i-1} (this is possible during a grace period — see Page 74) for the propagation, and contacts an active core AS to obtain a new certificate. Although AS *B* cannot sign objects (e.g., PCBs or path segments) with the new TRC_i, it replicates this TRC among its PCB and certificate infrastructure.

Finally, AS *A* receives the PCB, and although the latest AS entry is signed based on TRC_{i-1}, *A* learns about the new TRC from the previous entries. AS *A* verifies and updates the TRC similarly to how the previous ASes did.

Path servers learn the new TRCs from the path registrations and lookups, which are conducted on demand. For instance, if *A*'s beacon server registers a

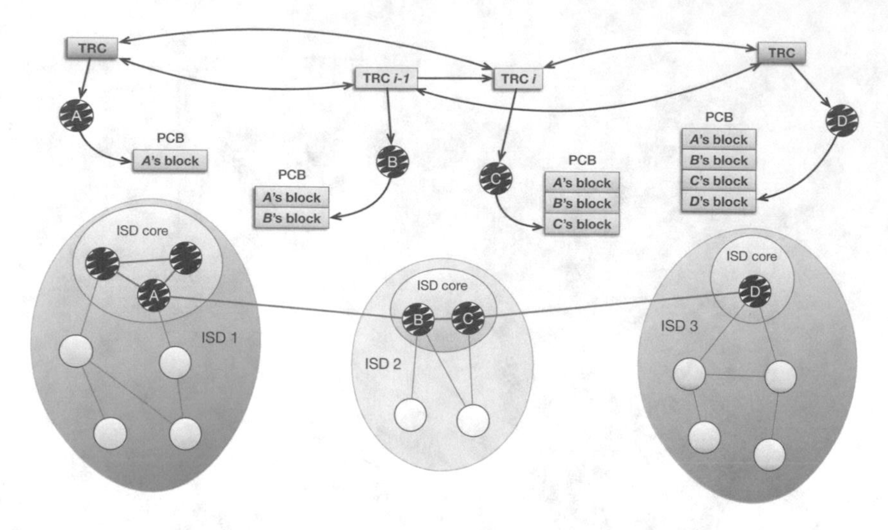

Figure 4.5: Example of core beacon authentication.

path segment that was built upon the new TRC, a receiving path server asks the beacon server to send this new TRC. However, after path servers learn about new TRCs, they still accept (during a grace period) registrations signed on the basis of an old TRC.

Inter-ISD Beaconing

An example of inter-ISD beacon authentication is presented in Figure 4.5, where a core PCB is disseminated from AS *A* towards AS *D*. In our scenario an intermediate ISD contains two core ASes (i.e., *B* and *C*) that have TRCs with different versions (such a situation can happen for example due to dissemination delays within an ISD core).

First, a beacon server in AS *A* initiates beaconing by creating the first AS entry. The entry is signed with *A*'s AS private key, and the corresponding public key is included within *A*'s AS certificate, which in turn can be validated with the TRC of ISD 1 (denoted TRC^1). *A* sends the PCB to AS *B*, which is in another ISD. AS *B* uses TRC^2_{i-1} and verifies the PCB. If TRC^1 or *A*'s certificate is missing, *B*'s beacon server asks *A* to provide that. AS *B* verifies *A*'s entry based on TRC^1, and verifies whether TRC^1 is cross-signed by TRC^2_{i-1}, which is currently used by AS *B*. Then, *B* creates and signs a PCB including its entry and disseminates the PCB to AS *C*. AS *C* verifies the PCB, validating *A*'s and *B*'s entries. As *B* used an old TRC, *C* checks whether the grace period is not violated. *C* also verifies whether TRC^1 is cross-signed with TRC^2_i, which is currently used by AS *C*. Next, *C* appends its entry and signs the PCB with its own AS key, which can be verified based on TRC^2_i. Then, the PCB is forwarded to ISD 3. AS *D* receives the PCB and obtains all missing (if any) TRCs and

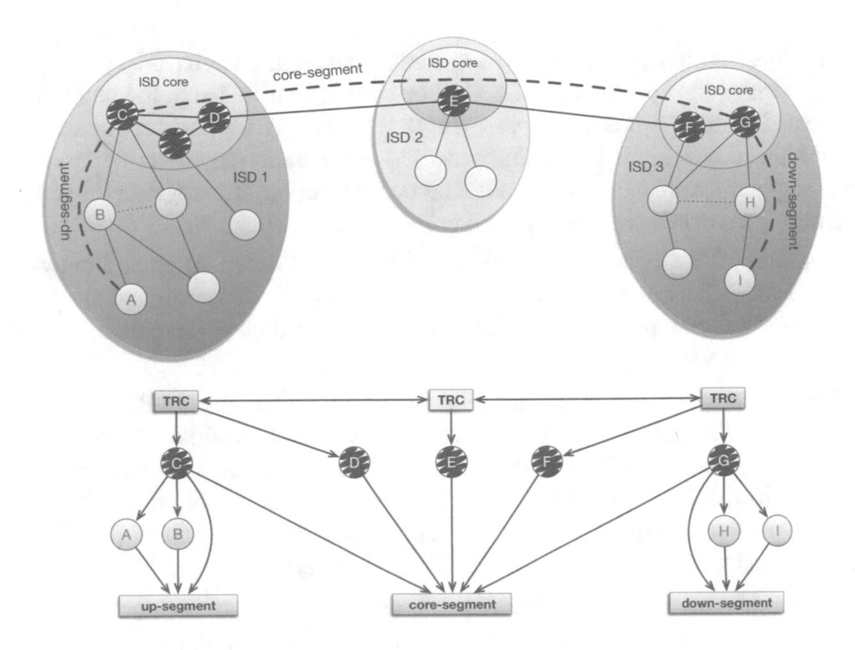

Figure 4.6: Example of path-segment authentication.

certificates from the sender (i.e., AS *C*). As ISD 1 is not a neighbor of ISD 3, *D* must verify that there is a chain of trust from TRC^3 to TRC^1. To this end, *D* verifies cross-signatures between TRC^3 and TRC^2_i, TRC^3 and TRC^2_{i-1}, TRC^2_{i-1} and TRC^1. Additionally, *D* verifies that the update from TRC^2_{i-1} to TRC^2_i was consistent.

Path-Segment Authentication

An example scenario of path-segment authentication is presented in Figure 4.6. The topology consists of three ISDs, where an end host from AS *A* wishes to connect to an end host from AS *I*.

The first step is the beaconing process. In our example, the following path segments are created:

- an up-segment between source AS *A* and core AS *C*,
- a core-segment between core AS *C* and core AS *G*, and
- a down-segment between core AS *G* and destination AS *I*.

AS *C* starts beaconing by sending a new PCB, which eventually is received by *A*'s beacon server (an example of this process is presented on Page 78). AS *A* receives and verifies the PCB, and converts the PCB into an up-segment. To this end, the AS creates its own AS-level information, appends it, and finally signs it (using the current TRC and *A*'s certificate). This path segment is registered with a local path server as an up-segment. During the registration, the path server verifies the up-segment. If a TRC or certificate is missing, the path server requests it from the beacon server. (Note that new TRCs and certificates are disseminated during the beaconing process.) After a successful verification, the path segment from *A* to *C* is saved. The beacon server also registers this path segment (as a down-segment) with a core path server (e.g., in AS *C*).

Similarly, AS *I* creates and registers the up- and down-segments between ASes *G* and *I* (the up-segment is registered at the local path server and the down-segment at the core path server). The TRC(s) and certificate(s) are disseminated (if needed) accordingly.

The core-segment is created in a slightly different way. A core PCB, disseminated from *G* to *C*, is validated by *C*'s beacon server. To validate it, the beacon server needs to have TRCs of ISD 2 and ISD 3. To create a core-segment, *C* appends to the PCB its own AS-level information, and then signs the PCB. Finally, the core-segment is registered with a local path server (i.e., path server of *C*), which will request any missing TRC(s) or certificate(s).

After paths are created and registered, end hosts can successfully conduct path lookups. In our example, an end host from AS *A* asks its local path server for a path to AS *I*. The path server does not have such a path segment cached, so a core path server in AS *C* is contacted. The core path server likewise does not have the path segment, hence the core of *I*'s ISD is queried. The core path server of AS *G* returns the *G–I* down-segment (which was registered by *I*'s beacon server) to the core path server in *C*. *C*'s path server can query *I*'s path server for TRCs and certificates used to authenticate the *G–I* down-segment (e.g., TRC^3 or *I*'s certificate can be queried if *C*'s path server does not already have it). The down-segment is then verified, and returned to *A*'s path server. Along with this down-segment, a core-segment (*C–G*) is added. *A*'s path server tries to validate the returned path segments. To this end, the TRCs TRC^2 and TRC^3, and all certificates involved in path segment authentication are needed. More specifically, TRC^2 is required to establish a chain of trust between TRC^1 and TRC^3 (note that TRC^2 cross-signs these TRCs). At the end of the lookup process, *A*'s path server returns the core-segment and down-segment to the end host, accompanied with the *A–C* up-segment. Finally, the end host obtains the path segments (and asks for missing TRCs or certificates), and verifies all paths and trust chains between TRCs.

4.2.5 SCION Control Message Protocol (SCMP)

Security of a control message protocol is essential for the security of higher-level protocols. For instance, the Internet Control Message Protocol (ICMP) does not provide any form of authentication. Consequently, the Internet transport protocols (such as TCP) suffer from attacks caused by maliciously generated ICMP packets [99]. However, providing security for control protocols is particularly challenging, as control packets are often created and processed by routers; thus the authentication and verification process has to be highly efficient.

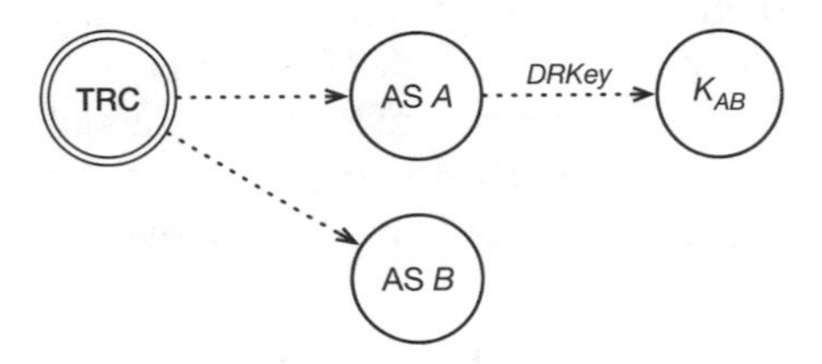

Figure 4.7: Verification of an SCMP message authenticated via MAC.

SCION provides a framework for authenticating SCMP packets, and two distinct authentication schemes are available. The first scheme is symmetric and based on the DRKey protocol (see Section 12.5 on Page 291). In this scheme, SCMP packets are protected by their sources and verified by their destinations. A symmetric key used to authenticate SCMP packets is derived from a secret symmetric key local to the AS, and is exchanged securely using AS certificates, as illustrated in Figure 4.7. In the second scheme, SCMP packets are signed in batches by border routers, and receiving end hosts verify signatures by using the corresponding AS certificates. More details on the authentication of SCMP are provided in Section 7.6.3 on Page 156.

4.3 Name Authentication

The goal of the name authentication infrastructure is to authenticate bindings between SCION addresses and human-readable names. To this end, name resolution responses have to be authenticated — for which SCION relies on a DNSSEC-like infrastructure. However, to achieve a higher level of security than DNSSEC, the validity of name resolution keys is additionally asserted by end-entity certificates.

4.3.1 Name-Resolution Key Infrastructure

SCION defines its own name resolution protocol, called RAINS, which is described in Chapter 6. RAINS provides an authentication infrastructure that is similar to DNSSEC [13], with one main difference that is important for authentication purposes. Assertions about names in RAINS are explicitly bound to an *assertion context* (see Section 6.3.3 on Page 109), which defines the chain of signatures used to verify the validity of a given assertion. Each ISD has its own native isolation context, at which signature chains for names looked up from within that ISD are rooted. This is analogous to the current DNSSEC tree with one root per ISD (instead of a single global root).

The root keys for each ISD's *native isolation context* are included and distributed with TRCs, as opposed to software updates in DNSSEC. Through the TRC cross-signing framework (see Section 4.2.1) clients can resolve names in other (than native) isolation contexts. RAINS assertions are natively signed (replacing DNSSEC's `RRSIG` resource record type), and delegation occurs via a digital signature, as opposed to via name-server redirection. For the sake of simplicity, we leave further details aside for now and refer the reader to Chapter 6.

Within a native isolation context, the chain of trust is tied to the domain namespace hierarchy. For instance, a domain '`ethz.ch.`', which wishes to sign its name resolution records, first has to obtain a signature for its key from its parent in the namespace hierarchy, i.e., from the '`ch.`' domain. Similarly, `ch.` has to have its key signed with the root key (i.e., by the root domain '`.`'). Figure 6.3 on Page 110 shows such chains for various assertion contexts. The name resolution root key is contained in the TRC.

Key updates in RAINS are similar to those in DNSSEC, i.e., every change of a domain's name resolution key affects the superordinate and subordinate domains according to the namespace hierarchy of the domain. Namely, the domain's parent has to sign the domain's new key, and then the domain has to re-sign the keys of each child. RAINS allows multiple keys to be valid for delegation to a zone at once; operational practices for overlapping validity can reduce the potential for disruption of verification during a key rollover.

To improve security of the standard DNSSEC trust model, we supplement RAINS authentication with our highly secure end-entity PKI. More specifically, domain public keys are additionally asserted by a special end-entity certificate. We call this special certificate a *subject certificate policy* (SCP); its details are presented below in Section 4.4.3. SCPs are certificates issued by multiple trusted parties (namely, certification authorities — CAs) and used by domains to govern their public-key certificates and secure connections. There is a unique active SCP per domain, and SCPs are published as RAINS assertions.[1] The domain asserts its own name resolution key by signing this key with its SCP's private key and publishing this signature within its name resolution zone. The public key of this certificate is also published as a RAINS assertion for cross-verification.

4.3.2 Validation of Name Resolution Entries

As part of a successful name lookup, the obtained assertions are validated as follows: (a) The signature chain is validated according to the assertion context for the entry; by default, for globally significant names, this chain is rooted at the current TRC (i.e., the TRC trusted by the user). (b) Once the assertion is

[1] SCPs are accompanied by proofs from the end-entity PKI that they are logged and fresh. If a domain does not have an SCP, the domain publishes an assertion with an absence proof.

validated, the last step is to verify the domain's SCP and the additional signature over the domain's name resolution key. (c) The client verifies whether the SCP belongs to the correct domain and whether it is correct (i.e., signed by the required number of trusted CAs and asserted by a trusted log — more on SCP validation can be found in Section 4.4.3). (d) The SCP is validated according to the TRC trusted by the user. (e) Finally, the client verifies whether the domain's name resolution key is signed with the SCP's key. This is the last check which ensures that name resolution responses are authentic. As the SCP is used in TLS connections, it can be stored by the client, so the client does not have to fetch it again over the potential TLS connection with the domain. In the case when the domain does not have an SCP, the domain asserts in RAINS a proof that claims that there is no SCP registered for this domain.

Note that, as in the previous cases, inconsistencies may occur while TRCs are updated. For instance, a client can have a new TRC while some name resolution entries (signed with an old one) are cached locally. However, the inconsistencies influence the entry validation only when the new TRC changes the root name resolution key. In such a case, the validation can be conducted using an old TRC as long as it is compliant with the specified grace period (see Equation 4.1 on Page 74).

Example. Consider an address lookup for the following name:

'`simplon.inf.ethz.ch`'

Assume the name is iteratively resolved, within the native isolation context, by a query server without any existing state. The query server performs the following steps:

1. Retrieve valid public keys for the naming root in the native isolation context from the TRC, and cache them for the limit of their validity.
2. Issue a query for the delegation key for the name '`ch.`' in the native isolation context using an intermediary server storing root assertions discovered through service anycast (see Section 7.5 on Page 153).
3. Verify the signatures on the resulting assertions against the root public keys in the TRC, and cache the delegation keys for '`ch.`' for the limit of their validity.
4. Issue a query for the delegation for the name '`ethz.ch.`' in the native isolation context.
5. Verify the signatures on the resulting assertions against the stored keys for the '`ch.`' zone, and cache the delegation keys for '`ethz.ch.`' for the limit of their validity.
6. Issue a query for the delegation for the name '`inf.ethz.ch.`' in the native isolation context.

7. Verify the signatures on the resulting assertions against the stored keys for the '`ethz.ch.`' zone, and cache the delegation keys for '`inf.ethz.ch.`' for the limit of their validity.
8. Issue a query for addresses for '`simplon.inf.ethz.ch.`' in the native isolation context.
9. Verify the signatures on the resulting assertions against the stored keys for '`inf.ethz.ch.`'
10. Additionally verify the signatures on the resulting assertions against the SCP key for '`inf.ethz.ch.`', if available.

Note that while the iterative verification of the delegation chain requires each of the public keys to be available and the signatures thereon verified, all of the queries can be issued in parallel.

4.3.3 Name Consistency

Each ISD maintains its own root zone for name resolution. This root zone contains delegations to the authority for each TLD. Each TLD authority then serves and authenticates assertions about second-level names, and so on. In the typical case, for a given TLD, each ISD will delegate to the same authority, and the chain of signatures will be identical beyond the ISD root signature of the TLD authority. There are two important deviations from the typical case:

- **Isolated TLD:** An ISD may delegate authority for a TLD otherwise not present in the root zone, creating a TLD that is available only in that ISD. Care must be taken with isolated TLDs, since they may lead to conflicts between ISDs, which can only be resolved through nontechnical processes (see Section 3.5 on Page 51).
- **Isolated subordinate TLD authority:** An ISD may delegate authority for a TLD to a registry other than the globally recognized registry for that TLD. This isolated subordinate authority tracks changes to the globally recognized registry, providing additional vetting of assertions about second-level domains and potentially declining to include those that are used for network abuse (e.g., malware domains).

The mechanism for ensuring consistency in a multiple-root environment using RAINS is described in detail in Section 6.5 on Page 116.

4.4 End-Entity Authentication

The TLS protocol is used globally to secure online communications (HTTP and SMTP communications, in particular). TLS enables the creation of end-to-end encrypted and authenticated channels. To authenticate entities (usually identified by domain names), TLS relies on certificates, which can be obtained from hundreds of geographically and administratively distinct CAs. In the traditional TLS PKI, a single CA can issue a certificate for any domain and bogus certificates can go unnoticed for long time periods due to a lack of transparency. Given that the security of the majority of web-based financial and commercial transactions relies on TLS, one would hope that its security is commensurate with its widespread acceptance and use. Unfortunately, although CAs wield significant power in the TLS ecosystem, their trustworthiness has recently been tarnished by several events. Operational mistakes, social-engineering attacks, and governmental compulsion have resulted in the issuance of fraudulent certificates for many high-profile sites [166, 167, 225]. In these cases, adversaries can impersonate domains to clients by performing active man-in-the-middle attacks, intercepting secure connections and stealing potentially sensitive information. Software vendors also hold significant power in the TLS ecosystem, since they manage the list of CA certificates that serve as the roots of trust.

For the SCION end-entity authentication PKI, we leverage a combination of ARPKI [23, 24] and PoliCert [235], which address the above issues and provide provable security guarantees. ARPKI provides the basis for a transparent and resilient infrastructure, while PoliCert allows domains to specify policies governing the use of their certificates to achieve additional security objectives and address the shortcomings of previous systems. The system we present is deployable (in an incremental manner) both inside and outside of SCION. In the inside case, however, SCION allows roots of trust to be defined at the ISD level, through TRCs, instead of being distributed independently by different software vendors. SCION allows the scope of these trust roots to be limited to ISDs, unlike current systems that rely on global trust.

4.4.1 Background

For the issuance of illegitimate certificates to be detected, the operations of CAs need to be transparent. Google's *Certificate Transparency* (CT) framework [149] proposed the use of append-only public logs to provide CA accountability. The goal is to make all issued certificates visible in order to alert domain owners and clients of any possible misbehavior. CT creates a system of *public logs*, which maintain a database of observed certificates issued by trusted CAs. The log can then provide evidence that it contains a given certificate, and the proof can be checked by clients during the TLS handshake. Additionally, the log is publicly auditable so that any party can fetch proofs of presence or consistency. However, CT has several drawbacks. Specifically, CT's main goal is to detect suspicious behavior, and thus it does not actively protect clients from ongoing attacks if an adversary successfully registers a bogus certificate at a public log [167]. Nevertheless, public logs similar to those used by CT can be employed to build systems such as ARPKI and PoliCert to provide strong security guarantees.

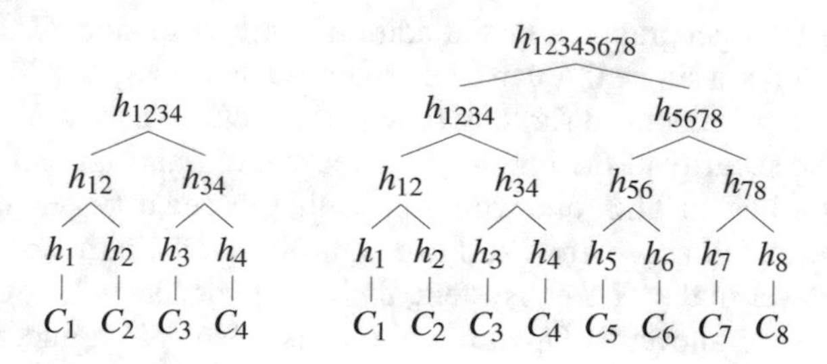

Figure 4.8: Example of appending entries in chronological order to a Merkle hash tree. The tree on the left-hand side represents the initial state of the log, and the tree on the right-hand side represents the state of the log after addition of certificates C_5–C_8.

Merkle hash trees* are generally used to implement public logs. Typically, hash trees are binary trees in which leaves contain certificate-related data, and non-leaf nodes contain the hashes of their two child nodes [174]. This structure can be leveraged to efficiently prove that a leaf is part of the tree. Only one node per level is needed in a *proof of presence*; hence, the proof size is proportional to the tree height, which is $\mathcal{O}(\log_2(n))$ for n leaves. If leaf nodes are ordered (e.g., lexicographically), the tree can additionally provide *proofs of absence*. If nodes are appended chronologically, then the tree can also provide *proofs of consistency* showing that the tree is indeed maintained in an append-only manner. Proofs are based on tree roots, which are lightweight cryptographic representations of the log at a certain point in time.

Example. Two Merkle hash trees are shown in Figure 4.8. It is possible, for example, to prove to someone who holds h_{1234} and $h_{12345678}$ that the tree on the left-hand side was extended (without removing any existing entry) to obtain the tree on the right-hand side, by providing h_{5678}.

4.4.2 Problem Definition

In this setting, the adversary's goal is to obtain a valid certificate and the corresponding private key for a domain that is not owned by the adversary (e.g., in order to obtain secret information through a man-in-the-middle attack). To this end, the adversary can either directly compromise the server in question to obtain its private key (in which case the PKI does not play any role in the attack; it is the server administrator's responsibility to protect the private key) or the adversary can try to produce a new valid certificate for that domain by compromising a number of trusted entities. However, for a PKI to satisfy any nontrivial security property, we assume that the adversary cannot compromise all entities. We also assume that the network is not trusted and, therefore, that the adversary can eavesdrop, modify, and insert messages at will. The main properties we seek are the following:

- **Compromise Resilience:** Unless more than a threshold number of trusted entities are compromised, it should be impossible for an adversary to impersonate a domain by forging a certificate or policy that would be accepted by clients.
- **Balanced Control:** All parties should be able to contribute towards determining whether or not a domain's certificate is valid, whether through signing information or specifying parameters for connection establishment.

4.4.3 ARPKI and PoliCert

Although ARPKI and PoliCert were initially developed as independent systems, they are compatible and are combined into the authentication infrastructure that we present in this section. We give a high-level description of these two systems and refer the reader to the academic papers for more details [23, 24, 235]. We start by listing the relevant entities in our infrastructure:

- **Clients** (usually browsers) can initiate TLS connections with servers in any domain. Depending on the authentication data provided by the server, a client can either accept the connection or display a warning/error message.
- **Domains** are identified by names unique within a given isolation context, and their servers respond to TLS requests from the clients. Domains authenticate themselves to their clients by presenting their certificate(s) and by using the corresponding private key(s).
- **Certification authorities (CAs)** are trusted entities responsible for issuing certificates for public keys that are associated with a domain. CAs must verify that the certificate requester is the legitimate owner of the domain in question. Clients must have access to a list of trusted *root CAs*, while *intermediate CAs* are certified by other CAs. CAs are also responsible for monitoring logs to detect their misbehavior.
- **Log servers** maintain a tree-based record of valid certificates and policies. Logs are able to prove that they behave in a consistent manner.

Subject certificate policies (SCPs) are central elements of the SCION end-entity PKI. An SCP contains parameters regarding the usage and validation of a domain's certificate, such as the list of CAs authorized to sign the domain's end-entity certificate. Each SCP has an associated key pair, and at a given point in time each domain can have only a single valid SCP. The policy private key is used to (a) sign the policy binding in a domain's certificate and (b) authorize certificate revocations and policy updates. Because the parameters in an SCP are bound to a domain's identity and policy key pair, we encode an SCP as a series of standard X.509 certificates (signed by distinct CAs), where each X.509 certificate authenticates the policy public key and lists the policy's parameters in an X.509 extension.

Let n be a security parameter that denotes the minimum number of parties (i.e., CAs and logs) that must be actively involved in asserting that an SCP has been registered. Then, an SCP must be confirmed and signed by at least $n-1$ CAs, and it must be registered by at least one log server to be considered valid. This n parameter is specific to each ISD and is defined in the TRC with the `ThresholdEEPKI` field.

Since domains are expected to only infrequently change their policy, SCPs are assumed to be stable (barring catastrophic events such as a weakness in a widely used encryption scheme). Therefore, we require that SCPs be valid for an extended period (of the order of months). Besides the end-entity PKI, SCPs are also used to certify RAINS keys (i.e., a policy key can sign the domain's RAINS key).

To provide some resilience against CA compromise, we use *multi-signature certificates* (MSCs), which allow multiple CA signatures to authenticate a single public key and require only a certain threshold of the signatures to be valid. Similarly to SCPs, an MSC is encoded as a series of standard X.509 certificates (signed by distinct CAs) authenticating domain D's public key. These standard certificates are followed by a special certificate that we refer to as a *policy binding certificate*. The policy binding certificate is signed with an SCP's private key controlled by domain D itself, and contains the current version number of D's SCP and an X.509 extension that lists the hashes of all certificates within the MSC. This allows the domain owner to change the certificates within an MSC. Because the policy binding can be generated by D independently of any CA, these changes can be made quickly. In order for an MSC to be considered valid, a threshold number of its certificates (defined in the policy) must be valid. An MSC with one certificate is equivalent to a regular certificate, but contains an additional policy binding certificate.

Log servers are highly available entities that monitor issued certificates, revocations, and policies. Each log maintains a *certificate tree*, which tracks certificates (MSCs); a *policy tree*, which tracks policies (SCPs); and a *consistency tree*, used to prove the append-only property of the log. The consistency tree contains all MSC and SCP registrations, updates, and revocations in chronological order. Additionally, upon each update the log appends the concatenation of the root hashes of the current certificate and policy trees to the consistency tree. Merkle hash trees allow the log server to produce efficient proofs that a leaf is present in or absent from the tree. These proofs can demonstrate that a certificate is logged, not revoked, and compliant with all applicable SCPs. To avoid frequent updates to the trees and thus to the proofs, objects are batch-added periodically (e.g., every hour). The update frequencies of log servers are public information, allowing clients to query them after each update or as needed. When an object is accepted for insertion into a tree, the log server schedules it and returns a signed *receipt* with the time at which the object is guaranteed to be present in the log's database. Log servers are required to produce a proof for a specific entry (certificate or policy) on request, which certifies the current validity of that entry. A log server is also required to provide a proof of consistency by showing that its database has been correctly extended from a previous version.

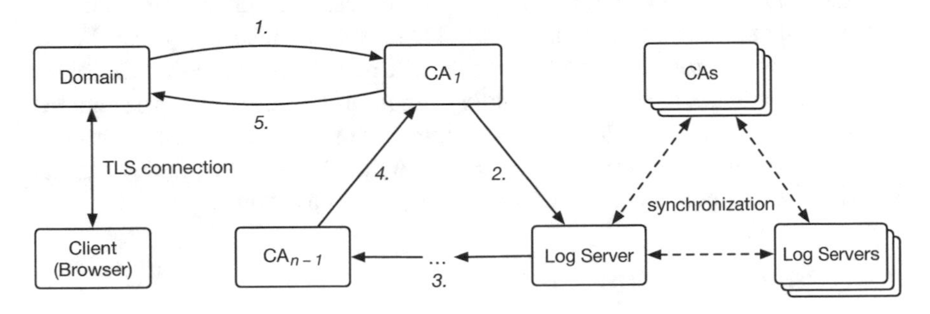

Figure 4.9: Overview of the SCP registration process.

SCP registration. At a high level, the SCP registration process (see also Figure 4.9) works as follows:

1. The domain creates a registration request along with a list of CAs that will confirm the registration. The request is sent to the first CA in the list.
2. The request is passed to the log, which performs verifications, synchronizes with other logs and CAs, and returns a receipt.
3. The receipt is then sent to the second CA, which checks that the registration was carried out correctly and passes it to the next CA.
4. The first CA receives the receipt and performs similar verifications.
5. The domain receives the log receipt confirmed by $n-1$ trusted entities.

Thereafter, the domain can create a key pair with which it will authenticate itself to clients. Then, the domain obtains standard X.509 certificates from authorized CAs (specified in the policy) and combines them, along with a policy binding (signed with its policy private key), into an MSC. The MSC is registered and confirmed by multiple entities, similarly to the SCP registration. A TLS connection can then be established.

After a successful SCP or MSC registration, the log returns a registration receipt promising that the certificate or policy will be added to its database within a certain amount of time. This registration receipt can be used as a short-term confirmation that an SCP or MSC is in the log, but proofs are more commonly used for this purpose. To successfully establish a connection to the domain, the client needs proofs that the policy is registered, as well as proofs that the MSC is registered and not revoked. While anyone can request such proofs from a log, they should be periodically retrieved by the domain and stapled to the MSC and SCP. To request a proof, the domain sends a request containing a hash of its MSC. The log uses this hash to locate the appropriate leaf node in its certificate tree and generates a proof of presence or absence. The log also produces a proof of presence for the domain's policy, as well as a proof that the policy and certificate trees' root hashes are the most recent ones. Additionally, proofs are confirmed by $n-1$ entities. The domain can then pass these proofs and hashes on to the client. There is also a possibility that the log does not have a proof for an SCP or MSC. It may be the case that the MSC, SCP, or both do not have a corresponding log proof because the log has not yet updated its database to reflect a registration. In this case, a registration receipt from the log suffices as a proof of presence so that domains that newly register a certificate and policy can begin serving customers as soon as possible.

4.4.4 Security Discussion

We now conduct an informal security discussion of our end-entity authentication infrastructure. We assume that a domain D has correctly registered its policy and certificate at the logs. We consider an adversary who is able to capture trusted entities of the system and whose goal is to impersonate the domain.

First, we observe that an adversary without the private key corresponding to a valid SCP for D cannot create a valid MSC for D and thus cannot impersonate it. Constructing an MSC requires a policy binding. Because the policy binding must be signed with D's policy private key, an adversary without that key cannot create any valid MSC. An adversary can either try to obtain that key directly from the domain or produce a fake SCP, but this second option would require compromising n entities. Even if we assume that the adversary has access to the original policy's private key, then the adversary cannot impersonate D without compromising at least a threshold number of D's trusted CAs. This is due to the MSC validation process, which requires a valid MSC to contain at least a number (specified in the policy) of valid certificates. An adversary who has compromised the required number of trusted CAs and D's policy private key can impersonate D by creating a malicious MSC and serving it to clients. However, to mount this man-in-the-middle attack the adversary must receive confirmations (a registration receipt or log proof) from the log. This requires registering the malicious MSC, which would make the fraudulent certificate publicly visible. The adversary could also attempt to update the SCP, but this would require compromising the number of CAs specified in the policy. MSCs by design remove single points of failure, which mitigates other threats such as too-big-to-be-revoked CAs [234]. If we assume that logs are not malicious, then the above attacks can be detected since the adversary's actions will become publicly visible. If we consider that logs might be misbehaving, gossip protocols [52] can be used as a last line of defense.

5 ISD Coordination 

LAURENT CHUAT, ADRIAN PERRIG,
RAPHAEL M. REISCHUK, PAWEL SZALACHOWSKI

In this chapter, we describe how ISDs are discovered and how they coordinate with each other, especially when a new ISD is created. The goal is for each ISD to have a list of all other ISDs — specifically, an identifier and a description for each ISD along with the roots of trust that enable authentication. An authority could create such a list and distribute it, but this would conflict with SCION's goal that each ISD can operate independently and communicate with other ISDs without any globally trusted entity. A global authority could also introduce a kill switch to take down parts of the Internet, so instead we present a decentralized approach.

In short, when an ISD joins the SCION network, its neighbors must announce it through dedicated beacon extensions. This prevents situations in which two new ISDs are created with the same identifier, as other ISDs have time to detect whether an ISD is being created under a misleading identity (intentionally or not). As ISD coordination is realized through beaconing, operations are transparent and accountable. To cover the possibility that conflicts might happen despite those preventive measures (due to negligence or malicious intent), we propose mechanisms to detect and resolve such conflicts. We also describe how ISDs are globally identified and how ISD descriptions (i.e., short textual representations of the entity running the ISD) can be specified and updated.

Chapter Contents

A. Perrig et al., *SCION: A Secure Internet Architecture*, Information Security and Cryptography, https://doi.org/10.1007/978-3-319-67080-5_5

5.1 Motivation and Objectives

As we describe in other chapters, core beacons (Section 7.1.3 on Page 127) and TRCs (Section 4.2.1 on Page 68) can be used to construct authenticated core-segments across ISDs that already know of each other. However, we have not described how new ISDs are announced or how conflicts are prevented when two ISDs claim the same identity but have different TRCs. New ISDs will indeed be created over time, and they should not be required to exchange keys — out of band — with *all* other ISDs before joining the network. Also, our assumptions regarding the honesty and cooperation of participants differ for the intra-ISD and inter-ISD case. Inside an ISD, we can expect ASes to collaborate, as they share a geographical, political, or organizational framework. At a larger scale, however, we cannot expect all ISDs to fully trust anyone, so we want to avoid a situation in which any single entity is responsible for the management of the whole SCION network. For these reasons, we have devised a process (described in this chapter) to announce, discover, and identify new ISDs without any central authority. But first, we discuss existing solutions and show why they are insufficient for our purposes.

Public-key certificates could be attributed to each ISD, but this would require that a trusted authority takes part in the initial issuance process, and such an authority could impersonate the entities for which it is responsible. Although an impersonation attack would eventually become visible, it cannot be prevented. Moreover, an attack can not only be caused by an ill-intentioned authority, but can also be the result of a key compromise or an honest administrative mistake. Self-certifying identifiers [7, 180] initially appear to be a promising direction, but on closer consideration, the difficulty of revoking or updating keys (to recover from key loss or compromise, for example) renders their use impractical in this context.

Blockchains are often presented as a way to achieve global consensus and implement a fully decentralized database. As decentralization and coordination on a global scale are the main properties we seek here, it may seem that we could build upon such a technology to achieve our goals in the context of ISD coordination, but there would be drawbacks. A blockchain is computationally expensive to maintain, peers need to be able to communicate with each other (which we cannot assume here, since we are defining the architecture upon which communication will be based), and consensus, when achieved through a proof-of-work system, also raises issues: with the majority of computing power, one can effectively manipulate the blockchain [200].

An absolute consensus is not needed to attribute identifiers to all ISDs. Consensus might not even be possible or desirable on a global scale. Instead, the system we describe tolerates ISDs having different views of the world, but encourages them to find an agreement and deters misbehavior through transparent operation. Besides, the information upon which ISDs must agree is minimal: only the binding between a globally unique identifier and an ISD (represented by a set of keys) must be determined, so we expect inconsistencies to be extremely rare.

Given that we employ a decentralized approach, different ISDs may use different methods, and this allows migration to new approaches over time. At a high level, the approach we propose for ISD coordination consists of three phases. In the first phase, a new ISD is announced in advance by all its future neighbors, which allows detection of colliding identifiers and gives administrators time to contact each other to resolve potential conflicts. In the second phase, the neighbors of the new ISD propagate their final announcement. During the last phase, if conflicts still exist, then each ISD can set some rules and pick which entity is to be trusted for building its own list of ISDs.

5.1.1 Potential Attacks and Undesirable Behavior

To provide insight into why ISD coordination is needed, we consider some unwanted situations. The difference between an attack and suspicious but benign behavior is often small; the scenarios we present can be the result of an error or malice, but we do not need to know the exact cause of abnormal behavior to prevent it from disrupting normal operations.

Identifier Squatting

The first type of undesirable misbehavior happens when an ISD is created with the same identifier as an already existing ISD. This situation is illustrated in Figure 5.1a and can occur in different cases:

- A newcomer ISD $1'$ is intentionally trying to replace ISD 1 by advertising the same identifier with a different path and different keys. To do so, it needs the cooperation of ISD 4, which is already connected to the network.
- ISD 4 is pretending to be connected to ISD $1'$ (although ISD $1'$ might not even exist) and is making an announcement in order to replace the existing ISD 1.
- ISD 2 and ISD 4 are both announcing — almost simultaneously — the creation of new ISDs (ISD 1 and ISD $1'$, respectively) with the same identifier, which results in a conflict.

The first two attacks are equivalent — from the victim's (i.e., ISD 1's) standpoint — and can be prevented by attributing an identifier to the first ISD who claims it. The result of doing so is that an attack cannot be successfully carried out as long as private keys are not compromised, since updating a TRC requires valid signatures. The third case, however, is more plausible and will be addressed in the remainder of this chapter, in particular, through early announcements.

Spurious ISDs

The second type of misbehavior concerns the situation in which many ISDs are created with new identifiers and arbitrary descriptions in a short period, as illustrated in Figure 5.1b. In that particular example, ISD 1's announcements might be legitimate, but the ISDs could also be fake and created with the intention of disturbing communication in the rest of the network. This situation is problematic, mainly for the following reason: an exhaustion or overcrowding of the identifier space is possible (since identifiers are positive integers). This could create conflicts and prevent someone who desires to create a new legitimate ISD from doing so.

To address this problem, the number of ISDs that any existing ISD can announce during a defined period should be limited. This makes an exhaustion attack infeasible if the creation period is chosen such that it takes a few days to create a single new ISD, and if the identifiers are represented with a sufficiently large number of bits. This safeguard slows down the ISD creation process, but this is desirable, since ISDs are the largest and most important structural unit in SCION. As such, ISDs are expected to take time to build and stabilize.

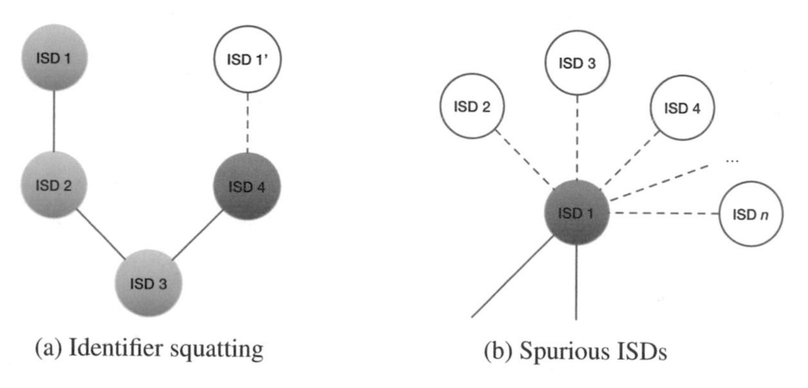

(a) Identifier squatting (b) Spurious ISDs

Figure 5.1: Examples of misbehavior in the context of ISD coordination.

Inappropriate Descriptions

The ISD descriptions could be misleading. Specifically, a new ISD could be created with an available identifier but a deceptive description in order to impersonate an existing ISD (e.g., a new ISD could be announced with the description "USA" when an ISD with the description "United States of America" already exists). Preventing inappropriate descriptions is more difficult than preventing colliding identifiers, because uniqueness is not sufficient: two descriptions might be highly similar without being identical. Therefore, the intervention of administrators is required to resolve such cases.

5.2 Announcing and Discovering New ISDs

In this section, we describe how new ISDs are identified, announced, and discovered.

5.2.1 Identifiers and Descriptions

ISDs are globally identified by unique positive integers. These identifiers are specified in the TRC and must be chosen with the help of neighboring ISDs that are already part of the network and know the list of existing identifiers. It is recommended that identifiers be picked in order, although this is not strictly enforced (as it would make the resolution of identifier collisions more complicated).

In addition to identifiers, all TRCs must also contain a human-readable description field that briefly describes the entity running the ISD (e.g., a country or a company name). Descriptions should not be misleading, and it is the neighboring ISDs' duty to verify that this condition holds before propagating an announcement. Descriptions can be changed through TRC updates (if neighboring ISDs approve the description change); identifiers, however, cannot be updated.

5.2.2 Announcements

A new ISD must be advertised to the whole network through a dedicated core beacon extension called an *announcement* that contains the TRC (version 0) of the new ISD, which contains an identifier and a description. Announcements containing invalid TRCs (e.g., with an insufficient number of signatures) must be ignored. Newcomer ISDs should collaborate with their neighbors to create their first TRC. Once the new TRC is created, it must be sent to all neighboring ISDs, so they can start generating announcements.

The advantage of this approach is that beacons (and thus beacon extensions) are signed by ASes from each ISD along the propagation path. Signatures constitute evidence that these ISDs have seen or generated the content of the beacon extension.

Early Announcement

During the early announcement phase, the announced TRC is valid but must not be used until a 7-day period has elapsed, and the TRC must come with a `quarantine` flag set to true.

Early announcements are produced by all the neighboring ISDs of the new ISD for at least one week and, as beacon extensions, are propagated through all core ASes of all ISDs. The recipients of an announcement set a timer to 7 days when the announcement is first received and they do not allow further steps (i.e., the final announcement) to take place until this timer expires. This gives some time to administrators to notice if another ISD is being announced under an existing identifier. As ISD creation is expected to be a fairly rare event, and as core PCBs propagate rapidly, unintentional identifier collisions are unlikely to occur. Nevertheless, if a collision happens, administrators can contact each other, agree to use different identifiers, and thus resolve the issue. By "different" here we mean that the new identifiers must be both different from the conflicting identifier(s) previously announced and different from each other.

When an announcement is modified in any way and propagated again, a new timer must be set by all receivers. ISDs are limited to making at the most five concurrent early announcements. This means that other ISDs only maintain a maximum of five running timers per source ISD. Also, an early announcement is only valid for 14 days. After that period, if the second phase has not started, an early announcement must be transmitted again to proceed.

Inappropriate descriptions should be detected by ISD administrators in this 7-day period, and the corresponding early announcements should be blacklisted to avoid the propagation of unwanted final announcements in the next phase.

During the announcement period, the new ISD must start building its own list of TRCs. For this to be possible, neighboring ISDs must provide an initial list and/or start forwarding core PCBs to the new ISD.

Final Announcement

A final announcement (still in the form of a beacon extension) must be generated by all neighbors of the new ISD and contain its initial TRC — with the quarantine flag set to false — when the 7-day period has ended. Because each ISD has its own timer, and because announcements take some time to propagate, final announcements should be generated and sent repeatedly 7 days after the first early announcement was sent. This constitutes the second phase of ISD creation. When a final announcement is received by a core AS, the corresponding TRC is added to the local list and propagated further, if the following conditions are met:

1. an early announcement containing the same TRC was received before,

2. that early announcement is not blacklisted,
3. the corresponding 7-day timer has expired,
4. the identifier specified in the TRC is currently not in use,
5. no other early announcement (excluding blacklisted ones) containing a TRC with the same identifier was received in the 7-day period.

If conditions 1–4 are not respected, then the final announcement is ignored. If condition 5 alone is not met, then a conflict resolution procedure must be initiated. When the final announcement has been propagated, the new ISD can be considered part of the network and start communicating with other ISDs.

The details of how early and final announcements must be validated and propagated by core ASes are specified in Algorithm 1. The isValid() function returns *true* if conditions 1–4 are respected by the final announcement, and hasConflicts() returns *true* if condition 5 is *not* respected.

Algorithm 1 Validating and Propagating ISD Announcements

```
1: data: ExistingISDs, EarlyAnnouncements, Timers, Blacklist
2: parameters: MinDays = 7, MaxDays = 14, MaxAnnouncements = 5
3: upon reception of announcement a do
4:     if a.quarantine = true then  // early announcement
5:         if not EarlyAnnouncements.contains(a) then
6:             for i ← 1 to MaxAnnouncements do
7:                 timer = Timers.get(a.sourceISD, i);
8:                 if timer.isNotRunning() then
9:                     timer.setTo(MinDays);
10:                    EarlyAnnouncements.add(a, timer, MaxDays);
11:                    break;
12:                end if
13:            end for
14:        end if
15:        Propagate(a);
16:    else // final announcement
17:        if a.isValid() then
18:            while a.hasConflicts() do
19:                ResolveConflicts();
20:            end while
21:            if not Blacklist.contains(a) then
22:                ExistingISDs.add(a)
23:                Propagate(a);
24:            end if
25:        end if
26:    end if
27: end reception
```

5.3 Local Resolution of Conflicts

After the first two phases (i.e., early and final announcements), which take at least one week, any remaining conflict is either intentional or the indication that ISDs failed to coordinate to avoid picking the same identifiers. We now describe how each ISD can resolve conflicts locally and we discuss measures that can be taken if misbehavior is observed.

In this context, we define a conflict as follows: a number n of ISDs has correctly generated announcements for a new ISD with identifier i during a given period, while a number m of other ISDs has correctly announced a different ISD (i.e., with a different TRC) with the same identifier i during an overlapping period, and at least one of the two corresponding final announcements is received.[1] There are several possibilities to resolve such a conflict:

- The decision can be based on the opinion of the majority (i.e., based on $\max(n, m)$), but this might not be possible (i.e., if $n = m$).
- Certain ISDs might be more trusted than others: by looking at which group (n or m) the most trusted ISD is in, a decision can be made.
- The conflict can be manually resolved on a case-by-case basis.

In case of conflict, it is up to each ISD to make a decision based on the above parameters. Conflicts may also concern similar descriptions, but such situations necessarily require human judgment in order to be detected and resolved. A conflict resolution procedure must result in incriminated early announcement(s) being blacklisted.

5.3.1 Conflict Resolution Policy

A policy can be specified to automatically resolve conflicts involving identical identifiers or descriptions. The policy must indicate whether the resolution should follow the majority of ISDs (if applicable), or instead use a complete list of ISDs ordered in terms of trust to make a decision. By default, new ISDs are placed at the end of the list and thus older ISDs are more trusted. However, administrators can arrange the list as they desire. Also, new ISDs are free to re-order the list they initially obtained. Alternatively, more elaborate trust metrics could be computed over time based on observed events.

[1] There might be even more conflicting groups, but we consider only two groups here for the sake of simplicity.

6 Name Resolution 

DANIELE ASONI, YIH-CHUN HU,
RAPHAEL M. REISCHUK, BRIAN TRAMMELL

While the path resolution process is necessary to turn a destination address into a set of paths, this is not sufficient for establishing communication between SCION-connected endpoints: we also need a way to turn an Internet name into a SCION address. As name resolution and path establishment are separate processes, with different timescales and triggered by separate events, we design a dedicated infrastructure that is optimized for each purpose.

We begin with an analysis of what a name resolution service is good for. At its core a naming service must provide a few basic functions, but in essence associate a human-understandable name with machine-understandable information. Although the Internet's Domain Name System (DNS) has been used and abused as a general-purpose distributed database, a useful Internet naming service need only provide information that is necessary to establish and maintain communication with an Internet-connected entity: addresses, namespace delegations, service information, certificates, and auxiliary information. There are two entities in this ideal naming service: (a) The *querier* is a client that wants to establish communication across the Internet with a named entity. That client uses the naming service to retrieve the necessary information. (b) The *authority* is an entity with the right to make assertions about names within parts of the Internet namespace. Before looking at the specific interplay between queriers and authorities, we start by discussing various possible resolution types.

Chapter Contents

A. Perrig et al., *SCION: A Secure Internet Architecture*, Information Security and Cryptography, https://doi.org/10.1007/978-3-319-67080-5_6

6.1 Background

6.1.1 Resolution Types

The assertions for name resolution are essentially mappings of various forms:

- **Name-to-address:** given a name, return associated addresses.
- **Name-to-name:** given a name, return equivalent names.
- **Name-to-service:** given a name representing a service, return a name and transport-layer ports for connecting to the service.
- **Name-to-certificate:** given a name representing a host or service, return an end-entity certificate representing the named entity, for authentication of a subsequent connection attempt with the named entity.
- **Name-to-delegation:** given a name representing a zone within the namespace, return the public key used to verify assertions in the zone.
- **Name-to-auxiliary-information:** given a name representing an organization-level zone within the namespace, return information about the zone and the organization behind it, analogous to the WHOIS service; as well as any restrictions on names in the zone (e.g., for confusability reduction).
- **Address-to-name:** given an address, return associated names, analogously to reverse DNS.

6.1.2 Properties of an Ideal Naming Service

We consider a set of properties of an ideal Internet naming service as background to selecting a design for SCION name resolution. A more in-depth discussion of these properties, enumerated below, is given in an IETF draft [239]. An ideal naming service must

- provide for names which are meaningful to human users;
- guarantee that different names are distinguishable by its users;
- allow for authority over names to be federated;
- allow a unitary authority for any given name to be transparently determined;
- operate without requiring trust in the operators of the name server infrastructure;
- provide for revocation of authority over a given name;
- allow assertions about names, and the nonexistence of a mapping for a name, to be unambiguously authenticated;
- provide for consistency, and predictability in the presence of changes to assertions about names, but

- allow for explicit inconsistency when necessary, and global transparency of this inconsistency;
- perform acceptably, in terms of availability, latency, and bandwidth efficiency;
- allow clients to specify tradeoffs between privacy and performance.

Since an Internet naming service is designed to provide information about Internet-connected hosts and services for the purposes of establishing a connection, note that assertion confidentiality (usually referred to in DNS literature in terms of zone enumeration) is a non-goal of our ideal naming service. If assertion confidentiality is required, an alternative service can be established that provides access control to the information that should remain secret.

6.1.3 Notation

Throughout the chapter, we make use of the terms, abbreviations, and resource record (RR) types defined in Table 6.1.

Term	**Definition**
Assertion	Mapping between a name and a property of that name
Shard	Set of assertions for some authority and context
Zone	Set of all shards and assertions for some authority and context
NCO	Naming Consistency Observer (see Page 116)
RR	Resource Record (fundamental data unit, see table below)
TLD	Top-Level Domain (such as `.ch` or `.com`)
ZK	Zone Key (key to sign assertions for the respective zone)
RZK	Root Zone Key (special zone key used for the root zone)

(a) Terms and Abbreviations

RR Type	**Content**
`A`	32-bit IPv4 address
`AAAA`	128-bit IPv6 address
`CNAME`	Canonical name for a given alias
`NS`	Responsible name server
`PTR`	Pointer to domain name (e.g., address-to-name mapping)
`SRV`	Service locator (locates service/protocol for a given domain)
`TLSA`	Certificate or public-key association (see DANE [61])

(b) Resource Record (RR) Types

Table 6.1: Notation used in the context of name resolution.

6.2 Name Resolution Architecture

We now turn our attention to designing protocols to provide this ideal naming service, which we call RAINS (a recursive acronym for "RAINS, another Internet naming service") [240].

> **Why not just use DNS?**
>
> We note that the DNS protocol as used in the present Internet, when deployed with the mandatory usage of DNS Security Extensions (DNSSEC) and one root per ISD, meets most of the properties of our ideal name system. Only explicit tradeoffs and explicit inconsistency are not well supported by DNS with DNSSEC. An initial approach to providing name resolution for SCION could therefore be to borrow DNS.
>
> Using DNS would have the following advantages:
>
> - It leverages an existing, widely deployed protocol, with which there is widespread operational experience.
> - It allows names for SCION-enabled nodes to be registered in the same name resolution system as the non-SCION Internet, which should make incremental deployment easier.
>
> It would also have some serious disadvantages:
>
> - DNS has no concept of explicit inconsistency or explicit tradeoff, especially for privacy.
> - Even with DNSSEC, DNS has poor operational security properties, specifically lack of query anonymity and vulnerability to abuse as an amplification attack vector.
> - Since DNSSEC would be mandatory for SCION RRs, SCION-enabled nodes could only use signed top-level domains (TLDs). Many country-code TLDs remain unsigned.
> - Support for extension mechanisms for DNS (EDNS0) and DNSSEC varies widely among stub and recursive resolvers, which negates the incremental deployment advantage above: lack of interoperability of the minimum DNS required for SCION and other DNS-supporting software and hardware would lead to difficult-to-debug issues.
>
> The final disadvantage is the most troubling, and led to our decision to build a new name resolution protocol for SCION.

The RAINS architecture is simple, and resembles the architecture of DNS. A RAINS server is an entity that provides transient and/or permanent storage for *assertions* about names, and a lookup function that finds assertions for a given *query* about a name, either by searching local storage or by delegating to another RAINS server. RAINS servers can take on any or all of three roles:

- **authority service**, acting on behalf of an authority to ensure properly signed assertions are made available to the system;
- **query service**, acting on behalf of a client to answer queries with relevant assertions, and to validate assertions on the client's behalf; and/or

- **intermediary service**, acting on behalf of neither but providing storage and lookup for assertions with certain properties for query and authority servers.

RAINS servers use the RAINS protocol, described in this section, to exchange queries and assertions. RAINS clients use a subset variant of the RAINS protocol (called the RAINS client protocol) to interact with RAINS servers providing query services on their behalf. RAINS protocol connections between servers are encrypted and authenticated. RAINS client protocol connections between clients and query servers are encrypted and optionally authenticated. In addition, the RAINS protocol provides object-level authentication. Section 6.4.1 provides details on bootstrapping trust using RAINS.

Authority service in RAINS resembles the role of authoritative servers in the present DNS. Query service resembles the role of recursive resolvers. Intermediate service resembles the role of caching resolvers. RAINS is therefore a drop-in replacement for the present DNS with better support for contexts and tradeoffs and with mandatory delegation and authentication by signature chain. As with DNS, a given RAINS server may play both the authority server and query server roles at any given time, depending on configuration. However, future implementations of RAINS could use other mechanisms for matching queries and assertions, and moving assertions to where they can be most efficiently matched with queries.

From the basic building blocks of these three services, any number of naming service architectures could be built. Within SCION, RAINS authority services are generally operated by TLDs (as isolation context root authority servers) as well as domain name registrants or ISPs acting on their behalf. Intermediate and query services are operated by ISPs and enterprise networks, and ASes make query servers available via service anycast (see Section 7.5 on Page 153).

RAINS also integrates into SCION's authentication infrastructure. End-entity certificates for named hosts can be stored in RAINS, and RAINS intermediary and query services support assertions signed via ARPKI (see Section 4.4).

There is an inherent tension between SCION's architectural principle of isolation and the need for a globally consistent namespace. RAINS on SCION resolves this by supporting *isolation transparency*. Queries and assertions can cross ISD boundaries, which is the basis of the Naming Consistency Observer (NCO) described in detail in Section 6.5 on Page 116.

So what is new in RAINS?

Though designed as a drop-in replacement, RAINS makes several radical departures from DNS as presently specified and implemented:

- Delegation from a superordinate zone to a subordinate zone is accomplished solely with cryptography: a superordinate defines the key(s) that are valid for signing assertions in the subordinate during a particular time interval. Assertions about names can therefore safely be served from any infrastructure.
- All time references in RAINS are absolute: instead of a time to live, each assertion's temporal validity is defined by the temporal validity of the signature(s) on it.
- All assertions have validity within a specific context. A context determines the rules for chaining signatures to verify the validity of an assertion. Within SCION, publicly available names within an ISD exist within that ISD's native isolation context. The use of context explicitly separates global usage of the DNS from local usage thereof, and allows other application-specific naming constraints to be bound to names; see Section 6.3.3.
- Explicit information about registrars and registrants is available in the naming system at runtime, combining the functionality of WHOIS with the naming service.
- Sets of valid characters and rules for valid names are defined on a per-zone basis, and can be verified at runtime.
- Reverse lookups are performed on a completely separate tree, supporting delegations of any prefix length, in accordance with classless inter-domain routing (CIDR) and the IPv6 addressing architecture.

6.3 Naming Information Model

Here we describe the information model for messages in the RAINS protocols. For simplicity of description, we omit details on error handling and parts of the information model necessary for protocol implementation and operation. The detailed protocol specification is in our IETF draft [240].

RAINS operates on two different basic types of information: *assertions* (Section 6.3.1) are mappings between a name and some property of the name, which can be grouped into *shards* and *zones* (Section 6.3.2) for performance and operational optimizations; and *queries* (Section 6.3.5) are expressions of interest about certain types of information about a name. RAINS matches queries to assertions that answer them.

6.3.1 Assertions

An assertion consists of the following elements:

- **Context**: name of the isolation context in which the assertion is valid. Section 6.3.3 provides more details.
- **Subject**: the non-qualified name about which the assertion is made. A non-qualified name is a local, not necessarily globally unambiguous identifier (e.g., 'foo'), which — in combination with the zone name (e.g., 'example.com') — yields the fully qualified (i.e., unambiguous) name (e.g., 'foo.example.com'). The domain name separator here is '.' and separates subject and zone.
- **Zone**: the name of the zone (e.g., 'example.com') in which the assertion is made.
- **Object**: the data associated with the name of the given type.
- **Type**: the type of information about the subject contained in the assertion. Each assertion is about a single type of data. Supported types include:
  - **Delegation**: the authority associated with the zone identified by the name (replaces the NS DNS record type for cryptographic delegation; see below).
  - **Redirection**: the authority servers for the zone identified by the name (analogous to the NS DNS record type).
  - **Address**: one or more addresses associated with the name, given an address family (analogous to the A and AAAA DNS record types).
  - **Service-info**: one or more layer-4 ports associated with the name, if the name identifies a service (analogous to the SRV DNS record type).
  - **Name**: one or more names associated with the name (analogous to the CNAME and the PTR DNS record types: a PTR-analog lookup is defined by the zone in which the lookup is made).
  - **Certificate**: an end-entity certificate representing the named entity, for authentication of a subsequent connection attempt with the named entity (analogous to the TLSA DNS record type).
  - **Nameset**: an expression of the set of names allowed within a zone; e.g., Unicode scripts or codepages in which names in the zone may be issued. An assertion about a subject within a zone whose name is not allowed by a valid signed nameset expression is taken to be invalid.
  - **Registrar**: a string identifying the registrar responsible for the appearance of a delegation within a zone, for TLDs that allow multiple organizations to modify their zones.
  - **Registrant**: a string containing information about the registrant of a zone within a TLD (analogous to the WHOIS service).
  - **Infrastructure-key**: a public key by which a RAINS server can be identified, for object security on RAINS messages.

  - **External-key**: a public key by which assertions in a zone can be verified outside the delegation hierarchy, e.g., via an SCP as described in Section 4.4.3.
- **Issued**: a timestamp at which the assertion was made.
- **Expires**: a timestamp after which the assertion is no longer valid.
- **Signature**: a signature generated by the authority, to authenticate the assertion. This signature covers all elements within the assertion except the signatures themselves. An assertion may have multiple concurrently valid signatures.

Issued and expired timestamps are always expressed in terms of UTC. Since the signature protects the timestamps as well, it is necessary to sign new assertions before old ones expire. At a single point in time, it is possible to have multiple active valid assertions with overlapping validity times for a given ⟨subject, zone, context, type⟩ tuple. The union of the object values of all of these assertions is considered to be the set of valid values at that point in time.

6.3.2 Grouping Assertions: Shards and Zones

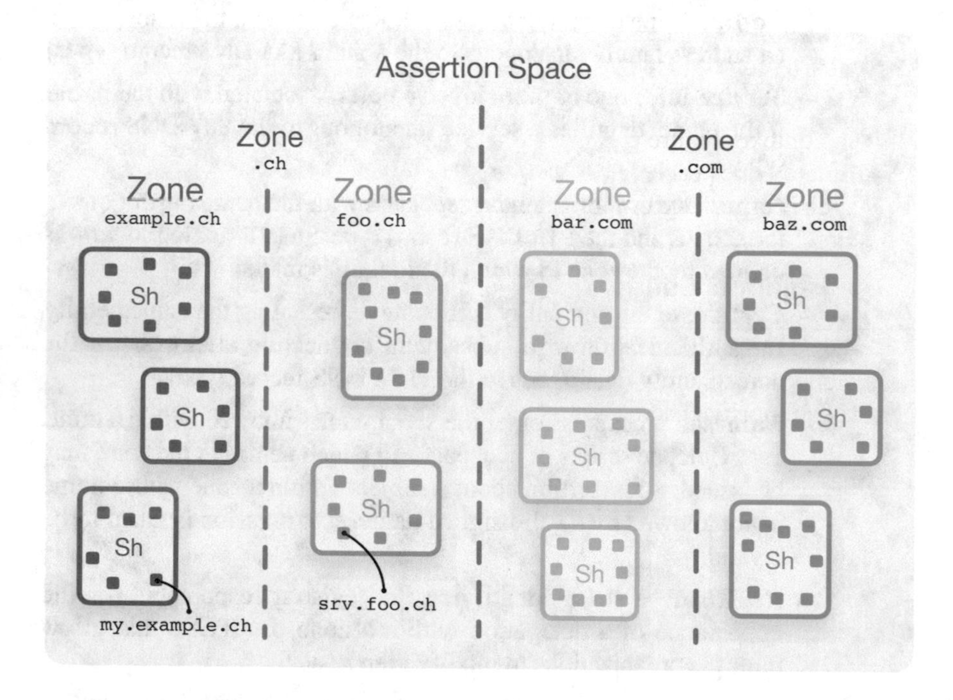

Figure 6.1: Hierarchical zones with shards and assertions.

An authenticated assertion with a valid signature provides a proof of existence of a name. Another mechanism is necessary to provide a proof of nonexistence; otherwise, malicious intermediary and query services could cause false negatives for queries by simply refusing to forward matching assertions. RAINS provides shards for this purpose.

A *shard* is a set of assertions for the same authority within the same context, protected by an additional signature over all assertions within the shard, which has the property that, given a subject and an authenticated shard, then either an assertion of a given type exists within the shard, or does not exist at all. We achieve this property by associating the shard with an exclusive *shard range* of names appearing in a shard: a shard with the range 'a' to 'b' contains all names in the zone and context that sort after 'a' and before 'b'. This property allows efficient verification of the nonexistence of an assertion for a given name at the query.

Example. Consider a zone containing the names `'aaa'`, `'aab'`, `'baa'`, `'cat'`, `'dog'`, `'nap'`, `'yyz'`, and `'zzz'`, as illustrated in Figure 6.2. This zone could be split into three shards: {`aaa, aab, baa`}, {`cat, dog, nap`}, and {`yyz, zzz`}. To ensure that a proof of nonexistence can be given for any name other than these eight using only one of these shards, the shard ranges overlap: the first shard has the range `null`–`cat`, the second the range `baa`–`yyz`, and the third the range `nap`–`null`. Note that names falling between the names in the shards can be disproved using either of the neighboring shards.

A *zone* is the entire set of *shards* and assertions for a given authority within a given context. Figure 6.1 shows two zones (`'.ch'` and `'.com'`) with two subordinate zones each. A zone may also contain assertions about the zone itself; this is especially useful for self-signing root zones.

6.3.3 Isolation Contexts

All assertions are held to be valid within an explicitly named *assertion context*. Assertion contexts are used to determine the validity of the signature by the declared authority. There are two broad uses for assertion contexts: *isolation*

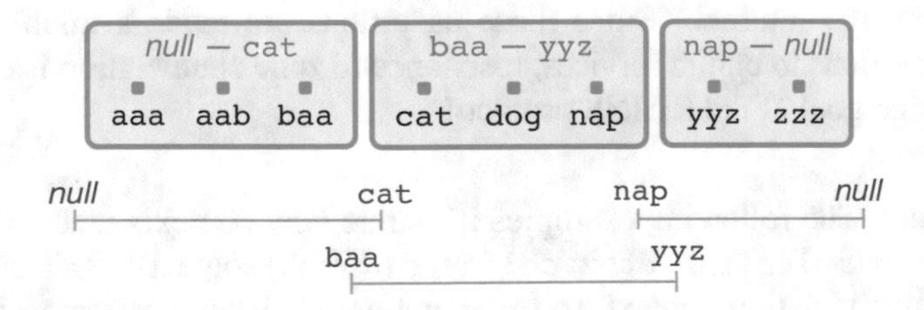

Figure 6.2: Eight assertions aligned in three shards with overlapping ranges.

and *local assertion*. Isolation contexts allow assertions and queries about an ISD other than that from which a query was made.

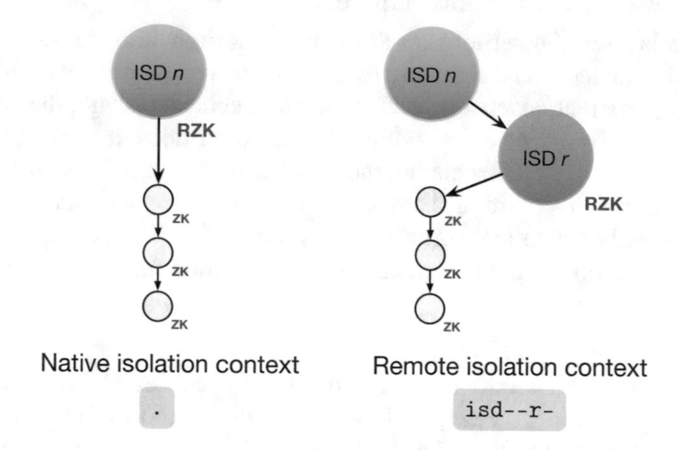

Figure 6.3: Isolation Contexts. RZK is the root zone key for an ISD; ZK is the zone key for a given zone.

There are two kinds of isolation context (as illustrated in Figure 6.3):

- **The native isolation context** is identified by the special context name '`.`'. Assertions in the native isolation context are signed by the authority for the subject name, with a signature chain rooted at the root authority for the ISD in which the assertion is made, such that the authority resides within that ISD (see also Figure 3.1 on Page 49).
- **A remote isolation context** is identified by the special context name '`isd--`*r*`-`', where *r* is the number of the ISD at which the context is rooted. Assertions in a remote isolation context are signed by the authority for the subject name, with a signature chain rooted at the root authority for the isolation domain identified by the context, such that the authority resides within that ISD. Remote isolation contexts can be used to make assertions about names as seen within other ISDs.

Assertions in an isolation context are intended to be publicly available throughout the Internet. Since these assertions are made available to support connections to public services, resistance to zone enumeration is explicitly not a design goal of the RAINS protocol.

Example. The following examples illustrate how contexts work. Consider the name '`simplon.inf.ethz.ch`' in the (default) context '`.`'. This context is the native isolation context, so the signature chain is determined from the name itself, rooted at the TRC for the current ISD. If the assertion is issued by an authority in ISD 33, the chain is as follows:

$$TRC_{33} \rightarrow RZK_{33} \rightarrow ZK_{\texttt{ch}} \rightarrow ZK_{\texttt{ethz}} \rightarrow ZK_{\texttt{inf}} \rightarrow \textit{assertion}$$

where TRC_{33} denotes the TRC of ISD 33, RZK_{33} its root zone key, and ZK the zone key for a given delegation.

Now consider the name '`simplon.inf.ethz.ch`' in the context '`isd--44-`' This is the remote isolation context for ISD 44, so the signature chain is again determined from the name itself, but rooted at ISD 44's TRC, as authenticated against the current ISD's TRC, as follows:

$$TRC_{33} \rightarrow TRC_{44} \rightarrow RZK_{44} \rightarrow ZK_{\texttt{ch}} \rightarrow ZK_{\texttt{ethz}} \rightarrow ZK_{\texttt{inf}} \rightarrow \textit{assertion}$$

Note that both ISD 33 and ISD 44 use the same authority for the top-level domain '`.ch`', but the verification path depends on the initial root of trust for each ISD. Other arrangements are possible; see Section 6.5 for more.

6.3.4 Local Assertion Contexts

Isolation contexts are useful for names pertaining to services made available to the Internet at large. The basic mechanism isolation context uses — providing an alternate signature chain to the root of a namespace — can be generalized. RAINS provides for *local assertion contexts* so that intentional inconsistency (often implemented in the current DNS) is transparent and can be authenticated.

A local assertion context is equivalent to a RAINS subject name designating the namespace within which the assertion is made. When a local assertion context is present on an assertion, the assertion is verified by following the delegation chain from the root through the names in the context before following the delegation chain for the name. Each context is then essentially an alternate root for a new namespace. While the same effect could be achieved simply by concatenating names together, separating this information into explicit subject name and context name allows the semantically meaningful part, which should be presented to the user (the subject name), to be separated from the namespace designator, which should be user-accessible but otherwise is a matter of system configuration.

Example: Split DNS. Consider an organization that places its workstations in their own top-level namespace. A workstation named `simplon` might carry the full name `simplon.workstations`. In the current DNS, this would be achieved through "split DNS", i.e., answering queries about the `workstations` zone only on certain networks. This arrangement, however, is operationally brittle and can lead to leakage of both queries and names beyond their intended scope.

To implement this split within RAINS, ETH Informatik could place assertions about its workstations in the local assertion context `isg.ethz.ch`, in essence creating a local root namespace containing the `workstations` zone. The signature chain for these assertions starts with the name components in the context before considering the subject name:

$$TRC_{33} \rightarrow RZK_{33} \rightarrow ZK_{\texttt{ch}} \rightarrow ZK_{\texttt{ethz}} \rightarrow ZK_{\texttt{isg}} \rightarrow ZK_{\texttt{workstations}} \rightarrow assertion$$

Note that the zone key for `workstations` above is local to `isg.ethz.ch`, unrelated to the zone key for the TLD `workstations`, if it exists.

Additional information can be placed in a context beyond the name of the local root. This additional information is separated from the authority part by a *context marker*, the special name `cx--`. Additional information in a context is used to group assertions signed by the same local root, and to provide a way to attach contextual information to queries.

Example: CDN zones. Consider a content delivery network (CDN) separating content into zones (data centers from which content is served) based on geography. It creates a local assertion context `some-cdn.com`, and places information about the zone in the additional context part: e.g., the local assertion context `zrh.cx--.some-cdn.com` names servers hosting content in a CDN's Zurich data center. A client could represent its desire to find content nearby by making queries in the `zrh.cx--.some-cdn.com`, `fra.cx--.some-cdn.com` (Frankfurt), and `ams.cx--.some-cdn.com` (Amsterdam) contexts. Note that, in this case, assertions in each of these content zones will be signed by the same delegation chain `.some-cdn.com`.

Local assertion contexts can be combined with remote isolation contexts, as well; here, the remote isolation is inserted into the signature chain before the name components in the context.

Example: Combining local and isolation contexts. Consider the name `example.com` within the context `zrh.cx--.some-cdn.com.isd--44-`, asserted within ISD 33. Here, the signature chain for the context is rooted at ISD 44's TRC, then follows the authority part of the local isolation context before looking for names in the root:

$$TRC_{33} \rightarrow TRC_{44} \rightarrow RZK_{44} \rightarrow ZK^{1}_{\texttt{com}} \rightarrow ZK_{\texttt{some-cdn}} \rightarrow ZK^{2}_{\texttt{com}} \rightarrow assertion$$

Here $ZK^{1}_{\texttt{com}}$ is the zone key for the top-level domain `.com`, while $ZK^{2}_{\texttt{com}}$ is a local key signed by `some-cdn.com`.

6.3.5 Queries

A query is a request for a set of assertions, shards, and zones supporting a conclusion about a given subject-object mapping. It consists of the following information elements:

- **Context:** the isolation context or local context in which responses will be accepted. A query may also name a special *any* context, signifying a willingness to receive information about names in any context available at the query server.
- **Subject:** the name about which the query is made; in contrast to assertions, the subject name here is fully qualified.
- **Types:** a list of the types of information about the subject that the query requests.
- **Valid-until:** a client-generated timestamp for the query after which it expires and should not be answered.
- **Token:** a client-generated token for the query, which can be used in the response to refer to the query.
- **Options:** a set of options by which a client may specify tradeoffs (e.g., reduced performance for improved privacy).

A response to a query consists of a message containing a set of assertions bound to the token supplied by the client in the query.

When used with the RAINS client protocol, the query server performing verification may sign the entire response; this is an assertion that the query server has verified the signatures from the appropriate roots, leaving the client only to verify the query server's signature on the whole response.

6.3.6 Registrar and Registrant Assertions

The registrant object type in the RAINS data model associates civil information about a name's registrant (organization or legal personality owning an entry under a top-level domain), and in essence integrates WHOIS into the naming service. The presence of a registrant object on a name identifies that name as a registrant-level domain, i.e., a name that exists due to a contractual relationship with a domain name registrar. This integration has two advantages: first, it provides authentication of WHOIS information. Second, it allows operational decisions to be taken based on WHOIS information.

The registrar object type identifies the registrar responsible for a given name's existence. This allows operational decisions to be taken based on the registrar, e.g., to block a registrar that is predominantly responsible for malware domains.

6.3.7 Augmented Assertion Authentication

To verify the authenticity of an assertion, a client or a query server can verify the signature against the delegation for the zone containing the assertion. The delegation assertion for that zone can be verified against the delegation from the zone containing it, and so on all the way back to the root delegation from the TRC for the isolation context. This delegation chain authentication has identical properties to the verification of an RRSIG in DNSSEC. It also has identical drawbacks: each level of delegation must be trusted in order to verify a name at the leaf.

RAINS provides for signatures by *external keys* on assertions, i.e., those outside the delegation hierarchy, to provide additional and/or parallel verification of the authenticity of the assertion. This facility, together with the certificate object type for storage of end-entity certificates, provides two-way integration between RAINS and ARPKI (see Section 4.3.2 on Page 84).

6.3.8 Address-to-Name Mapping

Information about addresses in RAINS is stored in a separate tree, indexed by address and prefix. An address assertion is similar to a name assertion, but is indexed by subject address as opposed to subject name, and the hierarchy of names is built upon delegation from less-specific to more-specific prefixes. Address assertions may only contain delegation, redirection, name, and registrant type objects.

Contexts are also available for address assertions, but the native isolation context may only contain assertions for SCION addresses within its ISD, remote isolation contexts may only contain assertions for SCION addresses within the remote ISD, and local contexts may only contain assertions for non-routable addresses within the address family (e.g., RFC 1918 [210] or unique local addresses (ULAs) [110]).

6.4 The RAINS Protocol

The details of the RAINS Protocol and the RAINS Client Protocol are specified in an Internet-Draft [240]), and consist of a relatively simple mapping of the information model in Section 6.3 to messages encoded in the Concise Binary Object Representation (CBOR) data format [39] that can negotiate operation over any underlying transport protocol that provides reliable, confidential, and authenticated message or stream transport. Our initial implementation experience with RAINS uses TLS over TCP.

RAINS is fundamentally a message-exchange protocol. A client sends queries to a configured query server (by default, this service is listening on an AS-level service anycast address), and expects responses. A query server may send queries to other query servers and/or authority servers to express interest in information about a given name. A query server sends assertions, shards, and zones to other servers in response to queries, or based on some other interest presumed by the query server.

Within SCION at present, query and intermediary servers are organized in a hierarchical cache. Each AS runs a service anycast query server, and queries that cannot be served out of that query server's cache are delegated to the service anycast query server of one or more upstream ASs. Query and intermediary servers may also recurse to the authority server for the zone, according to their configuration. However, the architecture of RAINS does not necessarily imply hierarchical caching: intermediary servers may connect to each other via a weighted distributed hash tree, for example, and authority servers may push assertions to intermediary and query servers without having been asked. This flexibility allows different networks to use different inter-server topologies for different performance tradeoffs.

6.4.1 Query Server Discovery and Bootstrapping Trust

When a RAINS client first connects to a network, it has no information about the available RAINS servers or the keys used to establish the authenticity of assertions it will receive from them. We assume that both the TRC and the address of a local RAINS query server are made available to a host during the host's initial configuration process. The TRC is available from the path dissemination process (see Section 4.2.3), and refers to the naming root key for its isolation domain. The address of the local RAINS server is provided at endpoint configuration time.

The TRC contains the public key for the naming root for the ISD's local isolation context. With this root, and the address of a local query server, the client can now begin using the query server for name resolution. Whether the client trusts the query server to verify the authenticity of names, or does the authenticity verification itself with the naming root key taken from the TRC, is a matter of the client's configuration: in general, clients will be configured to trust their "home" query servers, and optionally to perform verification of assertions received from local query servers on unknown networks.

Note that since the naming root key for an isolation domain is contained within that ISD's TRC, it cannot be forged by a malicious access network.

Clients may be configured to trust specific query servers other than the local query server. In this case, the client performs a name resolution for the name of its trusted query server using the local query server, verifying the signature chain itself. If the trusted query server is in a different ISD, it issues this query in the remote isolation context for that ISD. It then connects to the trusted query server, verifying the TLS certificate against a pinned certificate for that server.

Example. A mobile client, associated with `giant-enterprise.co.uk` and usually connected via AS 337 in ISD 2, roams on to AS 404 in ISD 7. During association with this network, it makes a service anycast query to find a query server in AS 337 in ISD 2 for the address of `rains-query.giant-enterprise.co.uk` in the remote isolation context `isd--9-`, without delegating assertion authentication to the anycast query server. It also queries for the certificate of `rains-query.giant-enterprise.co.uk`, if this is not pinned or otherwise available through ARPKI. Once it has the address and certificate of the trusted query server, it connects, authenticates the trusted query server using the certificate, and begins issuing queries, delegating authentication to this trusted server.

6.5 The Naming Consistency Observer (NCO)

Isolation, as noted, is a fundamental principle of the SCION architecture. At the same time, most users of naming systems expect *global consistency* in name resolution: even if a name does not resolve to a given address everywhere, the name should always point to the same service or content. "Owners" of names in the global namespace further expect *exclusion*: that their publication of assertions of a given namespace precludes other entities from publishing assertions about the same namespace. Global consistency and exclusion are impossible without a single global root of trust for naming, which runs counter to the principle of isolation.

To reconcile this conflict, RAINS provides *naming isolation transparency*. Entities connected to one ISD can observe name assertions in any remote ISD. An additional facility built on top of RAINS, the *naming consistency observer (NCO)*, provides continuous monitoring of inconsistencies among assertion signature chains in different ISDs, which ensures that any violation of global consistency and/or exclusion is publicly observable. The NCO operates on the principle of deterrence: since illegitimate behavior is made public, it should be rare.

Recall from Section 3.4 on Page 48 that, in the normal case, different ISDs have different root zones signed with different keys (derived from the ISD's TRC), but each root zone delegates to the same key for each TLD as shown in Figure 3.1 on Page 49. However, there are certain cases where an isolation domain might want to "edit" a TLD, by providing delegation to a different set of registrant-level domains (RLDs) (the level below the TLD, corresponding to organizations or other "owners" of names that have paid a registrar to place an entry in a name registry) than that provided by the primary operator of the TLD, or by failing to sign a delegation to a TLD. It does this by operating a shadow authority for that TLD, which may implement one of the following policies for each TLD or RLD:

- **New TLD adoption and TLD quarantine:** New TLDs created within a set of ISDs may be held over for some period of time while the TLD operator's practices are evaluated by the ISD.
- **RLD quarantine:** Newly created RLDs may only be available within an ISD's native isolation context after some period of time has passed, e.g., to ensure that they are not primarily used for abusive purposes (such as phishing landing pages or malware command and control).
- **RLD blacklist:** RLDs that are exclusively or primarily used for abuse can have their delegations from the TLD removed by an ISD after having this abuse demonstrated according to some policy followed by each ISD.

In any case, edits to TLDs are limited to either delegating to the RLD that the TLD operator delegates to, temporarily failing to delegate to an RLD (in the case of quarantine), or permanently failing to delegate to an RLD (in the case of blacklisting). Delegation to a different RLD key than that delegated to by the TLD operator — completely inconsistent naming — is treated as an error condition.

The purpose of the NCO is to make all such inconsistencies in native isolation contexts in different ISDs globally transparent, to detect and allow ISDs to correct unintentional inconsistencies, and to detect and repair error conditions. The NCO consists of a distributed service, run on servers in each ISD. For each ISD, it provides an authoritative view of that ISD's naming consistency policy (whether it performs RLD quarantine and what its timeouts are, its enumerated set of TLD delegations, and its enumerated RLD blacklist), and accepts RAINS delegation assertions sampled from various points in the infrastructure to compare against those policies. Sampled delegation assertions are shared with NCO servers in other ISDs, to be compared with those ISDs' views of the same delegations in their own native isolation context.

Sources of sampled delegation assertions include:

- Query servers can be configured to send a sampled set of delegation assertions used in verifying assertions on behalf of clients to the NCO server for their ISD.
- TLD authority servers can bulk-transfer delegation assertions to the NCO servers for each ISD on any change.
- RLD authority servers can send their public keys (in the form of a self-signed delegation assertion inside a local context) to the NCO servers for the ISD(s) providing their connectivity on any change.
- Specially deployed NCO clients can query for RLD assertions and report the resulting delegation assertions to designated NCO servers. The set of query targets can be derived from other sources of data available to the NCO.

Since inconsistencies uncovered by the NCO generally require human intervention and/or policy decisions to correct, the output of this process is made available via a RESTful API and a web front end, also operated by each ISD. Remedies for correction of intentional errors and undeclared inconsistency are a matter for inter-ISD coordination (see Section 3.5 on Page 51).

Since the NCO operates primarily through deterrence, it is not necessary that sampled delegation assertions cover every inconsistency with policy within some bounded time. It is enough that ISD operators know that edits they make to TLD delegations are visible, and that someone is watching. The sampling rate for sampled assertions at a query server should be selected to balance the tradeoff between the likelihood that maliciously transient inconsistency goes undetected with the overhead of sending assertions to the NCO.

7 Control Plane

SAMUEL HITZ, ADRIAN PERRIG,
STEPHEN SHIRLEY, PAWEL SZALACHOWSKI

In this chapter, we discuss SCION's control plane, whose main purpose is to create and manage path segments, which can be combined into forwarding paths to transmit packets in the data plane.

We first describe how path exploration is realized through beaconing, then we discuss the management of path segments (registration, lookup, and revocation), failure resilience, and the use of anycast to enable services to communicate with each other. We also show how SCION allows AS-level hierarchical anycast services to be built, and finally we describe the SCION Control Message Protocol (SCMP).

Chapter Contents

7.1 Path Exploration and Registration

In this section, we go into the details of generating and propagating path-segment construction beacons in SCION. We describe the control-plane format of a path, how beacon construction is initiated and propagated, and how ASes generate diverse paths.

A. Perrig et al., *SCION: A Secure Internet Architecture*, Information Security and Cryptography, https://doi.org/10.1007/978-3-319-67080-5_8

7.1.1 Path-Segment Construction Beacons (PCBs)

SCION introduces *path-segment construction beacons* (PCBs) to enable path exploration and registration. PCBs are used for intra-ISD and inter-ISD (core) path exploration, and contain topology and authentication information. They can include additional metadata that helps with path management and selection. Broadly speaking, a PCB represents a single path segment that can be used to construct end-to-end forwarding paths. Formally, a PCB is defined as

$$PCB = \langle\, INF \,\|\, ASE_0 \,\|\, ASE_1 \,\|\, \ldots \,\|\, ASE_n \,\rangle \tag{7.1}$$

where INF is an *info field*, and each ASE_i represents an *AS entry* that contains all information about a particular AS on the path segment represented by the PCB.

In the following, we describe all elements included in a PCB. The actual wire format of a PCB is presented in Figure 15.13 on Page 357.

Info Field (INF)

The first component of every PCB is the info field (INF), which provides basic information about the PCB. Specifically, the info field contains the following elements:

$$INF = \langle\, Flags_{INF} \,\|\, TS \,\|\, ISD \,\|\, SegLen \,\rangle \tag{7.2}$$

where $Flags_{INF}$ is used in the forwarding path to describe the type and the direction of the constructed end-to-end path, TS is a timestamp that denotes when the PCB's propagation started, ISD is an identifier of the isolation domain within which the beaconing was initiated, and $SegLen$ denotes the length of the forwarding path's segment (this field is set to 0 during the beaconing).

More information about the format of the info field is provided in Section 15.1.3 on Page 347.

AS Entry (ASE)

The complete information about an AS in a PCB is called an *AS entry* and is defined as follows:

$$ASE = \langle\, Meta \,\|\, \boldsymbol{HE} \,\|\, PE_0 \,\|\, PE_1 \,\|\, \ldots \,\|\, PE_m \,\|\, RevToken \,\|\, Ext \,\|\, \Sigma \,\rangle \tag{7.3}$$

The *Meta* field contains metadata describing the AS that generated a given entry. It contains ISD and AS identifiers (which together globally identify the AS), followed by the *TRCVersion* and *CertVersion* fields, which specify the TRC and certificate version number that the AS uses. It also signals what size of interface identifiers is used by the AS and the size of the *maximum transmission unit* (MTU) within the AS's network. Then, an AS entry consists of a single hop entry ***HE***, a list of optional peer entries PE_i, a revocation token *RevToken*, which enables revocation of any interface of the entry in an authenticated fashion (as we describe in Section 7.3), and optional beacon extensions *Ext*.

Each AS entry is signed with a private key that corresponds to the public key certified by the AS's certificate with version *CertVersion*. The corresponding signature Σ includes the PCB's metadata *INF*, the current AS entry ASE_i (without signature), and all previous AS entries in the PCB. Formally, the signature Σ_i of AS entry ASE_i in a PCB is defined as follows:

$$\Sigma_i = \mathsf{Sign}_K(INF \parallel ASE_0 \parallel ASE_1 \parallel \ldots \parallel ASE_{i\text{-}1} \parallel ASE_i') \tag{7.4}$$

where ASE_i' is the AS entry ASE_i without its signature, and K is the AS's private key (the corresponding certificate can be identified through the *CertVersion* field). Beacon extensions can contain unprotected fields, which are not included during the signature creation.

More information about the AS entry is presented in Section 15.3.1 on Page 357.

Hop Entry (HE)

A hop entry has the following format:

$$\boldsymbol{HE} = \langle\, InISDAS \parallel EgISDAS \parallel InIF \parallel EgIF \parallel InMTU \parallel \boldsymbol{HF_H}\,\rangle \tag{7.5}$$

where *InISDAS* is a concatenation of the ISD and AS identifiers of an ingress (i.e., the previous) AS, while *EgISDAS* identifies an egress (i.e., the next) AS. If a hop entry belongs to the first/last AS entry, then the ingress/egress ISD and AS identifiers are set to 0. The *InIF* and *EgIF* fields denote an identifier of the ingress and egress AS's interface, respectively, and the *InMTU* field specifies the MTU of the ingress interface. These fields help an end host to identify paths at the interface-level granularity and their MTUs. The last field $\boldsymbol{HF_H}$ is a hop field that includes the authenticated information of the ingress and egress interfaces.

To allow end hosts to explicitly select paths to reach other end hosts, the hop fields are propagated with the corresponding topology information to the end hosts (see below).

Details of hop entries are discussed in Section 15.3.2 on Page 358.

Peer Entry (PE)

Through the *peer entry*, an AS announces that it has a peering connection to another AS. Peer entries have the same format as hop entries, however, the first *PeerISDAS* pair identifies a peer AS (not an ingress AS):

$$PE = \langle\, PeerISDAS \parallel EgISDAS \parallel PeerIF \parallel EgIF \parallel PeerMTU \parallel \boldsymbol{HF_P}\,\rangle \tag{7.6}$$

The *PeerIF* and *EgIF* fields describe interface identifiers of the peer and egress ASes, and the *PeerMTU* field is an MTU value of the peer interface. Contrary to the hop field HF_H in a hop entry, the hop field $\boldsymbol{HF_P}$ in a peer entry authenticates the permission to use the peering between the *peer* and an egress interface.

More details on peer entries can be found in Section 15.3.2 on Page 358.

Hop Field (HF)

Finally, we introduce the *hop field* (HF), which is contained in hop entries and peer entries. A hop field is used directly in the data plane for packet forwarding: it specifies the incoming and outgoing interfaces of the ASes on the forwarding path. To prevent forgery, this information is authenticated.

A hop field encodes one of three cases for connecting adjacent ASes:

1. customer → provider: the egress interface connects the provider (who created the hop field) with its customer,
2. core AS → core AS: the hop field encodes information for the forwarding performed between core ASes,
3. peering links: the peer interface connects the AS (that created the hop field) with its peer AS over a peering link.

A hop field can be part of a hop entry or of a peer entry. We first discuss the case in which the hop field is contained in a hop entry. The hop field is then represented as follows:

$$\boldsymbol{HF_H} = \langle\, Flags_{HF} \parallel ExpTime \parallel InIF \parallel EgIF \parallel \sigma_H \,\rangle \tag{7.7}$$

where the $Flags_{HF}$ field describes the purpose of the hop field (thanks to this field, it is possible to encode forwarding cases other than the ones listed above, see Section 8.2), *ExpTime* defines for how long the hop field is valid (an expiration time of a hop field is an offset relative to the PCB's info field timestamp *TS*), *InIF* identifies the ingress interface (according to the direction of the beaconing), *EgIF* identifies the egress interface, and σ_H is a message authentication code (MAC) computed as

$$\sigma_H = \mathsf{MAC}_K(TS \parallel Flags'_{HF} \parallel ExpTime \parallel InIF \parallel EgIF \parallel HF') \tag{7.8}$$

where $Flags'_{HF}$ is the $Flags_{HF}$ field with only immutable flags set (see Section 8.1 on Page 162), HF' is the hop field of the previous AS (according to the direction of the beaconing) without its flag field included, and K is a local symmetric key, known only to the AS that creates the hop field.

In case the hop field is contained in a peer entry, the structure is slightly different:

$$\boldsymbol{HF_P} = \langle\, Flags_{HF} \parallel ExpTime \parallel PeerIF \parallel EgIF \parallel \sigma_P \,\rangle \tag{7.9}$$

The differences to the previous case are (a) replacing the *InIF* field with the *PeerIF* field identifying the ingress interface of the peering link, and (b) the authentication code σ_P, which is now computed as

$$\sigma_P = \mathsf{MAC}_K(TS \parallel Flags'_{HF} \parallel ExpTime \parallel PeerIF \parallel EgIF \parallel HF'_H) \tag{7.10}$$

where $Flags'_{HF}$ is the $Flags_{HF}$ field with only immutable flags set (see Section 8.1 on Page 162), and HF'_H is the hop field from Equation 7.7 without its flag field included. In other words, the verification of a peering link requires a locally generated provider-customer hop field.

More details on the format of hop fields is provided in Section 15.1.3 on Page 348.

7.1.2 Intra-ISD Beaconing and Path-Segment Registration

Paths in SCION are made available through the following two procedures:

1. *beaconing* (i.e., path exploration), which builds and propagates PCBs (from which path segments are created); and
2. *registration* of path segments to make them available to other entities.

The PCB generation process is initiated by each core AS, once per *propagation period*. The propagation of PCBs immediately follows PCB generation. When a PCB is received by an AS, its beacon server registers the contained path segment at the path servers, extends the PCB, and propagates the PCB further downstream. These steps are presented in Figure 7.1 on the next page. The propagation period is a parameter specified by each AS; its default value is 5 seconds in our current implementation.

Initiating Beaconing

Intra-ISD beacons are disseminated top-down (i.e., from core ASes to leaf ASes). Each core AS, through its beacon server, initiates the path exploration process by creating an initial PCB and propagating it downstream to each of its customer ASes. A core beacon server propagates a PCB to each customer. The process is repeated every propagation period. The beacon server inserts (among other information) the initial AS entry ASE_0 in the PCB. In this case, ASE_0's hop entry $\boldsymbol{HE}$ includes an initial hop field with ingress interface identifier set to $\bullet$ (indicating an empty value)

$$\boldsymbol{HF_0} = \langle\, Flags_{HF} \parallel ExpTime \parallel \bullet \parallel EgIF \parallel \sigma_0 \,\rangle \tag{7.11}$$

since $\boldsymbol{HF_0}$ represents the first hop and as such has no ingress interface (see Equation 7.7). We also use the empty value *null* for the previous hop-field entry:

$$\sigma_0 = \mathsf{MAC}_K(TS \parallel Flags'_{HF} \parallel ExpTime \parallel \bullet \parallel EgIF \parallel null)$$

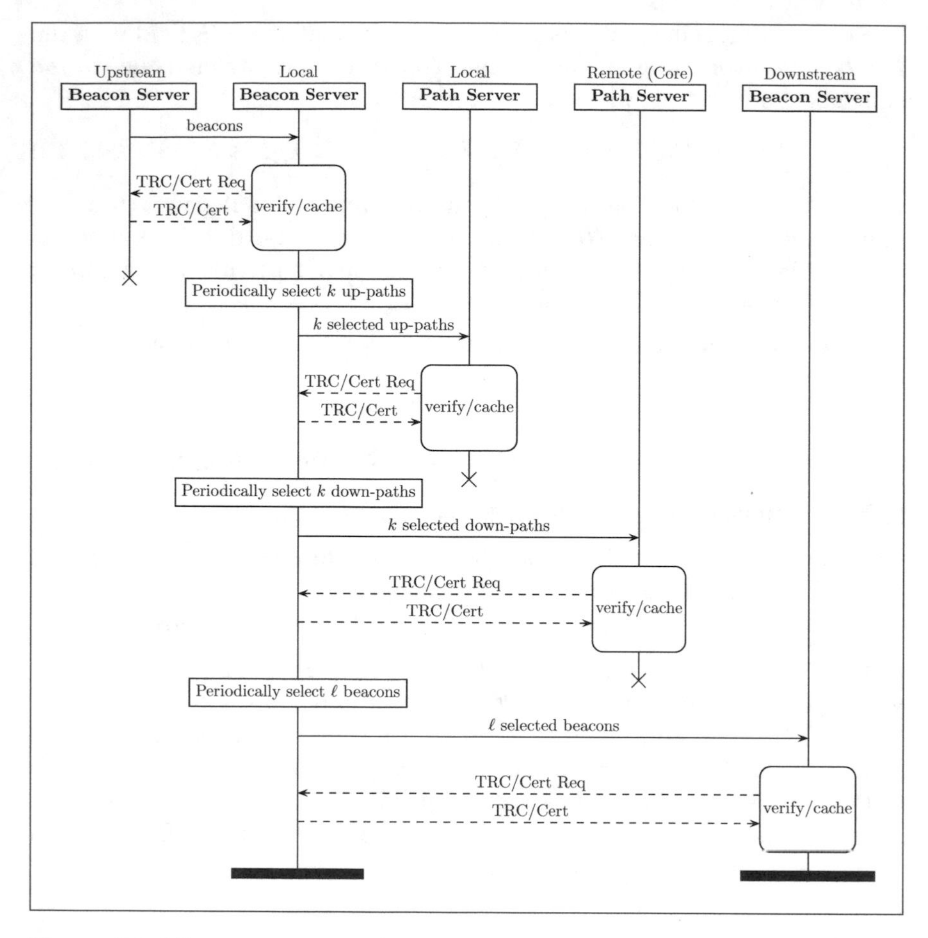

Figure 7.1: Message sequence chart illustrating the beaconing and path-segment registration process.

since no previous hop field exists (see Equation 7.10).

Using the combination of the info field's *absolute* timestamp *TS* and the hop field's *relative* duration *ExpTime*, each AS computes the *absolute* expiration time of the hop field. When the expiration time is exceeded, the hop field is considered expired and an AS's border router (the one assigned to *EgIF*) will drop packets with expired hop fields. The initial hop field denotes the beginning of a path and authenticates a forwarding decision for every packet that

- enters the AS through the interface *EgIF* and terminates inside the AS;
- originates from the AS and exits through the interface *EgIF*; or
- at this AS, switches to another path (which has to begin at this AS as well).

Finally, the beacon server signs the beacon and sends it to a border router (which corresponds to the *EgIF* identifier as specified in the hop field). The beacon server knows the mapping between interface identifiers and border router addresses from the AS discovery service (see Section 7.4.6).

PCBs are disseminated within packets addressed to the *beacon service*.[1] Initial PCB packets have to be processed differently from data packets as they do not contain full forwarding paths. To enable communication between two beacon servers in neighboring ASes a special one-hop path is created (see Section 15.1.4 on Page 351). The PCB is sent to the egress router, which then forwards it to the neighboring border router of the downstream AS.

Beaconing by Non-core ASes

The ingress border router of the downstream AS receives the PCB packet, detects that the destination is a SCION service address, and sends it to one of its beacon servers.[2] The beacon server verifies the structure and the signature of the PCB. The PCB contains the version numbers of used TRC(s) and certificate(s). It enables the beacon server to check whether it has the relevant TRC(s)/certificate(s); if not, it can be requested from the upstream beacon server, and then forwarded to a local certificate server. After the PCB verification is successful, the beacon server adds the PCB to its local database. The process is depicted in Figure 7.1.

Every propagation period (the time interval is configured by the AS), the beacon server selects the ℓ best PCBs from its database and continues path exploration by sending the PCBs to its downstream ASes (in our current implementation $\ell = 5$). PCB selection criteria are set according to local AS policies. The selection process is presented in detail below in Section 7.1.4.

For every selected PCB and for every interface that connects to a downstream AS, the AS creates a new PCB by adding a new AS entry. The AS entry includes an HF that authenticates the permission to send traffic between ingress and egress interfaces (see Equation 7.7), and HFs that authenticate forwarding between the peer interfaces and the egress interface (see Equation 7.9). (The AS can set an HF as *forward-only*, which denotes that the HF can be used only for transit, i.e., cannot be used to deliver a packet to the AS's end hosts.) The set of ℓ created PCBs are sent to the border router corresponding to the egress interface and forwarded to the downstream AS (see Figure 7.1).

---

[1] SCION introduces service addresses to address a service instance (with unknown actual address) in a remote AS. See details in Section 7.4.7 and Section 15.2.

[2] If there are several beacon servers in the AS, the PCB is sent to only one. The details are presented in Section 7.4.7 on Page 152.

Path-Segment Registration

Intra-ISD beaconing provides ASes with paths to communicate with their core ASes. To make paths accessible to their own and remote end hosts, the paths need to be published. Every time interval (called a *registration period*, determined by the AS, and set by default to 5 seconds in our current implementation), a beacon server selects two sets of path segments:

1. up-segments: to allow a local end host to contact core ASes, and
2. down-segments: to allow remote end hosts to fetch paths from core ASes towards a target AS.

An AS can set different selection policies for these two sets (see Section 7.1.4).

More specifically, in every registration period, beacon servers execute the following:

1. From the cached PCBs, select k PCBs that will be used as up-segments, and another k PCBs that will be registered as down-segments. As a default value, we use $k = 5$.
2. Remove all unprotected (i.e., non-signed) fields from the beacon extensions.
3. To every selected PCB, add a new AS entry with a final hop field of the following format:

$$\boldsymbol{HF}_H = \langle\, \mathit{Flags}_{HF} \,\|\, \mathit{ExpTime} \,\|\, \mathit{InIF} \,\|\, \bullet \,\|\, \sigma \,\rangle \tag{7.12}$$

   Only the ingress interface identifier is specified (i.e., *EgIF* is set to $\bullet$) since the path ends at the AS.

4. If the AS has peering links, for each peering link add to the AS entry a hop field of the following format:

$$\boldsymbol{HF}_P = \langle\, \mathit{Flags}_{HF} \,\|\, \mathit{ExpTime} \,\|\, \mathit{PeerIF} \,\|\, \bullet \,\|\, \sigma \,\rangle \tag{7.13}$$

   Only the peer interface identifier is specified (i.e., *EgIF* is set to $\bullet$) since the path ends at the AS.

5. Sign every selected beacon and append the computed signature. Such modified PCBs are then called *path segments*.
6. Register the resulting up-segments with a local AS's path server, and the down-segments with a core path server from a local ISD.

Unprotected fields of beacon extensions are removed for efficiency reasons (to reduce the size of the path segments). Path-segment registrations are sent as packets addressed to the path service (see Section 15.2 on Page 355). The format of the path registration message is presented in Section 15.4. Up-segments are registered at a local AS's path server, while down-segments are registered at a core path server from a local ISD. We note that the down-segment registration process is more complex since the core path server, which received a down-segment, has to replicate the segment among all core ASes within its ISD. Due to such replication, all core ASes can serve down-segments for all non-core ASes from the same ISD.

7.1.3 Inter-ISD Beaconing and Path-Segment Registration

The inter-ISD (or core) beaconing process is conducted only by core ASes in order to create core-segments, which enable two core ASes to communicate. The structure of inter-ISD beacons is identical to the structure of intra-ISD PCBs (see Section 7.1.1). However, the process of core beaconing differs slightly from the intra-ISD process. The main difference is that *every* core AS periodically initiates core beaconing by sending beacons to all its neighbor core ASes (not to its customers, as in the intra-ISD case). In inter-ISD beaconing the core PCB from each core AS is flooded to all other core ASes (forming a complete flooding tree), whereas in intra-ISD beaconing only PCBs originating from core ASes are disseminated along provider-customer links (forming a more limited distribution tree compared to core PCBs). For inter-ISD beaconing, our implementation sets the same default parameters as in the intra-ISD case (i.e., propagation and dissemination periods are 5 seconds long, and $\ell = k = 5$).

Initiating Core Beaconing

Inter-ISD PCBs (also referred to as *core PCBs*) are disseminated from every core AS to all other core ASes. Each core AS, through its beacon servers: (a) initiates the path exploration process by creating an initial core PCB and propagates it to all neighbor core ASes, and (b) propagates PCBs originated by other core ASes. The process is repeated in every propagation period (the period can be adjusted by every core AS, as before).

Among other information, the beacon server adds the following information to a core PCB: the current timestamp, the version of the used TRC and certificate, and the first AS entry, which contains only a single hop entry (peer entries are not added). This hop entry contains the ISD and AS identifiers of the current and the next ASes, and carries the hop field in the format presented in Equation 7.11. Similarly to the intra-ISD exploration process, the hop field denotes a beginning (or an end) of a path and authenticates a forwarding decision for every packet that

- comes from the interface *EgIF* and terminates inside the AS, or
- originates from the AS and exits through the interface *EgIF*, or
- at this AS, switches to another path (which has to begin at this AS as well).

Finally, the beacon server signs the PCB and sends it to the border router, which processes it similarly to the intra-ISD case (i.e., the PCB is finally passed to a neighbor beacon server). Note that the neighboring AS can be in the same or in a different ISD, and consequently, the ISD identifier included in the info field describes only the ISD of the PCB originator.

Beaconing by Core ASes

After an ingress border router passes a core PCB to a beacon server, the beacon server verifies the PCB, and similarly to the intra-ISD case, the beacon servers exchange TRC(s) and/or certificate(s) (if the TRC and/or certificate version has changed). As beaconing in the cores is based on flooding, it is necessary to avoid loops during path creation. A core beacon server avoids loops at both the AS and ISD levels as follows:

- it discards PCBs that include an AS entry created by itself,
- it discards PCBs that re-enter an already visited ISD.

Finally, the beacon server adds the PCB to its local database, as beacon servers collect PCBs to all seen ASes.

In every propagation period, the beacon server selects the ℓ best PCBs for every core AS from its database. PCBs are selected per unique core AS, as the goal of core beaconing is to have path(s) that connect every pair of core ASes. The selection criteria are set according to local AS policies, which are presented in detail in the next section. For every such selected PCB and for every interface that connects to a core AS, the beacon server creates a new PCB by adding a new AS entry. The AS entry includes only a single hop field that authenticates forwarding between ingress and egress interfaces (see Equation 7.7). The set of such created PCBs is sent to the border router corresponding to the egress interface and finally to the neighbor core AS.

Core Path-Segment Registration

The core beaconing process creates core AS path(s) to other core ASes. These paths have to be registered at local ASes' path servers so that local and remote end hosts can obtain and use them. In contrast to the intra-ISD registration procedure, there is no need to register core-segments with other ASes (as each core AS will receive PCBs originated by every other core AS).

In every registration period, a core beacon server

1. selects the k best PCBs towards each core AS observed so far, from the cached core PCBs;
2. removes all unprotected fields from the beacon extensions;
3. adds a new AS entry to every selected PCB with a hop field of the following format:

$$HF = \langle\, Flags_{HF} \parallel ExpTime \parallel InIF \parallel \bullet \parallel \sigma \,\rangle \tag{7.14}$$

(only the ingress interface identifier is specified (i.e., *EgIF* is set to •) since the core path ends here);
4. signs every selected PCB and appends the computed signature. Such modified PCBs are called *core-segments*;
5. registers the resultant core-segments with a local AS's path server.

7.1.4 Beacon and Path-Segment Selection

As an AS receives a series of intra-ISD or core PCBs, it must select the PCBs it will use to continue beaconing and to register path segments at path servers. A non-core AS must select (a) a subset of PCBs to propagate downstream, (b) up-segments to register at a local AS path server, (c) down-segments to register at a core path server. A core AS must select (a) a subset of PCBs to propagate to neighbor core ASes, and (b) core-segments to register at a local AS path server. Core ASes do not register core-segments at remote AS path servers, as due to core beaconing (see Section 7.1.3) all core ASes find a set of paths to all other core ASes.

The selection process is based on path properties (e.g., length, disjointness across different paths) as well as PCB properties (e.g., age, last transmission time). In this section, we describe the process by which an AS evaluates and selects PCBs. The beacon server of an AS maintains a data structure of received PCBs under consideration for downstream propagation and registration at path servers. Each AS can specify how PCBs are evaluated or eliminated from consideration through a local policy.

Although the policy-based selection process presented here enables a variety of path choices, ASes may need to express more sophisticated routing policies. In Section 10.9 we discuss how SCION can support routing policies fitting today's Internet business models.

Beacon Store

Each time a beacon server receives a PCB, it chooses whether or not the PCB will be stored as a candidate (i.e., under consideration for propagation and registration). To manage the set of candidate PCBs, the beacon server maintains a database of PCBs called the *beacon store*. The beacon store has a fixed capacity n and supports the following operations:

- add: add a new PCB to the beacon store if it complies with the selection policy. If the beacon store already contains n PCBs, remove the least desirable PCB.
- remove: remove a PCB from the beacon store.
- select: select a number (specified as a parameter) of PCBs.

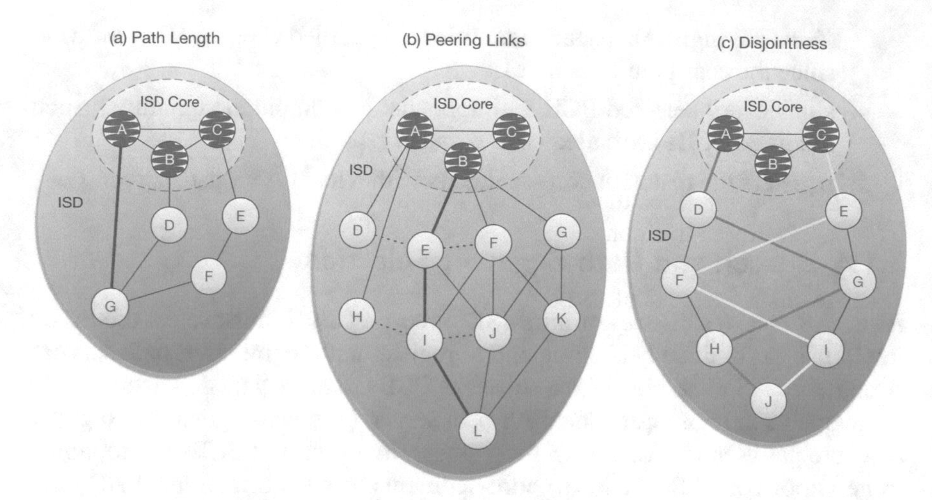

Figure 7.2: Example graphs to illustrate beacon and path-segment selection based on different path properties.

Through the above operations, the beacon store is thus implicitly responsible for applying the AS's selection policy (described below). In addition to storing PCBs, the beacon store also stores metadata for each PCB, such as when the PCB arrived at the beacon store and when it was most recently forwarded to a downstream AS.

Selection Properties

We propose a set of metrics that represent a range of desirable properties in a path or PCB:

- **Path length:** The first property we consider is *path length*. In this case, path length is defined as the number of hops from the originator AS to the local AS. This can give an indication of the path's latency (although there are many other factors affecting latency).

  In Figure 7.2a, we can see that AS *G* will receive the paths *AG*, *BDG*, and *CEFG*, which have lengths of 1, 2, and 3 hops, respectively. Based purely on length, *G* would prefer *AG* first, followed by *BDG* and *CEFG*.
- **Peering ASes:** We also consider *peering ASes*, defined as the number of peering ASes from all non-core ASes on the PCB. The number of peering ASes is important because a greater number of peering ASes on a PCB increases the likelihood of finding a shortcut using that segment.

  In Figure 7.2b, AS *L* receives seven distinct PCBs, all of which start at *B* and are three hops long. However, the number of peering ASes in these paths range from none (in *BGKL*) to four (in *BEIL*).

- **Disjointness:** Unlike other properties, the *disjointness* of candidate PCBs (illustrated in Figure 7.2c) is calculated relative to other PCBs and thus depends on PCBs that have been previously sent. We use the two following definitions of disjointness: paths can be *vertex-disjoint* (i.e., they have no common upstream/core AS for the AS the beacon store is in) or *edge-disjoint* (i.e., they do not share any AS-to-AS link). Both definitions are useful in this context: vertex-disjointness allows path diversity in the event that an AS becomes unresponsive, and edge-disjointness provides resilience in case of link failure.
- **Last reception:** The *last reception* of a PCB is defined as the time that has elapsed since the PCB arrived at the AS's beacon store. This metric is important because a short elapsed time indicates that upstream ASes found the PCB desirable and fewer catastrophic events (e.g., a failing link) can have affected the segment since it was propagated. Thus, older PCBs can be considered as more stable, thus more preferable. Because upstream ASes may propagate the same PCB multiple times, a beacon store may receive a PCB from its upstream AS that it has already received before. In this case, the beacon store simply updates the PCB's arrival time. On the other hand, new paths (never seen before) can also be desirable and should be propagated quickly to announce new paths.
- **Last transmission:** The time that has elapsed since the AS's beacon server last propagated the PCB must be taken into consideration. If the PCB has never been propagated downstream, then the beacon store assigns the PCB's last transmission a value of ∞. The last transmission of a path is important because it allows the beacon store to take into account paths that have not been propagated in a while and thus can improve the diversity of beacons transmitted downstream over time.
- **Feature support:** Beacon selection can be extended to support richer criteria, such as bandwidth reservations in SIBRA (see Chapter 11), consistent support for a certain SCION extension on a path, or support for a specific cryptographic algorithm, for instance.

Selection Policy

Each AS has a selection policy, which governs the storage and selection of PCBs at all beacon servers in the AS. In particular, a selection policy specifies the following:

- the maximum number n of candidate PCBs to store,
- the number k of up-path segments to register at a local path server each registration period,
- the number k of down-path segments to register at a core path server (specified only by non-core ASes) each registration period,

- the number ℓ of PCBs to propagate (downstream or to core ASes) each propagation period,
- a list of blacklisted ASes that must not appear in any PCB sent downstream or registered,
- a set of minimum and maximum allowable values for properties, and
- a set of weights representing the relative importance of the previously mentioned properties in evaluating and selecting PCBs.

Beacon policies are local to the AS and it might be in the commercial interest of the AS to keep them private.

Filtering Beacons

When the beacon server receives a PCB, the beacon store first checks the path against a series of filters defined by a selection policy. These filters check whether any ASes in the segment are blacklisted, and whether the path properties fall between the minimum and maximum allowable values specified in the selection policy. The latter type of filtering allows paths with certain undesirable properties, such as being longer than a threshold number of hops, to be ignored as a candidate PCB.

Selecting PCBs and Path Segments

The beacon store computes the overall quality of a PCB as a weighted sum, using the weights specified in the selection policy. Once it has computed the quality of all candidate PCBs, the beacon server selects the top-ranked PCBs. Time-based path properties, such as age and transmission time, must be recomputed when the beacon store selects PCBs. Disjointness is based on previous operations and must also be computed when PCBs are selected (i.e., every propagation or registration period).

7.2 Path Lookup

Path lookup is a fundamental building block of SCION's path management architecture. It enables end hosts to obtain path segments found during path exploration. End hosts can then construct end-to-end paths from a set of possible path segments returned by the path lookup process.

7.2.1 Requirements and Design Goals

We considered the following requirements and design goals that led to the design of SCION's path lookup infrastructure.

Low Latency

In the absence of a cached path at end hosts, a path lookup needs to be performed before a packet can be sent to a new destination. It is therefore performance-critical that a path lookup can be performed as fast as possible.

Effective caching is critical for the performance and scalability of path lookup, as it can decrease the latency of path lookups. To minimize the number of path lookups, path servers and end hosts should also cache paths for a short period of time to exploit the temporal locality of network destinations.

Scalability

Path lookup not only has to scale with respect to the number of users, but also to an increasing number of paths available in an ever-expanding network such as the Internet.

Caching can help with scalability with respect to an increasing number of requests. To ensure scalability with respect to the number of paths, the path lookup infrastructure can only contain a subset of all available Internet paths. It is also crucial that the amount of state needed to store and serve paths be as low as possible.

Availability

If the path lookup infrastructure experiences outages, end hosts might be unable to look up new paths, thus crippling the entire communication infrastructure. The path lookup infrastructure should therefore be distributed and replicated to guarantee high availability even when single parts of the system fail or are under attack, e.g., during a DDoS attack.

Cache Consistency

We argued that the use of caching is critical for path lookup with respect to performance, scalability, and availability. However, caching introduces consistency problems. If a cache delivers stale paths, then the performance of the path lookup and all upper layers are negatively impacted; the severity of this problem increases the more distributed the path lookup infrastructure is.

Security

In terms of security, the following properties are critical for the path lookup infrastructure to function properly in the presence of an attacker.

First, end hosts should be able to verify the authenticity of paths they receive from path lookup, i.e., that path segments were registered by the true destination and have not been altered since registration. This prevents an attacker from tricking an end host into using a fake path (similar to cache-poisoning attacks in DNS [177]).

Second, a path should only be removed from the path lookup infrastructure with proper authorization (apart from expiration). Otherwise, an attacker could disconnect an AS from the Internet by repeatedly revoking all paths to that AS.

Third, not all paths should be public. While path servers facilitate the retrieval of paths, it should be possible to distribute paths out of band directly to potential senders. SCION supports non-registered (or hidden) paths, which can serve as an important ingredient in DDoS attack defense.

7.2.2 Path Lookup Process

End-to-end communication is enabled by a combination of up to three path segments that form an end-to-end path. The goal of the path lookup process is to provide a source end host with diverse path segments and at least one set of *connecting* path segments, i.e., path segments that can be combined towards the destination by simply joining their corresponding endpoints. Depending on the location of source and destination end hosts, the path lookup process differs slightly.

Source and Destination from Non-core ASes

A source end host initiates a path lookup by issuing a path request, containing the destination ISD and AS identifiers, to a local path server. The local path server then forwards the request to one of the core path servers, using an up-segment that was previously registered by the beacon server (if the lookup succeeds, the local path server will append this up-segment to the corresponding response). At this point there are two possible scenarios:

1. The destination is in the same ISD as the source. In this case the core path server knows the down-segments to reach the destination and returns up to k segments to the local path server.
2. The destination is in a different ISD than the source. In this case, the core path server requests the down-segments from a core path server in the destination ISD (using a core-segment), before returning them to the local path server.

In both cases, the first core path server (the one requested by the local path server) returns up to k core-segments, which connect its AS and the ASes that originated the down-segments. If a down-segment originates in the core path server's AS, then the core-segment is not required as the up- and down-segments directly connect. However, it is guaranteed that if path lookup succeeds (i.e., an end host receives a set of path segments), then there is at least one set of connecting path segments; thus the end host is able to build a forwarding path.

The local path server then returns up to k up- and k down-segments (and optionally up to k core-segments — if required) to the source. The up-segment used for querying the core path server is included in the response. If the source wishes to communicate through the core and the received core-segments are unsatisfactory, then additional core-segments may be fetched (by asking another core AS). Depending on the received segments, there are different ways a source can combine them to create an end-to-end path. We describe these options in detail in Section 8.2.

End hosts and path servers accept path segments only when they are verified. This verification may require contacting the server that sent the given path. The details of this process are described in Section 4.2.3.

Example. An example of the entire path lookup process is depicted in Figure 7.3. In this example, we assume that the desired paths are not yet cached in the path servers. First, an end host from AS $(1,10)$ (ISD 1, AS number 10) that wishes to contact a host from AS $(2,23)$ contacts its own local path server, requesting path segments connecting source AS $(1,10)$ with the destination AS $(2,23)$. The local path server, using an up-segment, contacts a core path server inside the local ISD (i.e., AS $(1,1)$), requesting path segments from the core path server's AS $(1,1)$ to the destination AS $(2,23)$. (Note that the local path server postpones this request if it has no up-segment.) The core path server of $(1,1)$ takes any core-segment to an AS from the destination ISD 2, and queries this AS's path server. (This request is postponed until a core-segment is available.) In our example, the path server in AS $(2,2)$ is asked about down-segments of the destination AS $(2,23)$. We emphasize that the down-segments of AS $(2,23)$ do not have to originate from AS $(2,2)$, they can originate from any other core AS from ISD 2. As soon as the core path server from ISD 2 has appropriate down-segments, up to k of them are returned to the core path server in AS $(1,1)$, which verifies the path segments (it can query the origin core path server for certificates or TRCs if locally cached information is missing or outdated). Next, this core path server has to find up to k core-segments between its AS $(1,1)$ and ASes that originated the received down-segments. At least one such core-segment has to be found, otherwise the path server waits for it. Then, down-segments with the corresponding core-segments are returned to the local path server of AS $(1,10)$ (which verifies the path segments as well). The local path server adds up-segments to the set of obtained paths, adds additional core-segments (if they are cached) connecting up- and down-segments, and sends the entire response to the end host. The up-segment to the core AS $(1,1)$ has to be within this response. Finally, the end host verifies the received path segments.

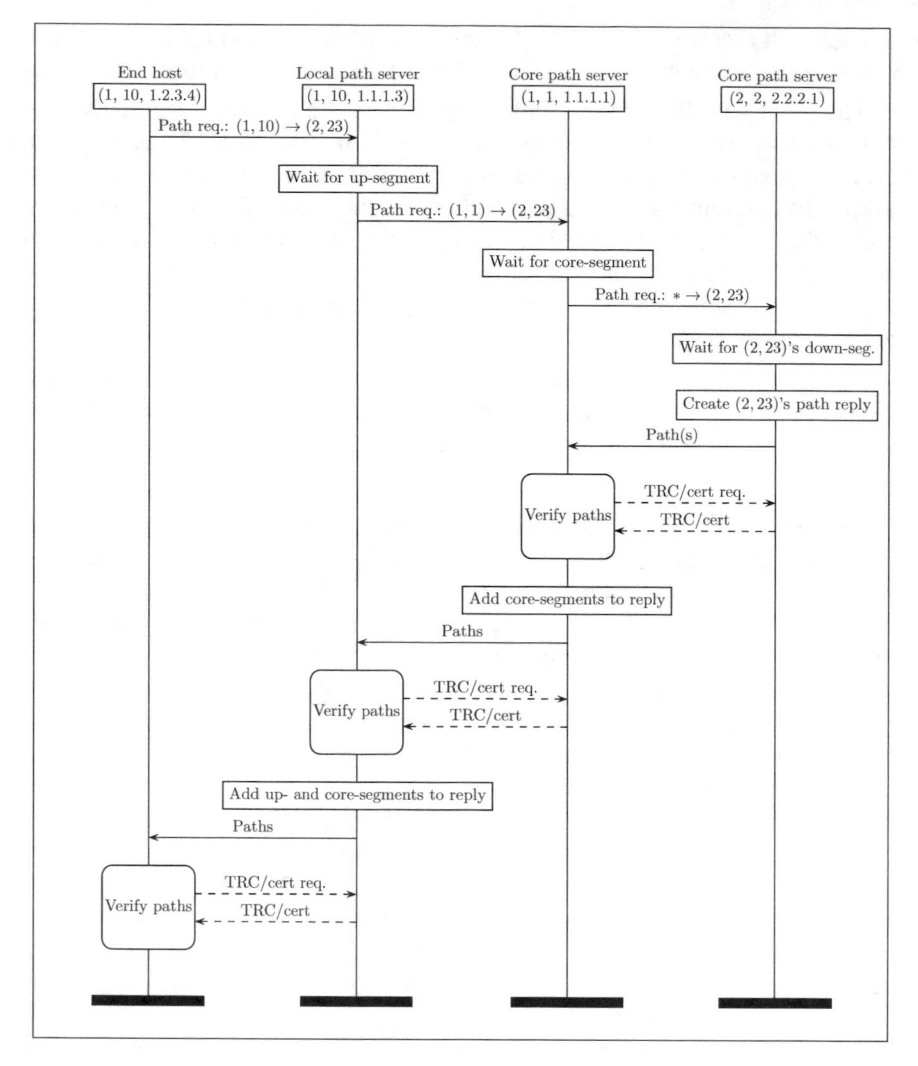

Figure 7.3: A path lookup example.

Source and/or Destination from Core AS(es)

When source and destination end hosts reside in the core, the queried path server returns up to k core-segments towards the destination. When the source is within a core AS, while the destination is within a non-core AS, the source receives up to k down-segments, and up to k core-segments between the source AS and the originators of the down-segments (to guarantee that there exist connecting path segments). Similarly, for a source in a non-core AS and the destination within a core AS, the source is provided with sets of up to k up-segments, and up to k core-segments between the requested core path server's AS and the destination AS. As in the previous cases, received path segments are verified by the receivers.

7.2.3 Caching

To reduce path lookup latency, path servers form a hierarchical caching infrastructure. Every path server internally maintains a cache of path segments received during path lookup. There are three events that trigger the removal of a path segment from a path server's cache:

1. The cache fills up completely. Path servers use the *least-frequently-used* replacement strategy to replace path segments if the cache completely fills up.
2. A path segment expires. Each path segment contains an expiration time (up to 24 hours) after which a path server evicts the path segment from its local cache. The expiration time of a path segment is the minimum of all hop-field expiration times contained in the path segment.
3. A path segment is explicitly revoked. Path revocation is covered in detail in Section 7.3.

Using a least-frequently-used replacement strategy ensures that the most frequently requested path segments are kept in a path server's cache. This is especially important for down-segments toward popular destinations or core-segments frequently involved in transit. Similarly, end hosts also cache obtained path segments.

7.2.4 Path-Segment Authenticity

Path segments are signed in the same way as beacons, i.e., by every AS on the path. Each path server (and end host) can verify a path segment regardless of its origin. By tying together path segments with information required for their verification (i.e., certificates and TRCs), we decouple verification of a path segment from the path server that delivers the segment during path lookup. Such an approach provides availability of the authenticity verification, as path segments can be freely distributed throughout the entire path lookup infrastructure. The details of the authentication process are described in Section 4.2.3.

7.2.5 Non-registered Path Segments

Public services typically want their servers to be reachable by as many hosts as possible. In these cases, maintaining an up-to-date set of path segments for that service's AS achieves this goal. However, certain use cases require services to be accessed only by authorized senders. While authorization can be achieved at the application layer, denial-of-service attacks may exhaust resources, preventing the data from reaching the authorization application or overwhelming the application so that it cannot process all requests. In these cases, it would be beneficial to make the path segment available only to specific authorized senders, and not allow attackers or unauthorized parties to even establish a connection to the service.

Non-registered (or hidden) path segments fulfill this need. Instead of registering the path segment to a path server, the path segment is communicated out of band (e.g., in person, via secure messaging, or posted encrypted to a public site) to authorized senders. Consequently, only authorized senders may then begin to use that path segment for communication.

7.3 Secure Path Revocation

In this section, we describe the SCION path revocation mechanism, which addresses the problem of removing faulty or undesired path segments from the path infrastructure. In SCION, path segments must be revoked, i.e., removed from path servers, in two cases:

1. due to changes in routing policies, i.e., *proactively*;
2. due to a link failure on the path, i.e., *reactively*.

The first case is usually not time-critical and can be addressed through expiration timestamps on path segments in conjunction with ASes ceasing to advertise these paths. An AS can always unregister its previously registered path segments in the core path server, which prevents end hosts in ASes that do not have a cached copy of the path from using it. However, cached copies will still be usable for as long as the path is valid. We assume that an AS is committed to a path segment it registers for the entirety of its lifetime.

For reactively revoking a path segment due to a link failure on the path, time plays a critical role; the faster a faulty path segment can be revoked, the fewer sources will try to make use of the faulty path segment and the quicker the system will converge to a state without stale (non-functioning) path segments. Thus, a path revocation system needs to be tuned to rapidly and efficiently remove faulty path segments.

Efficiency and scalability. To ensure scalability and also to prevent denial-of-service (DoS) attacks by malicious entities in the network infrastructure, it is critical to achieve low computational, storage, and bandwidth overhead. Thus, a revocation must not require involved network elements to keep an excessive amount of state or to generate a large number of additional messages within the network.

Additionally, a revocation must be efficiently verifiable to prevent overwhelming verifiers through many (possibly forged) revocations. Finally, revocations should be short, to minimize communication overhead.

Security. Path revocation is designed to remove a path segment from the path infrastructure. Therefore, the system needs to prevent unauthorized or malicious parties from revoking path segments. The system must thus ensure that revocations are authentic, i.e., only the operator of an interface should be able to revoke that interface.

Finally, it must be impossible to replay a recorded revocation with the effect of removing a valid path segment (resistance against replay attacks).

7.3.1 Design

The main task of the revocation system is to rapidly remove cached copies of path segments containing a failed link. The first design decision to make is whether the revocation system should be active or passive. An example of a passive design is the time-to-live (TTL)-based expiration of cached DNS records [178]. While simple in design, passive revocation suffers from conflicting goals: on the one hand, path segments should be cached for as long as possible, but on the other hand, failed path segments should be removed from caches as quickly as possible. TTL-based revocation cannot simultaneously achieve both goals.

With active revocation, long path segment retention can be achieved, while also enabling fast removal of failed path segments. The main design decision for active revocation lies in a suitable choice for revocation dissemination. In the following, we describe the salient features of SCION's path revocation system.

Our design is motivated by the observation that a cached but unused path segment does not have to be removed, because a faulty path segment will be detected with usage, which can trigger removal. While using a stale path segment leads to some overhead to detect the failure and recover from it, immediately removing it from path servers is not critical. We can exploit this to create a system with a loose consistency requirement, i.e., as long as a failed path segment is not used, then there is no point in expending effort to revoke it, but once a path segment is used, then the system revokes it, thus benefiting others who may want to use it at a later point.

Another design aspect is the granularity of revocations. Our revocation scheme works on the granularity of interfaces. To revoke a path segment, an AS simply revokes the interface corresponding to the failed link, i.e., its end of the link. This way, only a single revocation message is needed to revoke all path segments that contain a failed interface.

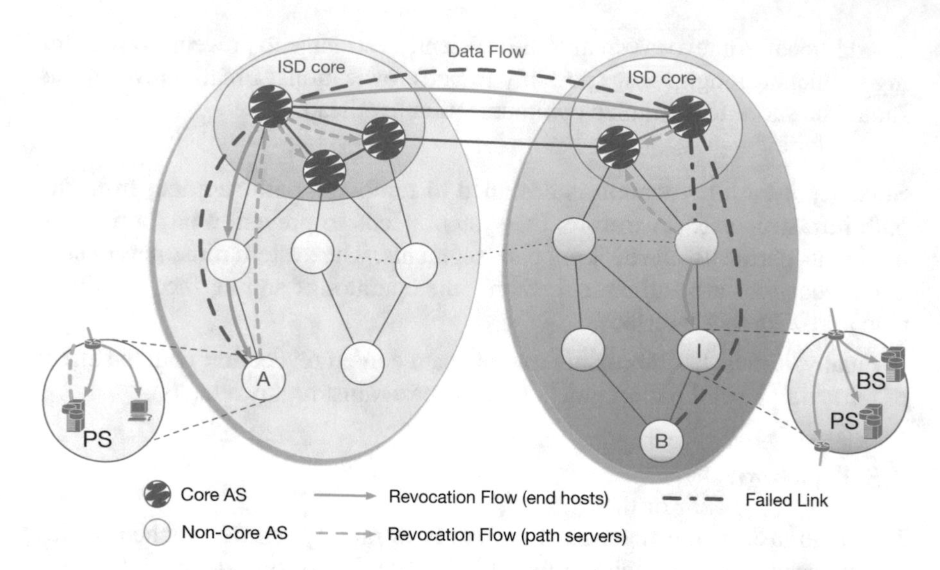

Figure 7.4: Overview of the path revocation system.

Figure 7.4 depicts an overview of SCION's revocation system:

- Whenever an AS needs to revoke an interface, the beacon server of that AS informs each border router about the revoked interface; thus a border router is always aware of all the revoked interfaces within its AS. Additionally, the beacon server sends the revocation to a core path server in its ISD (green arrows in the orange ISD).
- Whenever a packet with a forwarding path containing the interface ID of a revoked interface arrives at a border router, the border router issues a SCION control message protocol (SCMP) packet containing the revocation that is sent back to the sender along the reverse direction of the forwarding path contained in the packet header (blue arrows).
- The ingress border router in the AS of the source forwards the SCMP packet to the source, and additionally to the local path server. The local path server verifies and processes the revocation, and forwards it to a core path server in its ISD. That core path server then forwards the revocation to all other core path servers in the ISD (green arrows in the yellow ISD).
- If the SCMP packet travels downstream (away from the ISD Core), then border routers in that ISD downstream of the failed link send an additional SCMP packet to the local beacon server (AS I and AS B). This is to prevent beacon servers from disseminating beacons containing a failed link.
- End hosts receiving a revocation can verify it and immediately switch to (or request) an alternative path.

Algorithm 2 Initiate revocation at a beacon server

```
 1: procedure IssueRevocation(IF_x)
 2:     revMsg ← BuildRevMsg(IF_x)
 3:     for all router ∈ BorderRouters do
 4:         SendTo(router, revMsg)
 5:     end for
 6:     SendTo(CPS, revMsg)
 7:     RegisterNewDownSegments(CPS)
 8:     SendTo(LPS, revMsg)
 9:     RegisterNewUpSegments(LPS)
10: end procedure
```

- To meet our security goal, we propose a lightweight and efficient authentication scheme that allows each AS to prove to anyone in the network that it is the owner of the revoked interface and thus authorized to perform the revocation. Replayability of revocations is limited to 10 seconds, the lifetime of a revocation. (More details will be given in Section 7.3.3.)
- Due to the short lifetime of revocations, each network element can keep a map of all processed revocations and thus it can easily drop duplicates. Each entry in this map needs be kept for at most 10 seconds.

7.3.2 Processing of Revocations

Beacon Servers

The beacon server keeps track of the state of all interfaces within its AS through periodic keep-alive messages sent between adjacent border routers. If a link or an interface to a neighboring AS fails, the beacon server initiates the following revocation process (Algorithm 2):

1. For a failed interface IF_x, the beacon server creates a revocation message by calling the build revocation message algorithm (Algorithm 5).
2. The beacon server then sends a status update to all border routers in the AS, informing them about the status of the interface and installing the revocation message to revoke the interface.
3. It then sends the revocation to the core path servers in its ISD together with a new set of down-segments.
4. Finally, the beacon server sends the revocation together with a new set of up-segments to the local path server.

A beacon server that receives a revocation for an upstream interface checks whether any of its currently registered paths are affected, and if so, immediately registers a new set of up/down-segments with the local and core path servers.

Algorithm 3 Process revocation at a border router

```
 1: procedure SendRevocation(pkt, IF_x)
 2:     revMsg ← GetRevMsg(IF_x)
 3:     SCMPPacket ← SCMPPacket(this.addr, pkt.src, ReversedPath(pkt))
 4:     SCMPPacket.payload ← revMsg
 5:     NormalForward(SCMPPacket)
 6: end procedure
 7: procedure ForwardRevocation(rev)
 8:     NormalForward(rev)
 9:     if FromLocalAS(rev) or AlreadySeen(rev) then
10:         return
11:     end if
12:     if ToLocalAS(rev) then
13:         ForwardTo(LPS, rev)
14:     end if
15:     if rev.ISD == this.ISD and FromUpstream(rev) then
16:         ForwardTo(BS, rev)
17:     end if
18:     AlreadySeen(rev) ← True
19: end procedure
```

Border Routers

Border routers perform different functions with respect to revocation processing, depending on their position on the path of the packet that triggers a revocation message (Algorithm 3):

- If the current or next hop interface of the packet's forwarding path is revoked within the local AS, a border router sends an SCMP packet containing the corresponding revocation back to the source host. To that end, a border router reverses the path of the packet that triggered the revocation message.
- An ingress border router in the AS of the source host forwards the SCMP revocation packet to the source host, and also sends it to the local path server.
- An ingress border router downstream of the failed link forwards the SCMP revocation packet toward the destination and also sends it to the local beacon server. This prevents beacon servers from disseminating beacons containing failed links.
- If none of these conditions apply, a border router simply forwards a revocation message along the path in its header.

Algorithm 4 Process revocation at a non-core path server

```
 1: procedure ProcessRevocation(rev)
 2:     if AlreadySeen(rev) then
 3:         return
 4:     end if
 5:     x, P, R_{T_{i-1}}, R_{T_{i+1}} ← ExtractProof(rev)
 6:     for all segment ∈ {seg ∈ PathSegments | rev.IF_x ∈ seg} do
 7:         revToken ← segment[rev.IF_x].token
 8:         if Verify(x, P, R_{T_{i-1}}, R_{T_{i+1}}, revToken) then
 9:             Remove(segment)
10:         end if
11:     end for
12:     if rev.src ≠ BS and rev.ISD ≠ self.ISD then
13:         SendTo(CPS, rev)
14:     end if
15:     AlreadySeen(rev) ← True
16: end procedure
```

Non-core Path Servers

Non-core path servers either receive revocations from the local beacon server or from a border router if the path server is in the same AS as the source host that triggered the SCMP revocation packet. The revocation message is processed as follows (Algorithm 4):

1. The path server first checks whether it has already received this revocation and if so, stops processing it.
2. The path server then verifies the revocation against the token included in path segments that contain the revoked interface (using Algorithm 6), to ensure that it was issued by the interface's owner. If the verification succeeds, the path server removes all path segments that contain the revoked interface. More details on revocation authentication can be found in Section 7.3.3.
3. Finally, the local path server forwards the revocation to a core path server if the revocation message originated from a remote ISD.

Core Path Servers

Core path servers receive revocations either from a non-core path server downstream, from a border router if the path server is in the same AS as the source host that triggered the SCMP revocation packet, or from another core path server in the same ISD. In any case, core path servers process the revocation in the same way as non-core path servers (Algorithm 4). Additionally, a core path server also forwards the revocation to all other core path servers in the same

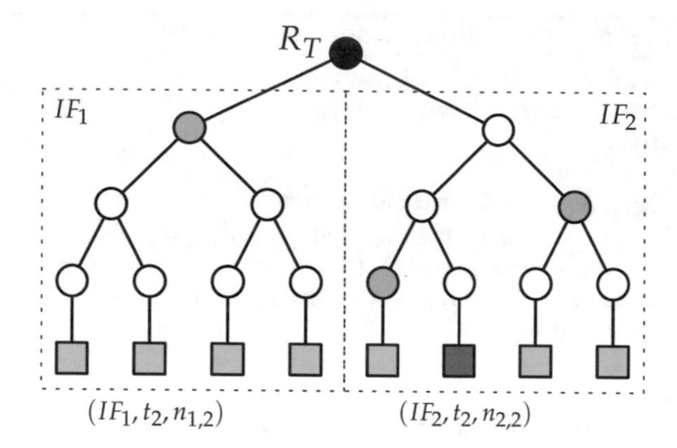

Figure 7.5: A hash tree for two interfaces with $T = [t_1, t_2, t_3, t_4]$. Squares contain the triples $(IF_x, t_i, n_{x,i})$. To revoke IF_2 in epoch t_2, an AS reveals $(IF_2, t_2, n_{2,2})$ (red square) along with the green nodes.

ISD, unless it received the revocation from another core path server in the same ISD.

End Hosts

Much like the path servers, end hosts can perform the `Verify`() procedure and if it succeeds they remove all path segments containing the revoked interface from their cache. Revocations received by an end host in response to a sent packet immediately allow packet retransmission on a different path without having to wait for a timeout.

7.3.3 Revocation Authentication

To efficiently authenticate revocations we designed an authentication mechanism based on hash trees [174]. The idea is the following: given a time interval T, divide T into m equal, smaller intervals $t_1, t_2, ..., t_m$. Each AS constructs a hash tree, whose leaves are of the form:

$$\mathcal{H}(IF_x \| t_i \| n_{x,i}),$$

where IF_x is the interface identifier, t_i is a time interval of T, and $n_{x,i}$ is a secret nonce. For each interface IF_x of an AS, there are m such leaves, i.e., in total the hash tree contains $n \cdot m$ leaves, where n denotes the total number of interfaces of an AS. Such a tree is only valid within time interval T. Figure 7.5 shows a graphical representation of such a tree.

With the authentic root R_T distributed to verifiers (see below), an AS can revoke an interface IF_x for a given epoch t_i by revealing $x = (IF_x, t_i, n_{x,i})$ together

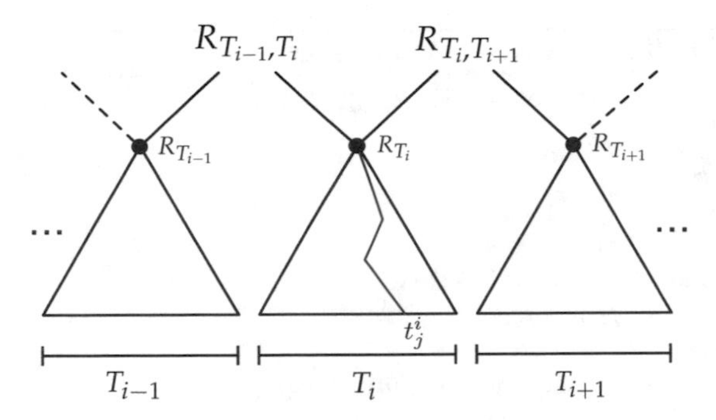

Figure 7.6: Hash trees for a time interval T_i are connected by introducing a new root node that connects two subsequent trees. To revoke IF_x at epoch $t_j^i \in T_i$, an AS reveals $(IF_x || t_j^i || n_{x,j}^i)$ together with the labels of the nodes of the hash tree rooted in R_{T_i} and the roots of the hash trees for T_{i-1} and T_{i+1}, $R_{T_{i-1}}$ resp. $R_{T_{i+1}}$ (red path).

with the hash values of the siblings of the nodes on the path from $\mathcal{H}(x)$ to the root R_T (see Figure 7.5 and Algorithm 5, lines 4–11). Let P be this set of hash values. Using x and P, a verifier can compute R'_T (Algorithm 6, lines 5–12). To verify a revocation, a verifier checks the freshness of the revocation by executing `VerifyEpoch` (Algorithm 7) and then ensures $R'_T \stackrel{?}{=} R_T$. An expired revocation message must be ignored to prevent replays of old revocations. When verifying the freshness of the revocation, a tolerance ε is added to account for imperfectly synchronized clocks as well as the propagation time of revocations.

Connecting Hash Trees

A hash tree, as described above, is only valid for time interval T, but clearly, revocation authentication needs to be possible over an arbitrary amount of time. To achieve this property, we consider time as an infinite series of time intervals T_i. To each T_i we associate a corresponding hash tree with root R_{T_i}. To achieve continuity between the time ranges, we propose connecting two consecutive trees with roots R_{T_i}, $R_{T_{i+1}}$ by making them the left and right subtree of a new root node $R_{T_i,T_{i+1}}$ (see Figure 7.6).

With this enhancement, we can present our complete scheme. During beaconing, an AS adds $R_{T_i,T_{i+1}}$ (the revocation token) to the PCB, assuming the path expires in T_{i+1}. Selecting T to be at least as long as the longest lifetime of a PCB, then in epoch $t_j^i \in T_i$, there exist only PCBs containing either R_{T_{i-1},T_i} or $R_{T_i,T_{i+1}}$. Thus, to revoke an interface IF_x at $t_j^i \in T_i$, the AS reveals $(IF_x || t_j^i || n_{x,j}^i)$ together with the hash values of the nodes of the hash tree rooted in R_{T_i} and the roots of the hash trees for T_{i-1} and T_{i+1}, i.e., $R_{T_{i-1}}$ and $R_{T_{i+1}}$.

Algorithm 5 Building a revocation message.

1: **procedure** BUILDREVMSG(IF_x, $R_{T_{i-1}}$, $R_{T_{i+1}}$)
2: $x \leftarrow (IF_x \| t^i_j \| n^i_{IF_x,j})$
3: $m \leftarrow \mathcal{H}(\mathrm{x})$
4: **for** $l \leftarrow 1..\mathrm{height}(T_i)$ **do**
5: $P \leftarrow P \,\|\, \texttt{sibling}(m)$
6: **if** `IsLeftSibling`(m) **then**
7: $m \leftarrow \mathcal{H}(m \,\|\, \texttt{sibling}(m))$
8: **else**
9: $m \leftarrow \mathcal{H}(\texttt{sibling}(m) \,\|\, m)$
10: **end if**
11: **end for**
12: **return** $(x, P, R_{T_{i-1}}, R_{T_{i+1}})$
13: **end procedure**

Verification proceeds similarly. First, a verifier checks the freshness of the revocation. Then, the verifier computes R'_{T_i} and checks whether:

$$\mathcal{H}(R_{T_{i-1}} || R'_{T_i}) \stackrel{?}{=} R_{T_{i-1},T_i} \quad \text{or}$$

$$\mathcal{H}(R'_{T_i} || R_{T_{i+1}}) \stackrel{?}{=} R_{T_i,T_{i+1}}.$$

Note that an AS, at any point $t^i_j \in T_i$, only needs to store the hash trees with roots R_{T_i} and $R_{T_{i+1}}$. From the previous hash tree, only the root $R_{T_{i-1}}$ needs to be stored.

To provide a conservative estimate of the size of a revocation message, we set $T = 24$ h and each epoch $t^i_j = 10$ s. Thus there are $m = 24 \cdot 60 \cdot 6 = 8{,}640$ epochs in T. Additionally, we set $n = 10{,}000$, which is considerably more than the maximum number of links to neighboring ASes for any AS in today's Internet [77]. In total, the entire hash tree contains about 86 million leaves and thus has a height of 28, i.e., a revocation message has to include an additional 28 hash values to enable verification.

7.4 Failure Resilience and Service Discovery

The path infrastructure is a fundamental piece of the SCION architecture whose availability is crucial for basic communication. In this section, we describe how we achieve high availability for services that are part of the path infrastructure. We note that ASes can use other techniques than the ones we describe here, but the default strategies below are sufficient to provide high availability.

The control-plane infrastructure is based on a *consistency service* that provides the following primitives:

- a *distributed database* that allows entities connected to the service to share information,

Algorithm 6 Verifying a revocation message.

1: **procedure** VERIFY(x, P, $R_{T_{i-1}}$, $R_{T_{i+1}}$, y)
2: **if not** `VerifyEpoch(x)` **then**
3: **return** False
4: **end if**
5: $R'_{T_i} \leftarrow \mathcal{H}(x)$
6: **for** p **in** P **do**
7: **if** `IsleftSibling(p)` **then**
8: $R'_{T_i} \leftarrow \mathcal{H}(p \| R'_{T_i})$
9: **else**
10: $R'_{T_i} \leftarrow \mathcal{H}(R'_{T_i} \| p)$
11: **end if**
12: **end for**
13: **if** $\mathcal{H}(R_{T_{i-1}} \| R'_{T_i}) == y$ **or** $\mathcal{H}(R'_{T_i} \| R_{T_{i+1}}) == y$ **then**
14: **return** True
15: **end if**
16: **return** False
17: **end procedure**

Algorithm 7 Verifying an epoch.

1: **procedure** VERIFYEPOCH(x)
2: $e \leftarrow$ `CurrentEpoch()`
3: **if** $e == x.t^i_j$ **or**
4: ($e == x.t^i_j - 1$ **and** `TimeSinceEpoch()` $< \varepsilon$) **then**
5: **return** True
6: **end if**
7: **return** False
8: **end procedure**

- a *leader election* to elect an entity that acts as a master, and
- a *group membership primitive* to discover which instances are currently alive.

In the current SCION implementation, Apache ZooKeeper [11] provides the above primitives, though any software providing the three primitives listed above can be used to implement the consistency service.

7.4.1 Beacon Service

The path exploration process within an AS relies on the availability of a beacon server. In order to prevent a beacon server being a single point of failure, the AS can run multiple, coordinated beacon server instances.

All beacon server instances in an AS connect to the consistency service and appear as group members. An instance that gets disconnected from the service for any reason will no longer appear as a group member. Upon joining, each instance tries to become a leader, which makes the instance a *master beacon server* until the leader terminates (e.g., due to failure or shutdown). When the leader terminates a new election process takes place.

When a border router receives a new PCB, the router finds a running beacon server instance (as described in Section 7.4.7) and forwards the PCB to that instance. The beacon server instance can then share the PCB with the other instances by writing the PCB into a specified location of the distributed database. All beacon server instances watch this location and copy any new PCBs into their caches. Every new beacon server instance populates its cache with PCBs from the distributed database.

Once per propagation and registration period, the master beacon server initiates the beaconing and path-segment registration processes, respectively. If the master beacon server fails, a new master is elected and the new master starts the beaconing and path registration.

Although we assume that network partitions in an AS or failures of the consistency service are unlikely, the beacon servers can also handle these failures. If a beacon server instance is disconnected from the consistency service, it will initiate the beaconing and the registration at a planned interval. While this approach may cause several beacon server instances to simultaneously propagate PCBs, it guarantees that the beaconing and path-segment registration processes can continue even under catastrophic failures.

7.4.2 Path Service in Core ASes

We now describe how we achieve high availability for the path service in core ASes. Similarly to beacon servers, we deploy multiple coordinated instances of path servers that elect a master (via leader election) and share a distributed database containing registered down- and core-segments. However, due to load, the core AS path servers cannot replicate all registered path segments as beacon servers do with PCBs. Despite this limitation, each core AS must be able to respond to queries for down-segments. We thus propose a two-level replication scheme to meet these requirements: non-master path servers cache registered down-segments and replicate them only with a master path server (that caches all seen down-segments).

Down-Segment Registration

When an AS registers a down-segment at a core AS with replicated path servers, registration proceeds as follows:

1. A down-segment is sent to a running path server instance as determined by the group membership protocol. The instance is selected randomly by the last border router as described in Section 7.4.7.
2. The path server instance verifies and registers the down-segment.

3. If the path server instance is not a master, it forwards the segment to the master path server instance of that AS.
4. The path server instance forwards the path segment to all other ISD core ASes. Each AS processes the segment in the same manner.

Each ISD core AS thus has a master path server instance that keeps all registered segments, and a copy of the database is also distributed among the running path server instances in the AS.

Down-Segment Request

A down-segment request is handled differently depending on the destination's location. Regardless of the destination's location, a path server instance in an AS is randomly selected as described in Section 7.4.7.

If the destination is within the local ISD then the request is handled as follows:

- If a down-segment to the requested destination exists in the path server's local cache, the server responds with the segment.
- If no down-segment to the destination exists in the path server's local cache and the server is not the master instance, the path server instance asks the master instance of the AS, which responds to the query. The answer is cached by the non-master server and used to serve the initial request.

If the requested destination is in a remote ISD, then the request is handled as above, except the path server forwards the request to a path server in the remote ISD rather than to its local master path server instance. The remote path server then processes the requests using the above steps.

Core-Segment Registration and Request

Core-segments are only registered with path server instances within a core AS and are not sent to other ASes. The core-segments are replicated among path server instances based on their destination. In particular, if a core-segment's destination is an AS in the local ISD, then the segment is stored in the distributed database where each path server instance within that core AS can access it. If a core-segment's destination is in a remote ISD, the core-segment is cached at the path server and forwarded to the master path server instance of the core AS.

A core AS's path server instance handles a path request for a core AS (and thus for a core-segment) as follows:

- If a core-segment for the AS exists in the local cache, return it.
- If not, and the target AS is from the local ISD, wait for the appropriate core-segments (timing out after a waiting period).

- If the target AS is within a remote ISD, query the local master path server for a core-segment, timing out after a set waiting period.

Failure of a Core Master Path Server

When a core master path server fails, the following procedure is executed:

- A new master path server is elected from the group members (i.e., from all active path servers within the AS).
- All non-master path servers from the AS send the following to the new master:
  - their locally cached down-segments (i.e., down-segments from the local ISD), and
  - their locally cached core-segments to ASes in remote ISDs.

For efficiency reasons, the number of replicated paths per destination AS is limited. Note that the distributed database stores all core-segments that originated within the local ISD, so there is no need to send those path segments to the new master.

7.4.3 Path Service in Non-core ASes

In non-core ASes, path server instances join the consistency service and access the distributed database, but do not participate in master election.

Up-segment registrations are handled by all path servers in a non-core AS and are fully replicated through the distributed database. Since up-segments are accessible to all path servers in a non-core AS, a path request can be handled by any path server. By default, paths to remote ASes (core- and down-segments) are only cached by path servers that have received them (i.e., there is no replication of these path segments for scalability reasons).

If a path server is disconnected from the consistency service, it serves the requests as usual, but for all new up-segments obtained, an attempt is made to synchronize them via direct communication with all remaining path servers (known from the discovery service — see details in Section 7.4.6).

7.4.4 Certificate Service

The certificate service in both core and non-core ASes has a similar architecture for high availability to that of the path service in non-core ASes, in that the instances do not participate in master election. New TRCs and certificates are replicated across all instances via the distributed database, providing all servers with the same view of TRCs and certificates. Thus each certificate server instance can serve TRC and certificate requests independently.

If a certificate server is disconnected from the consistency service, it serves the requests as usual, but attempts to replicate new TRCs and certificates via direct communication with other certificate servers.

7.4.5 Inactive Interfaces

PCBs (through their AS entries) should reflect an accurate state of the network within an AS, and thus interfaces that are down should not be added to PCBs. To achieve this accuracy, every AS implements an interface failure detection mechanism. In a nutshell, every border router periodically sends a *keep-alive* message with the respective interface identifier to its neighboring router, which propagates this message to all the beacon servers in its AS. The interval between these keep-alive messages is known in advance, allowing an AS to detect that it has missed a keep-alive message. After a threshold number of missed messages, a master beacon server can consider the interface inactive; such interfaces will no longer be added to new PCBs. An AS can also revoke an inactive interface from all paths that contain information on the interface, as described in Section 7.3.

7.4.6 Service Discovery

Both infrastructure elements and end hosts need to be able to find instances of services they require for their operation. To facilitate this, a SCION AS runs a *discovery service.* The discovery service gathers information from several sources and exposes it in a standard format in a standardized set of URLs (presented in Section 16.3 on Page 374).

The discovery service exports two views of the information: a *full view* intended for infrastructure servers and routers, and a *reduced view* for end hosts. The AS can make a policy decision on which part of the infrastructure is visible in the reduced view, e.g., the entries for beacon servers may be excluded.

The main source of information for the discovery service is the consistency service employed by the AS. The discovery service connects to the consistency service and reads the membership information created by the group membership primitive. In this way, the discovery service obtains the list of instances of a given service — IP addresses and ports — and updates the exported information accordingly. If the consistency service detects that a server has failed, it is removed from the corresponding group. Since this change is visible to the discovery service, it can then update the dynamic view it exports.

Additional information, such as the addresses and ports of border routers and the MTUs of links, is configured statically for the discovery service, since this information will rarely change and is typically gathered from additional configuration files.

Finally, the discovery service augments all the exported records with a timestamp of the last update. This timestamp can vary between services, but not service instances. That is, there is one timestamp for all listed certificate servers, another for all path servers and so on. The discovery service also exports a TTL for the information it provides.

As a fallback, the discovery service exports static versions of both the full and reduced views. This information is exported on a different path of the URL, so clients have discretion in when to switch between the dynamic and static views. The static view will typically have a longer TTL than the dynamically generated view.

All views (static or dynamic, full or reduced) are signed by the discovery service with the AS's private key (the same key that is used for signing control-plane messages). The minimal information that end hosts have to be provided with is an address of a discovery server.

If an end host discovers that for a given service there are no instances listed in the dynamic view, it can choose to use the content of the static view. If both are empty, it can either fall back to a copy it has cached earlier, or use a static configuration that was provided by other means. It can also choose to switch to a different discovery service instance it knows about. Note that the information provided by a discovery service will typically include all the discovery service instances an AS wants to be used by end hosts or infrastructure elements.

If a discovery service instance has stale information (i.e., the TTL has passed with no updates), it must still export this stale information. The decision what to do with stale information is entirely up to the client.

7.4.7 Service Instance Selection

In order to facilitate control-plane anycast communication, SCION introduces a dedicated service-addressing scheme. For instance, a beacon server that wishes to register segments with a remote AS's path service does not have to know the actual address of a remote path server. Instead, the SCION service address of the path service suffices, so that the SCION border router in the remote AS can select an alive instance of the service to deliver the packet to.

To implement this primitive, all border routers, through the discovery service, keep lists of alive instances for all supported services within their ASes. These lists are frequently updated by the discovery service. When a border router detects a packet addressed to a supported service, an instance of the service is selected pseudo-randomly, and the packet is sent to the instance. To support connectionless protocols, the selection process has to be deterministic, such that two consecutive packets sent from the same application to the same service are delivered to the same instance of that service.[3]

[3] In the case of TCP connections, only the first packet (i.e., `SYN`) is addressed to a given service.

7.5 AS-Level Anycast Service

SCION deploys anycast as a standard communication model for control-plane requests. In essence, the service anycast system provides a service-oriented communication infrastructure, enabling a request to be routed to the nearest server. Due to the hierarchical nature of caching infrastructures, service lookups should progress through subsequent servers at increasing levels in the hierarchy. SCION intrinsically supports this primitive by embedding an up-segment in the anycast request, further enabling the requester to specify which of the ASes on the path to the core should invoke the anycast primitive to establish whether an internal service can answer the request. This flexibility endows the SCION service infrastructure with powerful primitives to implement a variety of services, without introducing dedicated service-oriented stacks and layers [186, 256]. In this section, we describe this infrastructure in more detail.

By default, all control-plane requests are sent within a requested AS as anycast packets through a SCION service destination address. Within an AS, server instances of a given service are discovered and tracked by the consistency service, catalogued by the discovery service, and exported to all SCION border routers through a server list (see the previous section). Moreover, SCION introduces a separate anycast mechanism that works at the AS level. It allows an anycast request to a service's server to be sent to any intermediate AS on a path to the ISD core.

Figure 7.7 presents several examples of SCION anycast requests. Each service that is accessible via anycast has a dedicated anycast address that is globally registered via an organization like IANA. Additionally, service discovery (or anycast routing) has to be implemented within an AS (as described in Section 7.4.7).

Case 1 in Figure 7.7 is the standard request to a server in the core (e.g., a path request). The requester, using an up-segment to the ISD core, sends the request, which traverses the ASes toward the core. The last border router on the path sends an anycast packet to a pseudo-randomly selected server that implements the service. The server responds to the requester using the reversed path from the request packet.

For some services it is preferred (mainly for efficiency reasons) to send a request to servers within intermediate ASes instead of contacting ISD core servers, to benefit from hierarchical caching. To satisfy this requirement, SCION introduces service anycast that can be targeted to specific intermediate ASes. The mechanism is implemented by a dedicated extension (see Page 354) and can be enabled by an end host that wishes to request intermediate servers. The end host simply marks a hop field that corresponds to a requested (intermediate) AS as an anycast hop field. The bit informs the border router of the selected AS that this packet should be forwarded to the service's server within the AS.

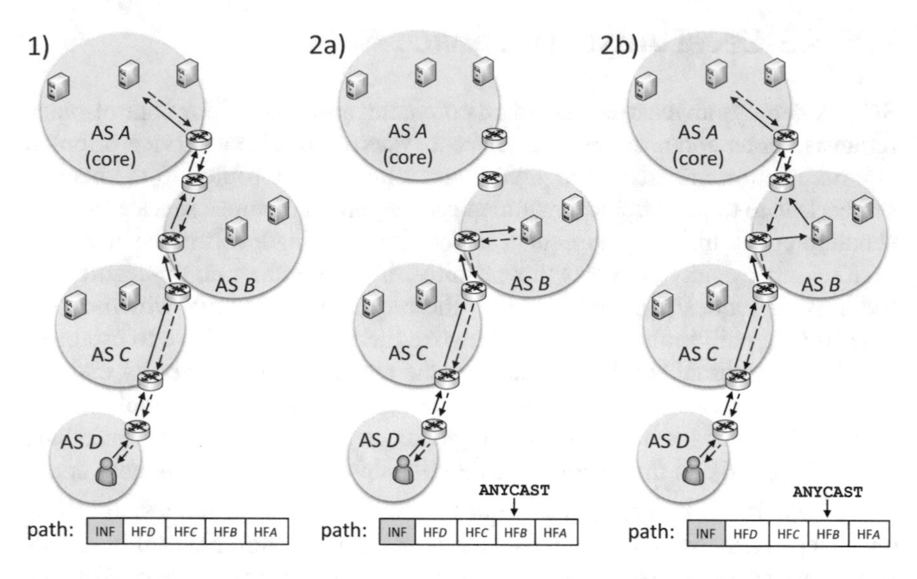

Figure 7.7: Examples of service anycast, where solid lines indicate a request and dashed lines indicate a response. Case 1 shows a request to a server in the ISD core. Cases 2a and 2b demonstrate a request where either a server inside AS *B*, or a server in the ISD core is requested to respond, respectively. In Case 2a, the server in AS *B* responds directly. In Case 2b, the server in AS *B* forwards the request to the server in the core. The path header diagram indicates the HFs that are marked as anycast; in this example only the HF corresponding to AS *B* is set as anycast in Cases 2a and 2b.

After the packet is received by the server, it can handle the request in several ways. The processing logic depends on the contacted service, but we distinguish the cases where the intermediate contacted service's server

- can serve the request, and respond to the requester reversing the path (Case 2a in Figure 7.7);
- cannot serve the request, and passes the request upstream sending it to the next border router (Case 2b in Figure 7.7);
- can serve the request partially, and respond to the request in the packet payload, but passes the request upstream (Case 2b in Figure 7.7).

Only the hop field that is processed by the ingress router of the AS that should handle the anycast is marked as anycast.

The SCION AS-level anycast service enables design and implementation of services that leverage a hierarchical caching infrastructure to minimize latency (e.g., content distribution), or services that need to perform an action by every AS on the path (e.g., on-path key agreement). Additionally, through the beacon extension mechanism ASes can announce which services they support.

7.6 SCION Control Message Protocol (SCMP) 

The SCION Control Message Protocol (SCMP) is analogous to ICMP in the current Internet and provides the following functionalities:

- *Network diagnostic:* allows debugging tools such as the SCION equivalents of `ping` or `traceroute` to be built.
- *Error messages:* signal problems with packet processing or inform end hosts about network-layer problems.

The SCMP protocol is the first instance of a secure control message protocol in a network infrastructure we are aware of. The main challenges include scalable Internet-wide key distribution and highly efficient generation of authentication information at line speed. In this section, we describe the design, goals, and use cases of SCMP. Low-level details, such as packet headers, are presented in Section 15.6.

7.6.1 Goals and Design

SCMP must be flexible as it is used for many purposes in various applications. For instance, (a) some SCMP messages are processed by intermediate routers on the path, while other messages are end-to-end, (b) there are various types of SCMP messages (for various types of diagnostics or network errors), and (c) the messages can influence different parts of the SCION stack (such as the transport protocol or the beacon selection mechanism).

SCMP packets can carry either error messages or non-error messages. One basic rule of SCMP is that an error packet should never generate another SCMP packet (to prevent loops), thus border routers must be able to efficiently check whether a packet is an SCMP error message. To this end, each SCMP packet contains a mandatory and easily accessible SCMP *extension header* (see details in Section 15.6.1 on Page 363). The extension header indicates whether the packet should be processed by every router on the path (i.e., *hop-by-hop flag*), and whether the packet contains an SCMP error message (i.e., *error flag*).

An SCMP packet has a simple SCMP *layer-4 header* that contains the length of the carried SCMP message, describes its class and type, and contains the message creation timestamp.

Finally, SCMP packets contain an SCMP *payload*, which carries the actual content of the message, necessary for interpreting a given message class and type specified in the SCMP layer-4 header. In particular, it can contain information about the SCION packet that triggered the SCMP message.

SCMP is implemented by network devices and end-host stacks. Usually, SCMP packets are generated in response to a SCION data packet (that triggered an SCMP message). As the SCMP packet has to be delivered back to the initiator, it contains the reversed path and address from the initial packet. SCMP packets also contain information to identify the source application (such as a layer-4 header).

7.6.2 Supported Message Classes and Types

SCMP supports messages of the following generic classes:

- **Forwarding:** errors that can happen during packet forwarding or delivery. This class contains message types that represent problems such as end-host unreachability (e.g., unreachable ports), network issues (e.g., MTU exceeded), or administrative decisions (e.g., destination denied).
- **SCION common header:** errors that can be found during basic packet parsing; for instance, types such as wrong packet or header length, incorrect path pointers, or invalid address type.
- **Path:** errors related to the processing of the packet's forwarding path. This class can signal problems such as expired hop field, revoked interface, invalid interface, or wrong MAC.
- **Extension:** errors that can happen while processing SCION packet extensions (e.g., unsupported extension or too many extensions).
- **General:** messages that do not fall into any other class; for instance, types such as echo request/reply and traceroute.[4]

Besides generic messages, SCMP is also able to handle specific errors of SCION extensions such as SIBRA (see Section 15.6.3 on Page 365).

7.6.3 Authentication

All SCMP packets are authenticated, thus it is infeasible to perform attacks analogous to ICMP-based attacks on TCP/IP [99]. To the best of our knowledge, SCMP is the first control message protocol to provide an authentication property.

SCION provides two means of SCMP authentication, using symmetric or asymmetric cryptography. The methods can be used interchangeably, and they both deploy a SCION packet security extension (see details in Section 15.1.4); consequently, they protect the entire SCMP packet (not only its payload). The symmetric authentication method uses AS-level keys to compute a message authentication code (MAC) — while this approach offers high speed and scalability, the disadvantage is that only the destination AS infrastructure or the destination end host can verify the SCMP message. The asymmetric authentication mechanism is based on digital signatures, enabling any AS on the path and end hosts to verify the SCMP message using the appropriate public key.

[4] SCION implements its own (more verbose) version of traceroute (note that in SCION the forwarding topology is known by the source). Details can be found in Section 15.6.3 on Page 365.

However, the disadvantage of the asymmetric approach is the much slower speed for signature generation.

SCION border routers take an active role in creating SCMP packet authenticators. If an SCMP packet is generated by a border router, the router decides which authentication approach to use. For SCMP packets generated by end hosts, the end host decides how the packet is protected, by setting a chosen option for the SCION packet security extension for this packet. The extension indicates an authentication method, and — if asymmetric authentication is used — the first border router on the packet's path authenticates the packet.

Symmetric Authentication

The first method of SCMP authentication leverages symmetric cryptography. In this method, SCMP packets are authenticated by the router that generates them, and verified by the final SCION border router on the path (i.e., the border router of the SCMP message's receiver). Thus, this method provides AS-level authentication, i.e., the receiving AS can be sure that the packet was indeed created by the sending AS.

To efficiently create authenticated SCMP messages, we use the DRKey protocol (as described in Section 12.5 on Page 291). The DRKey protocol lets each router derive a symmetric key for the receiving AS or end host on the fly. This symmetric key will be used to compute a MAC of the SCMP message. Upon receiving an authenticated SCMP message, the receiver AS has most likely already cached the verification key from a previously verified SCMP message (originated from the sending AS). If the key is not available, the receiving AS contacts the AS that generated the SCMP message and engages in a key exchange protocol to fetch the current MAC key.

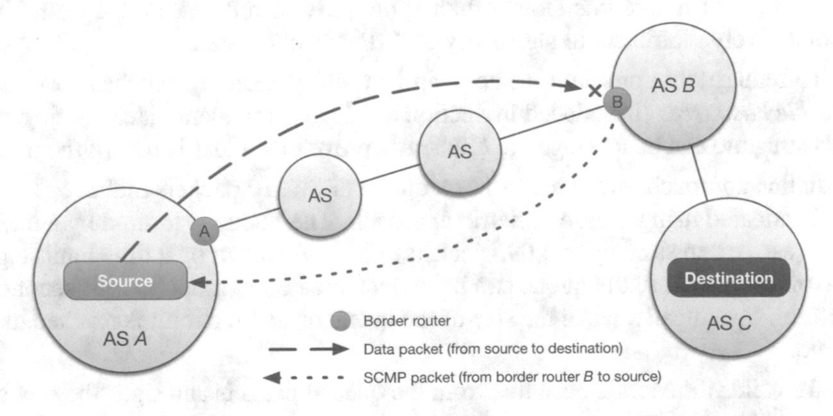

Figure 7.8: An example of SCMP authentication using MACs.

Example. In Figure 7.8, the source host from AS A sends a data packet to the destination in AS C, but forwarding fails (e.g., due to an expired hop field) at the ingress interface of AS B's border router BR_B. The router creates an SCMP message, which describes the problem and uses the source address and the path from the original data packet to build an SCMP packet destined for the source. When the packet is created, then BR_B (a) derives a key for authenticating packets destined for the source based on the key shared with AS A, (b) computes a MAC over the packet and puts it into the packet, and (c) sends the SCMP packet back to the source. If the source has a key shared with AS B, then the packet is verified and (on success) delivered. If the key is not established, then the certificate server is contacted and queried for the missing key. The certificate server derives the requested key from a shared secret between AS A and AS B and returns it to the source, which can then verify the SCMP packet.

The main advantage of this approach is that the authentication process is extremely efficient. A router can efficiently derive a key for any AS and use it to authenticate packets to this AS. Moreover, it does not need coordination within an AS. If all routers within an AS share a secret SCMP key, each of them can locally and efficiently re-create the SCMP key using the DRKey protocol for any destination AS, without any additional communication and without storing any per-AS state.

Asymmetric Authentication

The second form of authentication is based on digital signatures, which provides a stronger security property than symmetric authentication — SCMP packets are again authenticated by routers, but can be verified by any entity including other on-path entities. However, even fast digital signature schemes are a few orders of magnitude slower than symmetric primitives, and it would be prohibitively inefficient to sign every SCMP packet created.

To remedy this problem, routers sign SCMP packets in batches. We use *Merkle hash trees* (introduced in Section 4.4.1) to implement batch signing, as this structure can be leveraged to efficiently prove that a leaf is part of the tree.

In this approach, every router has a queue of SCMP packets that need to be authenticated using the asymmetric approach. The queue is limited by a fixed size (e.g., it can store up to 4,096 packets) and is restricted by a time limit (e.g., the oldest packet in the queue can have been created at most 20 milliseconds before). Specifically, when the size of the queue or its time limit is reached, the router

1. builds the Merkle hash tree from the queued packets and signs the root of the tree (using the same key used by beacon servers to authenticate PCBs and path segments);

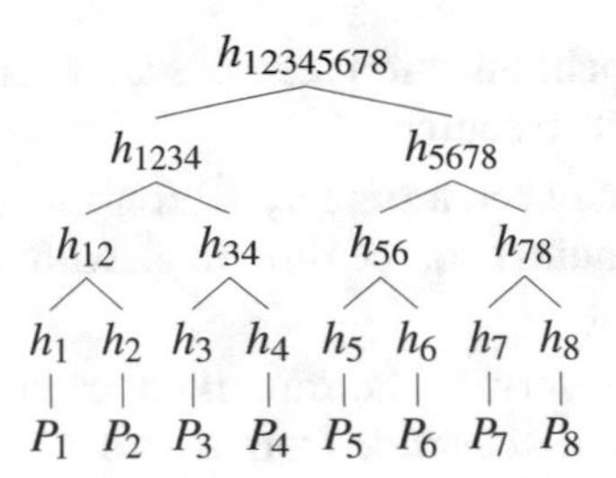

Figure 7.9: Queue of the SCMP packets and the corresponding Merkle hash tree.

2. creates a proof for each packet that belongs to the tree (see Figure 7.9, where packet P_4 is authenticated via values h_3, h_{12}, h_{5678}, and the signed root value $h_{12345678}$);
3. extends each packet by its proof and sends all packets towards their destination; and
4. clears the queue.

Although this scheme introduces higher overhead (proofs are longer than MACs and the signing is less efficient than MAC creation) it has several advantages. First, the packets can be verified by any entity, not only by the destination AS or end host. Moreover, in this scheme, a source host does not need to conduct any certificate or key lookup. The connection initiator that receives an SCMP packet can immediately verify it, as it already has the required certificate (the SCMP is sent via a forwarding path derived from the signed path segments). If the certificate is missing, however, an end host can obtain it from its local certificate server. Finally, the digital signature offers non-repudiation, a stronger property than authentication offered by the MAC.

7.7 Time Synchronization

A standard assumption in security protocols is synchronized time. SCION protocols also rely on this assumption. It is required that end hosts, servers, and border routers are synchronized with at least second-level precision, although some protocols may still work effectively when time is less precisely synchronized.

To provide reliable time information, SCION proposes a time synchronization framework as follows.

1. Each core AS runs a public time synchronization service that is accessible to anyone inside its ISD.

2. The core time synchronization services are stratum-1, i.e., they synchronize with a stratum-0 source.
3. A non-core AS may run a time synchronization service, synchronized with at least a stratum-1 source time (e.g., with its core time synchronization service).
4. The infrastructure (servers and routers) of each AS is synchronized with a time synchronization service from a core AS of its ISD.
5. Finally, end hosts should be synchronized with a core (or, if possible, with a local) time synchronization service.

The time synchronization service has its own service address. As the default protocol we propose the use of the Roughtime protocol [101]. Roughtime is a novel protocol that provides higher security than currently deployed time synchronization protocols (such as NTP [175]). Every response from a Roughtime time server is signed, and the protocol allows misbehavior on the part of time servers to be cryptographically proven.

8 Data Plane 

Christos Pappas, Adrian Perrig, Raphael M. Reischuk, Stephen Shirley, Pawel Szalachowski

In this chapter, we discuss the SCION data plane. The purpose of the data plane is to forward packets containing a SCION header. In SCION, inter-domain forwarding decisions are encoded as a sequence of hop fields (HFs), which encode AS-level hops augmented with ingress and egress interfaces.

Two important aspects of the SCION data plane are *HF integrity* (to prevent forgery or alteration of HFs) and *efficiency* (to enable high-speed processing). SCION provides a data plane that, despite its secure operation, is more efficient than the current Internet infrastructure in several aspects: processing time, router complexity, scalability to large networks, and energy consumption. In particular, our investigations suggest that the cryptographic verification of HF information can be made faster and more power-efficient than the longest-prefix matching by current routers. (The power efficiency is discussed in Chapter 14.) The absence of inter-domain routing tables improves scalability. Finally, the implementation of cryptographic functions is well understood today, and can lead to simple router implementations, helping to reduce the complexity of current routers.

In this chapter, we discuss, among other things, the format of hop fields, how path segments are combined to create forwarding paths, and how routers compute a forwarding decision.

Chapter Contents

A. Perrig et al., *SCION: A Secure Internet Architecture*, Information Security and Cryptography, https://doi.org/10.1007/978-3-319-67080-5_8

8.1 Path Format

We start with the description of the data-plane path format used in SCION. A path in this format is called a *forwarding path* and is placed directly into a SCION header. It is considered by all SCION border routers on the path to make forwarding decisions. To determine when to terminate the packet, border routers check the destination address, which is also present in the SCION header.

In contrast to the verbose control-plane path format (see Section 7.1), the forwarding-path format includes minimal information that is needed to forward packets. The rationale behind this design is that the path construction operation is infrequent (compared to forwarding), and SCION's control plane offers path transparency so that end hosts obtain detailed path information when they compose paths. However, only a fragment of this information is needed for packet forwarding, which in turn is a very frequent operation and thus needs to be highly efficient. Roughly speaking, a forwarding path is created once per connection, and then it is processed by each border router on the path, for every packet sent. For local communication (i.e., within an AS) a forwarding path is not necessary (i.e., the path within the SCION header is empty).

A path in the data-plane format (i.e., a forwarding path) can be defined as a concatenation of at most three lists of hop fields, which are extracted from an up-segment, a core-segment, and a down-segment, respectively. Each list of hop fields is optional, but hop fields have to be inserted into a packet in the correct order (i.e., hop fields from a down-segment cannot precede hop fields from a core- or up-segment, and core-segment hop fields cannot precede up-segment hop fields). An example of how a forwarding path is constructed from path segments is presented in Figure 8.1. The hop fields obtained from each path segment are prepended with an info field corresponding to the path segment, which includes the following information:

- a timestamp used for hop field freshness verification (each hop field of a given path segment is verified against the corresponding timestamp);
- the identifier of the ISD that initiated the propagation of the path;
- the length of a given segment; and
- $Flags_{INF}$, which describes the type and the direction of the constructed forwarding path with the following flags:
  - **UP:** describes a forwarding path's orientation (as forwarding paths are bidirectional, the orientation information is required for correct processing). When a packet travels in the direction of beacon propagation, the flag is set to *false*; otherwise it is set to *true*.
  - **SHORTCUT:** describes whether the constructed forwarding path is of *non-core AS shortcut* type. This flag is set when up- and

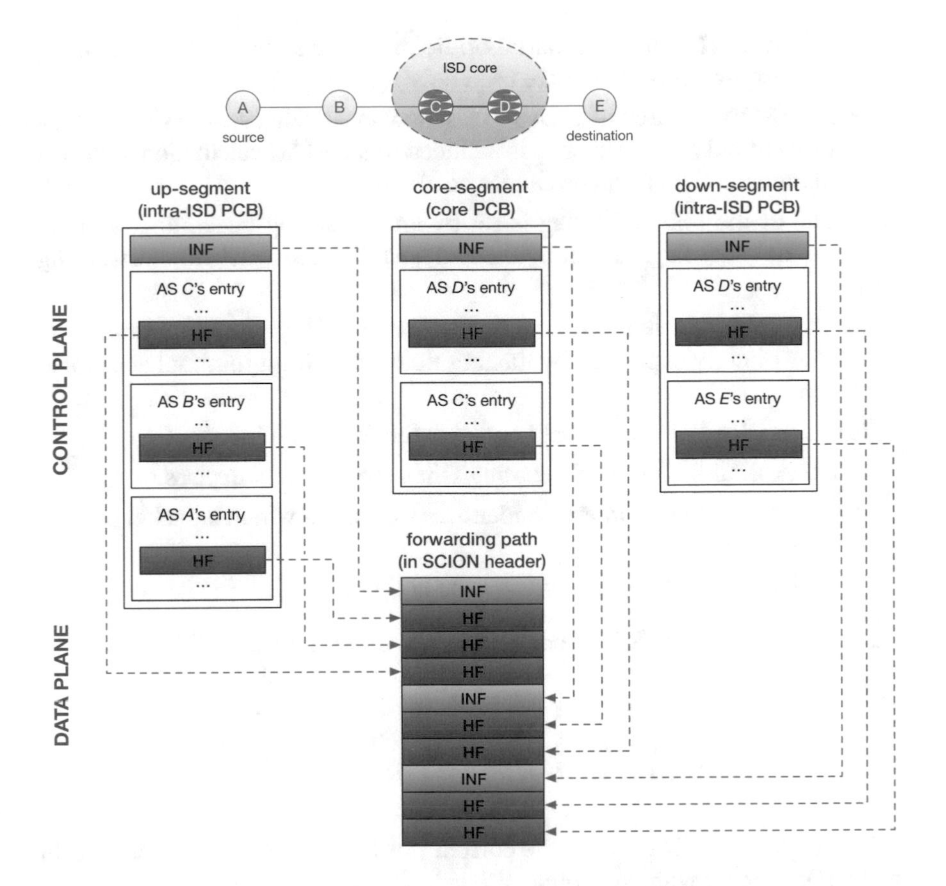

Figure 8.1: Example of the path construction.

down-segments have a common AS through which the forwarding path is constructed (see Section 8.2 and Figure 8.5).

- **PEER:** describes whether the constructed forwarding path is of *peering shortcut* type (see Section 8.2).

The values for these flags are set by an end host that constructs the forwarding path.

Similarly, each hop field has a $Flags_{HF}$ field that describes the purpose of the hop field. It contains the following flags:

- **XOVER:** used to signal that this hop field is at a cross-over point between hop fields from different path segments, and needs special processing. A border router that processes the hop field will advance to the next info field and thus switch to the next list of hop fields (i.e., hop fields from the next path segment).

- **VRFY-ONLY:** used to mark hop fields that are not used for making a forwarding decision, but are used only for MAC verification.
- **FWD-ONLY:** used by an AS to disallow local delivery of packets (i.e., a forward-only AS). This flag is included in the MAC calculation, thus it is immutable (i.e., its value cannot be altered).

The value of the `FWD-ONLY` flag is set by an AS during beaconing, and the values of the other flags are set by the end host that constructs the forwarding path.

The flag fields enable fast forwarding at border routers since the forwarding decision is directly expressed by the flag fields, requiring minimal additional validation.

Beside the flag field, a hop field contains the following information:

- the expiration time field denoting when the hop field expires,
- ingress and egress interface identifiers between which a packet is to be forwarded, and
- the MAC field that authenticates the hop field.

The forwarding path format can be described as follows:

$$\begin{array}{l} INF_{up} \parallel HF^{0}_{up} \parallel HF^{1}_{up} \parallel \cdots \\ \parallel INF_{core} \parallel HF^{0}_{core} \parallel HF^{1}_{core} \parallel \cdots \\ \parallel INF_{down} \parallel HF^{0}_{down} \parallel HF^{1}_{down} \parallel \cdots \end{array}$$

To forward SCION packets, the current position on the path is encoded in the SCION header with two pointers:

- **CurrINF:** points to the current info field,
- **CurrHF:** points to the current hop field.

Details on the forwarding path format are presented in Section 15.1.3. The next section describes how an end host transforms a combination of path segments into an actual forwarding path.

8.2 Creation of Forwarding Paths

In this section, we illustrate path combination, i.e., how an end host constructs a forwarding (end-to-end) path under different topologies. The forwarding path construction process is executed when an end host establishes a connection with another end host. Prior to the path *construction* process, the path *lookup* process returns a set of path segments that are later combined by the end host to reach the desired destination (see Section 7.2). The packet-forwarding process using the constructed forwarding paths is demonstrated by an end-to-end example (including updates of the `CurrINF` and `CurrHF` fields) in Section 10.8 on Page 223.

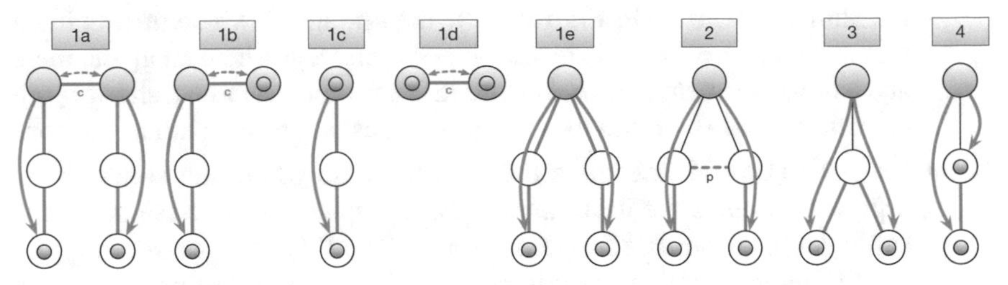

Figure 8.2: SCION forwarding paths created by an end host. The blue circles represent the end hosts; the shaded gray circles represent core ASes, possibly in different ISDs; blue lines without arrow heads denote hops of created forwarding paths; the dashed blue line denotes a peering link (labeled “p”); orange lines with arrows stand for PCBs and indicate their dissemination direction; dashed orange lines represent core beacons exchanged over core links (labeled “c”). All created forwarding paths in cases 1a–1e traverse the ISD core(s), whereas the paths in cases 2–4 do not enter the ISD core.

In the following, we assume that the source and destination end hosts are in different ASes, as end hosts from the same AS use an empty forwarding path to communicate with each other. We stress that although end hosts enjoy freedom in composing their forwarding paths, the possible combinations (i.e., the provided path segments) follow AS routing policies and rules such as the valley-free property (discussed in Section 10.9). The possible combinations are the following:

- **Communication through the ISD core (Case 1 in Figure 8.2):** We consider five scenarios in Case 1, grouped in two blocks as follows:
  - **Core-segment combination (Cases 1a, 1b, 1d):** the last AS of the up-segment is *different* from the first AS of the down-segment. This case requires a core-segment that connects the up- and down-segment. The final forwarding path then consists of the combination of up-, core-, and down-segments. The case works analogously if the two end hosts are in different ISDs.
  - **Immediate path segment combination (Cases 1c, 1e):** the last AS on the up-segment (ending at a core AS) is the *same* as the first AS on the down-segment (starting at the core AS). In this case, a simple combination of up- and down-segments creates a valid forwarding path.
- **Peering shortcut (Case 2 in Figure 8.2):** a peering link exists between the up- and down-segment. The extraneous path segments to the core

are cut off. We note that the peering link could also be traversing to a different ISD.

- **AS shortcut (Case 3 in Figure 8.2):** the up- and down-segments intersect at a non-core AS. This is the case of a *shortcut* where an up-segment and a down-segment meet *below* the ISD core. In this case, a shorter path is made possible by removing the extraneous part of the path to the core.
- **On-path (Case 4 in Figure 8.2):** in the case where the source's up-segment *contains* the destination AS, the up-segment of the source is sufficient to construct a forwarding path. The ISD core is again not part of the final path. If delivery is not permitted to the destination AS on the up-segment (as specified in the hop field), a down-segment needs to be combined with the up-segment, resulting in the AS-shortcut case discussed above.

In the following sections, we describe these cases in detail. We begin with the common case, in which two communicating end hosts are located in non-core ASes. We then describe communication between end hosts located in core ASes.

Notation

We use the following notation in this section: **HF** A_{BC} stands for a hop field that was generated by AS A, and that can be used for packet forwarding through AS A, with ASes B and C as preceding or succeeding ASes (resulting in the AS sequence BAC or CAB). If A is the first or last AS in a path segment, the absence of either the preceding or succeeding AS is indicated by the symbol $\bullet$. Arrows in the figures are oriented according to the direction of beaconing.

8.2.1 End Hosts in Non-core ASes

The common case for end-to-end communication is when two communicating end hosts reside in non-core ASes. Depending on the location of the two hosts, there are four possible types of path composition, which will be described in this section. We assume that after the path lookup process (see Section 7.2 on Page 132), an end host is provided with sets of up-segments, core-segments, and down-segments.

Paths Through the Core

It is always possible to construct forwarding paths that traverse core AS(es). However, this option is not preferred and is used as a last resort path since such a path is usually longer than its alternatives (e.g., peering or shortcut paths). In order to construct a path that traverses a core AS, the source needs an up-segment, a core-segment, and a down-segment, which are *connecting*, i.e., the up-segment starts where the core-segment starts/ends, and the core-segment ends/starts where the down-segment starts. Such a topology is presented in Figure 8.3. In the case where up- and down-segments originate from the same AS, a core-segment is not required. Note that a successful path lookup guarantees that there exists at least one connecting set of path segments.

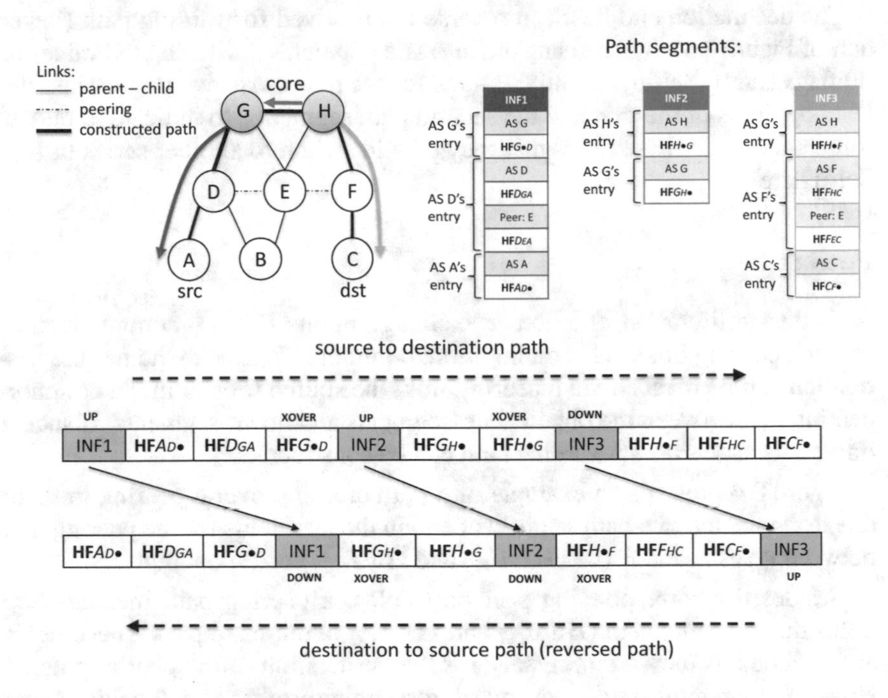

Figure 8.3: An example of a path traversing core ASes.

An example of a forwarding path traversing core ASes is presented in Figure 8.3. Combining the path segments is straightforward in this case: the source embeds a series of up-segment hop fields (from its own AS *A* to core AS *G*), two core-segment hop fields (from AS *G* to AS *H*), and three down-segment hop fields (from core AS *H* to the destination AS *C*) as a forwarding path.

Each series of hop fields is prepended with an *info field*, which encodes a timestamp used to verify the freshness of every corresponding hop field and information about the path type. The forwarding path goes through the core; thus it does not have either a shortcut or peer flag set. Info fields also contain information about the direction of the forwarding path (i.e., how interfaces within the hop fields should be interpreted and how hop fields should be verified). The direction depends on the propagation direction of a segment: The down direction (i.e., `UP` flag set to *false*) is set when the hop fields are listed in accordance with the direction of the beacon propagation. For the reversed direction the `UP` flag is set.

Hop fields within each series are ordered according to the order of traversed ASes. The source marks the last hop fields in each series with a dedicated *crossover* flag XOVER, to indicate that hop fields of another path segment start at this AS. Thus a border router that processes this hop field should switch to another hop-field series.

The destination end host can reverse the received forwarding path (lower half of Figure 8.3), by reversing the info and hop fields, switching the values of UP flags and by setting the XOVER flags for the new crossover points (i.e., the last hop fields of the first and the second path segments, to indicate to border routers that read these fields that they should switch to another series of hop fields here).

Peering Path

With the proliferation of Internet exchange points (IXPs), communication through peering links is becoming more common. To check whether the destination can be reached via a peering link, the source tries to find a common peering link between the obtained up-segments and down-segments. If such a link exists, a peering forwarding path can be constructed.

Figure 8.4 presents an example of a path crossing over a peering link. In this example, the two path segments contain the hop fields for the peering link between ASes E and F (i.e., **HF** E_{FB} and **HF** F_{EC}, correspondingly).

Besides the corresponding peer hop fields, a peering path includes hop fields that precede them (**HF** E_{GB} and **HF** F_{GC} in our example). These fields are included as they are necessary for the verification of the peer hop fields. However, to minimize traffic overhead, forwarding-irrelevant hop fields are not added to the path. In particular, a core-segment and its hop fields are not used to construct a peering path. In Section 8.3, we show an efficient algorithm to find a common peering link in two path segments.

In the example presented in Figure 8.4, the source specifies its intent to use the shortcut by setting a special flag (i.e., PEER flag) on the info fields. It also sets the direction (i.e., UP) flag of the hop-field segments. The hop fields **HF** E_{FB} and **HF** F_{EC} are marked with a *crossover* flag XOVER, which indicates that ASes corresponding to these hop fields need to process the hop fields in a special way. Specifically, border routers verify whether hop fields are created correctly, i.e., whether their ASes permitted the use of the mentioned peering link. For example, for a packet sent from the source to the destination, AS B first verifies whether the forwarding encoded within **HF** $B_{E\bullet}$ is allowed for the packet. To this end, a hop field **HF** E_{GB} of the parent is needed (as it was chained to **HF** $B_{E\bullet}$'s generation in the computation of the MAC field of **HF** $B_{E\bullet}$). When AS E receives the packet, the hop field **HF** E_{GB} is needed for the verification of **HF** E_{FB}, but is not used for the actual forwarding. Such hop fields are marked with a special VRFY-ONLY flag. The AS verifies that **HF** E_{FB}

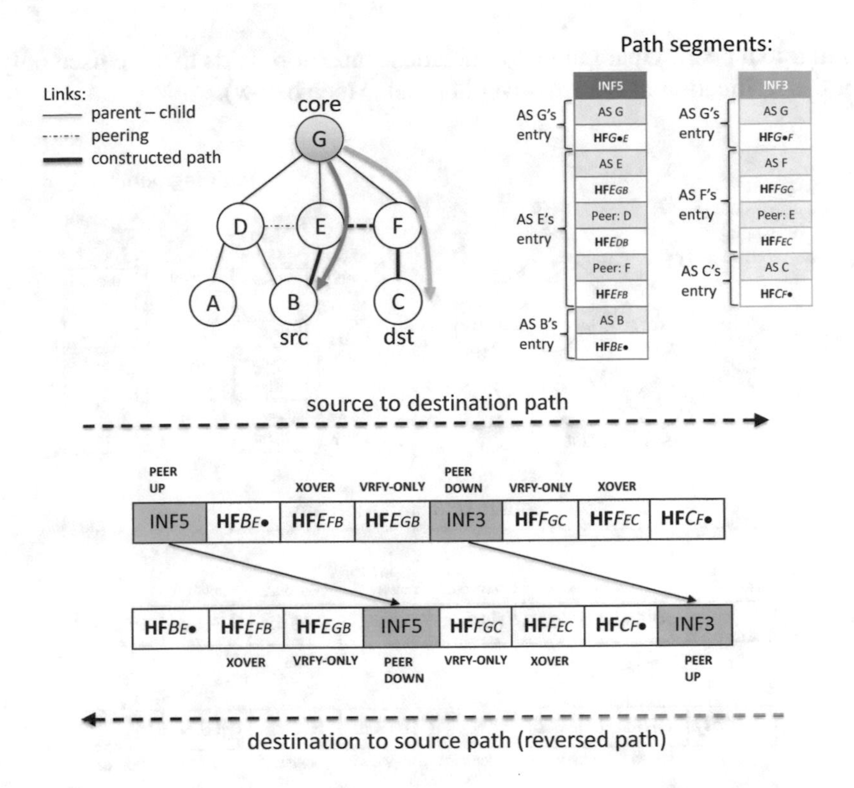

Figure 8.4: Path composition with a peering link.

was created correctly (i.e., that it is chained to **HF** E_{GB}), and forwards traffic to the peer AS.

A similar procedure applies to the processing of hop fields on the down-segment (i.e., AS F executes analogous steps). The reversed path is analogous to the original one, except that the UP flags are inverted and the info fields are moved.

Peering links are allowed between ASes from different ISDs and peering paths with such links are created in the same way as presented.

Shortcut Path (Common AS on Paths)

The up- and down-segments obtained may contain a common upstream AS. The example in Figure 8.5 shows such a case, where an end-to-end shortcut path can be constructed through AS D.

In this case, packets do not need to traverse the ISD core, but can be directly forwarded from the common AS D to destination AS B. A shortcut path is similar to a peering path in that it includes some hop fields that are used only for the verification of the crossover hop fields (see below).

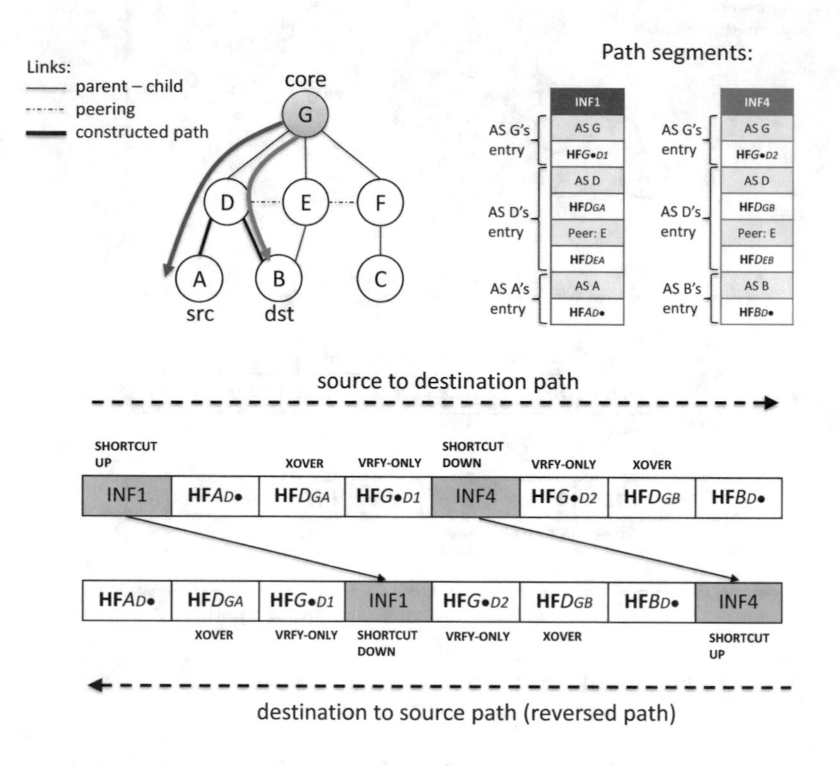

Figure 8.5: The path composition through a common AS.

To establish such a shorter path, the source specifies the path type as shortcut (i.e., it marks the info fields with the SHORTCUT flag), and marks the hop fields **HF** D_{GA} and **HF** D_{GB} as crossover points (i.e., it marks them with the XOVER flag). Similarly to the peering case, hop fields **HF** $G_{\bullet D1}$ and **HF** $G_{\bullet D2}$ are added only for verification purposes, and are marked with the VRFY-ONLY flag. The reversed path is analogous.

Destination AS on Path

In the last case, the destination is an AS on the up-segment. No forwarding path for this case should go through a core AS. Also, the up-segment (on which the destination is placed) alone should be sufficient to forward packets to the destination. However, the ingress router of the destination AS has to be informed that a packet can be terminated at the AS.

An example of such a path construction is presented in Figure 8.6, where the destination is within AS D, which is on the up-segment. In this case, the

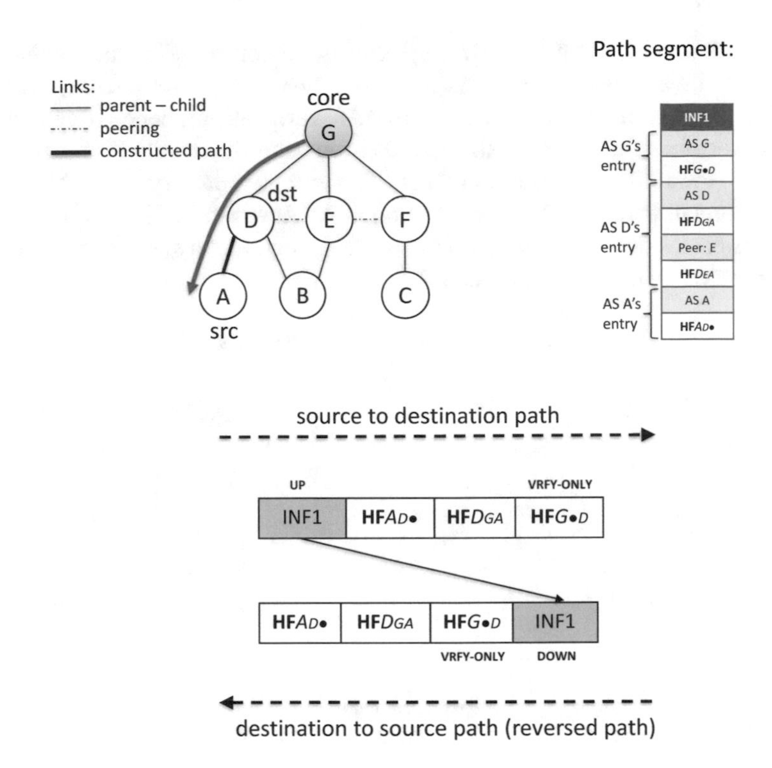

Figure 8.6: Path composition when the destination is contained in the up-segment.

source composes the series of hop fields up to the destination and sets AS D as the destination address. To enable verification of the **HF** D_{GA} hop field, the source additionally adds the next (upstream) hop field, and marks it with the `VRFY-ONLY` flag. The last router, i.e., the ingress router of the destination AS, verifies the hop field **HF** D_{GA} using the upstream **HF** $G_{\bullet D}$ hop field. If the verification succeeds and the destination AS matches the router's AS, then the packet is also allowed to terminate at the destination AS. In our example, **HF** D_{GA} is the last hop field processed, and **HF** $G_{\bullet D}$ is used for verification only.

The construction of the reversed path and the forwarding path where the source is on a down-segment are analogous.

An AS can disallow such a packet termination (i.e., it can create hop fields that can only forward packets between the AS interfaces, and that cannot deliver packets to the AS end hosts). This is done by setting the `FWD-ONLY` flag on the hop field during its creation (i.e., during beaconing).

An example of such a situation is presented in Figure 8.7. The topology and the path segment are identical to the previous example, except that AS D has set a FWD-ONLY flag on a hop field **HF** D_{GA} during beaconing. This means that AS D can only be a transit AS for this path, and a packet cannot terminate at this AS. This flag is immutable (i.e., used in MAC verification), hence if an end host tries to terminate a packet at the AS D (by setting the destination AS to D, as shown), the router will not deliver it. The end host can try to unset the FWD-ONLY flag, but then the MAC verification at AS D will fail. (To deliver the packet in this case a down-segment to D is needed, such that a shortcut path as described previously can be constructed.)

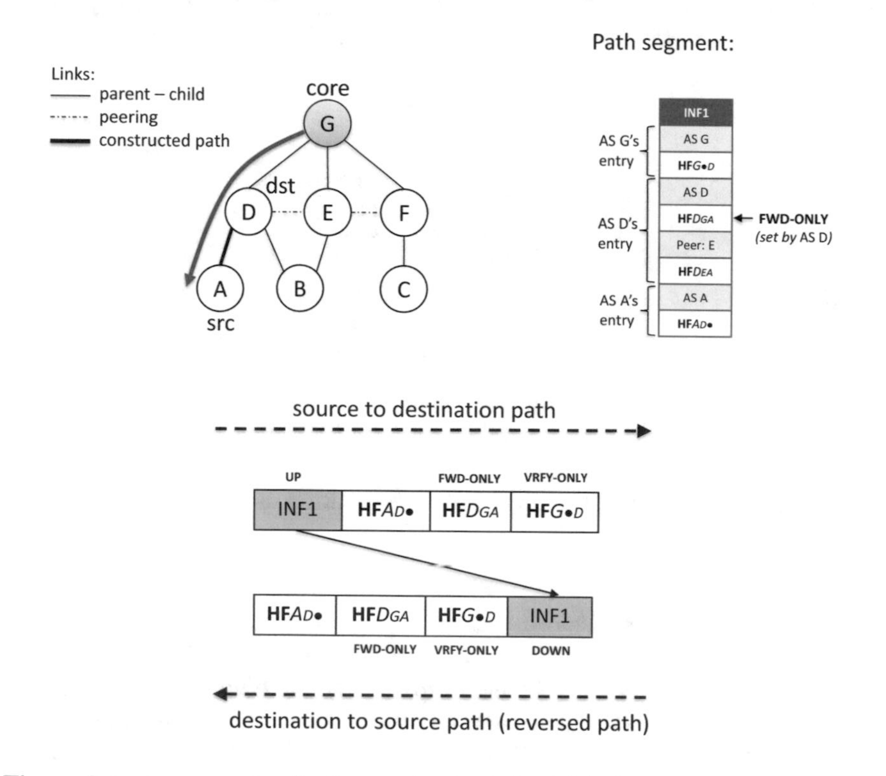

Figure 8.7: An example of a forward-only AS, which does not work to deliver a packet.

8.2.2 End Host(s) in Core ASes

We next describe the cases where at least one end host of an end-to-end connection is in a core AS. These cases do not involve peering links, common ASes of the involved path segments, or end-host ASes on the path. As a consequence, creating a forwarding path is less complex since it is simply a combination of entire path segments (which is similar to the path-through-the-core case in Section 8.2.1).

One End Host in the ISD Core

In the case where the source is in a non-core AS, but the destination is in a core AS, the following two options are possible.

First, the source can have a direct up-segment that originates from the destination AS (e.g., when AS *A* from Figure 8.8 wants to communicate with AS *G*). This is possible when the destination AS is a parent of the source AS, and the destination AS has propagated a beacon that was chosen by the source AS as an up-segment.

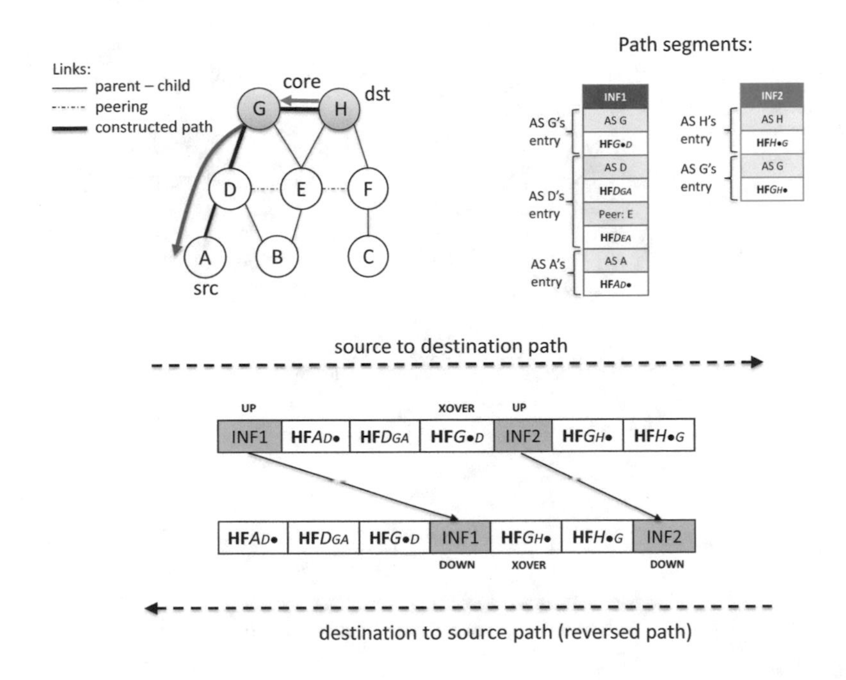

Figure 8.8: An example of path joining when the source is in a non-core AS and the destination is in a core AS.

Second, a direct up-segment to the destination does not exist; see again Figure 8.8 with source in AS *A* and destination in AS *H*. The source needs to obtain an up-segment to a core AS (e.g., to AS *G*), and a core-segment between that core AS and the destination AS *H*.

The forwarding path in this case is constructed as a special case of the path through the core from Section 8.2.1. The source just joins all hop fields of the up- and core-segments, and sets the crossover flag for the hop field **HF** $G_{\bullet D}$ (in order to inform a border router that, at this point, it should switch over to the hop fields of the next path segment).

Both End Hosts in the ISD Core

Path construction between two core ASes is straightforward since path propagation in the core guarantees that there is always a *direct* core-segment that connects two core ASes. Thus, the source simply obtains a core-segment and uses it directly.

An example of such a case is presented in Figure 8.9. The source just extracts the hop fields from the core-segment and concatenates them to construct a forwarding path.

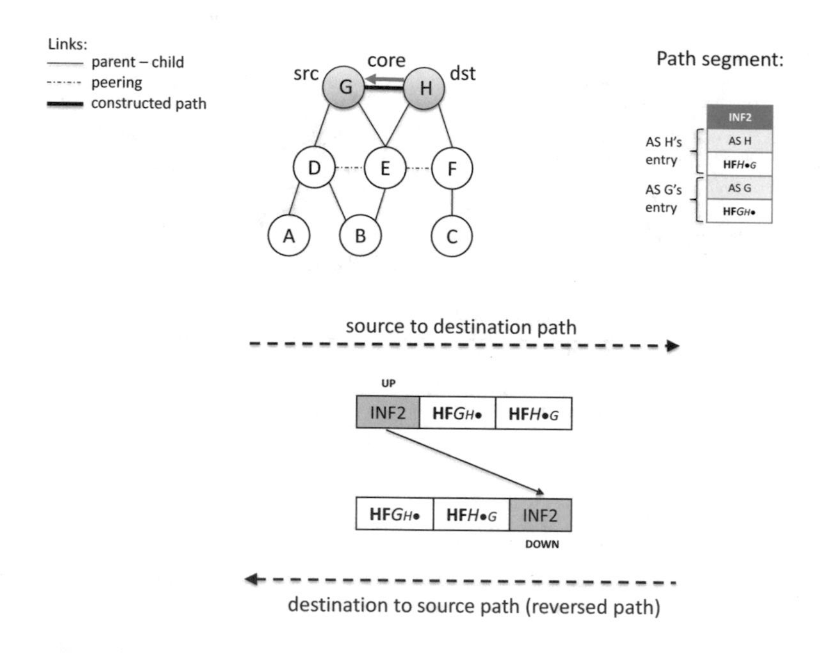

Figure 8.9: An example of a forwarding path between two core ASes.

8.3 Efficient Path Construction

In this section, we present an efficient path-construction algorithm, which allows an end host to find and build the shortest forwarding path towards a destination AS. The algorithm is executed by the end host once it has obtained the corresponding up-, core-, and down-segments to the destination AS.

A path segment consists of a list of ASes, where each AS is identified by its ISD and AS identifiers (ISD ID, AS ID) and a list of entries with ingress and egress interfaces. Thus, even if two paths contain the same list of (ISD:AS) tuples, they can differ at the granularity of interfaces (e.g., when two ASes share more than one link). Naively, an end host can try to join all possible combinations of up-, core-, and down-segments and select the best end-to-end path among them. However, exhaustive exploration is inefficient if path segments contain many peering links, or if there are many path segments available. We resolve this issue by designing an efficient path-construction algorithm that finds the shortest path(s) in terms of AS hops. The algorithm operates in two steps: the graph-construction step and the path-construction step.

8.3.1 Graph Construction

In the first step, the source host creates a weighted and directed graph, based on the up-, core-, and down-segments. Each path segment contains a list of (ISD ID, AS ID) tuples; each tuple uniquely identifies an AS. The graph is then constructed as follows:

1. For each up-segment of the source host's AS, the algorithm traverses the ASes starting from the source AS and creates a node in the graph for every new AS encountered in the up-segment. The algorithm adds a directed edge from the source AS to the encountered AS, annotated with the hop distance from the source AS. Furthermore, a path identifier is also used to annotate the edge.

   If the encountered AS already has a node in the graph, a new edge is added from the source AS node to the encountered AS node, and the edge is annotated with the same information as described earlier. Thus, multiple edges can exist between two nodes.

   If an AS has a peering link with another peer AS, a new edge is similarly added between the nodes for the source and the peer AS. In this case, the edge is additionally annotated as a peering link.
2. The same procedure is followed for each down-segment of the destination, with two differences. First, the direction of the edge is reversed, so that the edge points towards the destination AS. Second, peering links are not added, since a valid end-to-end path can traverse at most one peering link; if a peering link is traversed, it has already been added in step 1 of the algorithm.
3. For core-segments, the algorithm complements the graph as follows. First, it selects only the segments that connect the core ASes of the up-segments to the core ASes of the down-segments. Then, it traverses every AS in the selected core-segments and adds an edge from the core AS of the up-segment to the encountered core AS, similarly to the previous two steps. Also, it annotates the edge with the hop distance and with a core path flag.

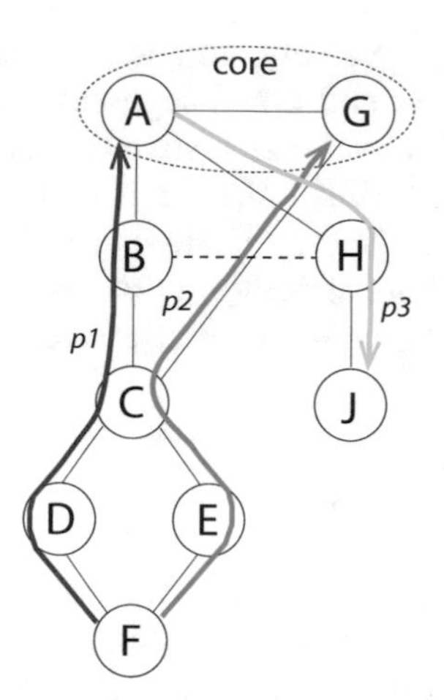

Figure 8.10: An example topology, with up-segments (*p1*, *p2*) and down-segments (*p3*) obtained by the source (placed in AS F) towards the destination (placed in AS J).

Example. We provide an example to explain how the graph is constructed. Figure 8.10 shows an example topology with the up-segments (*p1* and *p2*) of the source AS F, and the down-segments of the destination AS J; for simplicity, we omit the core-segments. The source host in AS F obtains the path segments *p1*, *p2*, and *p3* and constructs the graph in Figure 8.11, according to the procedure described earlier.

All outgoing edges from F point to an AS that is either on an up-segment of F or has a peering relationship with an AS that is on an up-segment of F (e.g., AS H). Note that the edges do not correspond to physical links (as shown in Figure 8.10) and that the weight of each edge denotes the hop distance from the source AS to the corresponding AS. The reason the graph is constructed in this way is that the source host should not combine valid path segments to create a new path segment. For instance, the combination $F \rightarrow E \rightarrow C \rightarrow B \rightarrow A$ is not valid, as it is a combination of *p1* and *p2*.

The same procedure is followed for the down-segments, but the direction of the edges is reversed. Thus, all edges point towards the destination AS (AS J in Figure 8.11).

The algorithmic complexity of the graph-construction step is linear with respect to the number of received path segments (whether up-, down-, or core-segments), since the algorithm processes every received path segment once. The intermediate nodes in the graph of Figure 8.11 can be added to a hash table for efficient lookup, i.e., lookup in constant amortized time, when they have to be looked up in order to add the corresponding edges in the graph; this is possible as the number of paths is relatively small.

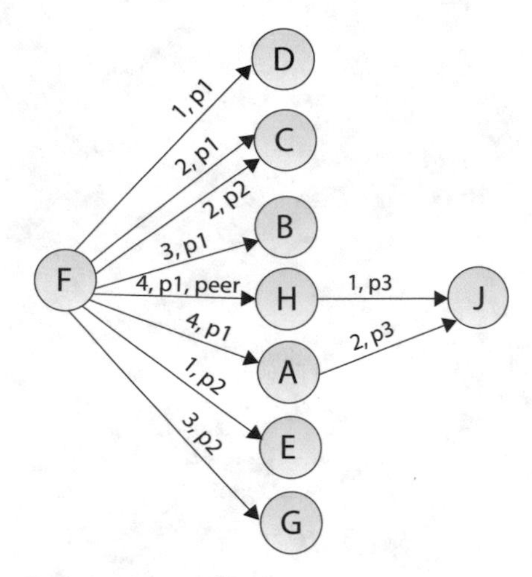

Figure 8.11: Constructed graph based on the up- and down-segments. Edges are annotated with the hop distance between the corresponding ASes, a path identifier, and a peering-link flag in case a peering link is traversed.

8.3.2 Path Construction

The construction of the graph turns path construction into a simple graph traversal problem. The source host can use existing algorithms to discover the shortest path, all shortest paths, or all paths to the destination. For example, using Dijkstra on the graph in Figure 8.11 yields the shortest path $F \rightarrow H \rightarrow J$, with a total cost of 5.

The paths that are discovered on the constructed graph do not correspond directly to physical paths. However, the edges of the constructed graph are annotated with all the required information so that the source host can discover the actual path and set up the hop fields. For example, the edges of the path $F \rightarrow H \rightarrow J$ inform the source host that the end-to-end path is formed by combining *p1* and *p3* and that it traverses a peering link.

Using Dijkstra for path construction yields an algorithmic complexity of $\mathcal{O}(|L| + |V| \log |V|)$, where $|V|$ is the number of distinct AS nodes in the received path segments and $|L|$ is the number of distinct links in the received path segments; recall that a link is uniquely identified by the interface identifiers and not by the AS identifiers. However, note that the graph-construction step generates a directed acyclic graph (DAG), which enables an even faster shortest-path algorithm: with topological sorting on the DAG, the single-source shortest distances can be calculated in $\mathcal{O}(|L| + |V|)$.

9 Host Structure

JASON LEE, ADRIAN PERRIG, PAWEL SZALACHOWSKI

This chapter introduces the host software components that enable applications to communicate via SCION. An overview is given in Figure 9.1.

The central point of the host structure is the *SCION dispatcher*, which handles all incoming and outgoing packets and interacts with the higher-layer protocols. The *SCION daemon* handles control-plane messages, e.g., obtaining paths to remote ASes. Currently, SCION supports the following transport protocols: TCP, UDP, and the *SCION Stream Protocol* (SSP). The last is SCION's native multipath transport protocol. UDP and SSP data is handled by the *SCION socket library*. Implementations of all three protocols provide an API similar to the Berkeley Socket API. Since the host structure is under active development, we expect many changes in this area over the coming years. For example, we are currently working on a SCION extension of the QUIC protocol [105].

Chapter Contents

9.1 SCION Dispatcher

The initial version of the SCION host stack follows two design choices, which affect how packets are handled: (a) the host stack runs over an IP/UDP overlay to communicate within an AS, and (b) all code executes in userspace.[1] Consequently, we designate a UDP port through which all SCION packets are

[1] In a future release of SCION, the functionality of the dispatcher will execute inside a kernel module.

A. Perrig et al., *SCION: A Secure Internet Architecture*, Information Security and Cryptography, https://doi.org/10.1007/978-3-319-67080-5_9

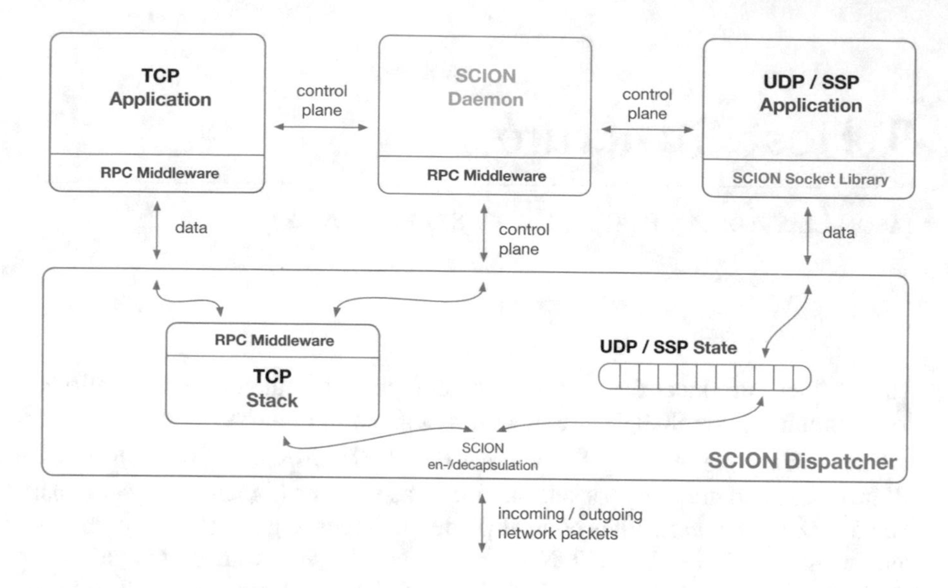

Figure 9.1: Overview of SCION's end-host software structure.

communicated and define a single process (within each host) that handles all incoming SCION packets.

The dispatcher performs two tasks: (a) handling encapsulation and decapsulation of the IP/UDP overlay and the SCION headers, and (b) interacting with the transport protocol stacks. That is, the dispatcher mediates between applications and transport protocol implementations to process incoming and outgoing packets.

More specifically, for outgoing packets, the TCP stack and UDP/SSP applications send their data along with the packet's metadata. The metadata includes a forwarding path and an address of the first SCION hop (i.e., border router) on the path. The dispatcher, upon receiving the data and its metadata, encapsulates the data and sends the packets to the specified border router. The path and the first-hop address are provided to applications by the SCION daemon, which has the required control-plane information from the discovery service (Section 7.4.6).

For incoming packets, the dispatcher decapsulates the overlay header and parses the SCION header to identify the transport protocol. The dispatcher processes packets differently for different transport protocols. All TCP packets are passed to the TCP stack; for UDP and SSP packets, the dispatcher parses the identifier associated with the incoming packet and delivers the packet to the application that is registered for that identifier.

9.1.1 Application-to-Dispatcher Communication

Communication between the dispatcher and applications is implemented via reliable Unix domain sockets. However, depending on the transport protocol, the communication between applications and the dispatcher differs.

TCP Applications

TCP applications and the dispatcher communicate through a *remote procedure call* (RPC) middleware which consists of interacting parts: (a) the applications side (to send calls and receive responses), and (b) the TCP stack (to receive calls, execute them, and send responses back).

For each application-level socket there is a corresponding Unix socket and native TCP state. For example, an application executing the `connect()` system call sends an RPC message to the RPC middleware of the TCP stack, which receives the message, executes the native TCP `connect()` call on the corresponding TCP socket, and returns to the application the results of the call (e.g., a success or an error code).

For connected sockets, i.e., after a successful `connect()` or `accept()`, the RPC middleware enters *pipe mode*, in which it simply passes the data between an application socket and the corresponding TCP socket.

UDP and SSP Applications

The interaction between the dispatcher and UDP/SSP applications is simpler than in TCP, as the UDP and SSP stacks are implemented within the SCION socket library. All UDP and SSP applications must register with the dispatcher to send and receive packets. The format of the registration message depends on the transport protocol used by the application.

UDP registration messages have the following format:

- addr represents the address bound by the application (its length depends on the type).
- SVC is an optional 2-byte field that specifies the service type to register for (e.g., beacon servers register for SVC 0).

SSP registration messages have the following format:

- flowID is an 8-byte identifier for the connection used by the application.
- addr is a variable-length address bound by the application.
- SVC is an optional 2-byte SVC address type to register for.

We emphasize that SCION addresses are triples: `(ISD,AS,ADDR)`, where `ADDR` is an address that has significance only within the host's AS. Because IP addresses need to be unique only within an AS, the ISD-AS information is needed in the registration message to correctly identify the receiver. For example, an application that registers for the address `(1,1,1.2.3.4)` should not receive packets with destination address `(2,1,1.2.3.4)`, even though both addresses may belong to the same host.

9.1.2 Outgoing Packets

TCP

When the TCP stack sends TCP segments to the dispatcher, they are passed along with their metadata. A segment's metadata includes the information required to encapsulate the segment into a SCION packet (i.e., source and destination addresses, a forwarding path, and optionally extensions). The TCP stack also passes the first-hop information, which is necessary to encapsulate the SCION packet within a UDP (overlay) packet. The dispatcher creates the SCION packet, then encapsulates it within a UDP packet, and finally sends it to the first-hop address.

UDP and SSP

Data sent by UDP and SSP applications is processed by the SCION socket library. When the library sends a chunk of data it is encapsulated within a SCION packet. Next, outgoing SCION packets are sent to the dispatcher along with the first-hop information only (forwarding path and optional extensions are already included within the SCION packet). Upon receiving a data packet from an application, the dispatcher parses the first-hop information, creates a UDP packet and sends the packet to the first-hop host.

9.1.3 Incoming Packets

The dispatcher receives incoming packets by listening on the SCION UDP port. Upon receiving a packet and decapsulating the overlay IP/UDP header, the dispatcher obtains a SCION packet. When a SCION packet arrives at the host, the dispatcher first checks whether both ISD-AS identifier and IP address indeed identify that host. If not, the packet is dropped. Then, the dispatcher parses the SCION header (see details in Section 15.1.1 on Page 343) to identify the layer-4 protocol. To this end, the *protocol number* (encoded in the `NextHdr` field inside the SCION common header, see Section 15.1.1) is used. Then, depending on the protocol, the data is passed either to the TCP stack or to the UDP or SSP stack.

TCP

If the TCP protocol (i.e., protocol number 6) is used, the SCION header is decapsulated. The obtained TCP segment is then passed to the TCP stack for further processing. Along with the segment, the following metadata is passed to the stack: source and destination addresses, forwarding path, and optionally extensions. The processed payload (if any) is delivered to an application's socket through the RPC middleware.

UDP and SSP

For the UDP and SSP packets, the dispatcher identifies the protocol-dependent *application identifier*. For UDP packets (protocol number 17), the application identifier is the port number. For SSP (protocol number 152), the application identifier is the 8-byte flow ID. If the flow ID has not been registered, the dispatcher will look for a wildcard entry that accepts all incoming flows on a given port number. The SCION packet is not decapsulated at the dispatcher, and is entirely passed to the listening application, where the SCION socket library further processes the packet.

9.2 SCION Daemon

The SCION daemon is a background process running on end hosts with the goals of (a) handling SCION control-plane messages, and (b) providing an API for applications and libraries to interact with the SCION control plane. More specifically, the SCION daemon implements the following services:

- **Path lookup:** provides path lookup functionality for host applications. The path lookup process is described in Section 7.2, and the path creation process is described in Section 8.2.
- **Name resolution:** provides name resolution functionality, i.e., a translation from a human-readable domain name to a SCION address. The details of name resolution are described in Chapter 6.
- **Trust management:** stores received certificates and TRCs, and checks their authenticity and consistency (when a new certificate/TRC is received). Certificates and TRCs can be provided to applications on demand.
- **Topology information:** provides information about the topology of the local AS. Topology information includes addresses of border routers (with their interface identifiers) and information on running services (e.g., RAINS or path servers). The topology information is obtained from the local AS through the discovery service (see Section 7.4.6). If an end host is multi-homed, the information is provided for all local ASes.

- **Extensions:** various SCION extensions and sub-protocols, such as SIBRA (Chapter 11) and OPT (Chapter 12), implement their control plane as part of the SCION daemon and extend its API.

To start, the SCION daemon contacts the discovery service(s) of its AS(es) (Section 7.4), parses the TRC(s) of its ISD(s) (Section 16.1), and optionally parses the corresponding certificate(s).

9.2.1 API

The SCION daemon exposes its API through a reliable Unix domain socket with a pre-defined path, which allows userspace applications to interact directly with it.

Request and Responses

We next describe the most important API calls with their expected results. To support multi-homed end hosts (i.e., residing in multiple ASes) each API request can be accompanied by an ISD and AS identifier to specify the AS whose infrastructure should be used to handle the request.

- **Path request:** Consider a process that wishes to obtain a forwarding path to a destination AS. In this case, it queries the SCION daemon with a path request specifying the destination.

  Upon receiving such a request, the SCION daemon attempts to build forwarding paths to the destination using locally cached path segments. If it fails, the SCION daemon contacts a local path server, which initiates the path lookup process (see details in Section 7.2 on Page 132). In the case of a successful path lookup, the SCION daemon returns a list of forwarding paths with their metadata. The metadata includes
  - the local address of the first border router on the path (in case the requester process encapsulates and sends a SCION packet on its own),
  - the maximum transmission unit (MTU) of the path, and
  - a list of interface identifiers that provide a path-requesting process with the information about the path's ASes (and their interfaces).
- **Name resolution:** Consider a process that needs to resolve a human-readable domain name to a SCION address (as described in Chapter 6). The request to the SCION daemon includes the domain name and an optional resolution context, while the response consists of a list of SCION addresses and additional optional information associated with the requested domain name.
- **Topology discovery:** Some processes might need topological information about the local AS(es). To this end, a process can send a topology

discovery request to the SCION daemon. The SCION daemon returns the topology information (see details in Section 16.3) including addresses of border routers and their interface identifiers, and information on running services.

- **TRC request:** To request a TRC, a process sends a request that specifies the requested version of the TRC for a given ISD. Without specifying the version, the most recent TRC (that the SCION daemon has) is requested. To serve the request, the SCION daemon first searches its local cache, and if absent, it contacts the local certificate server. We note that the SCION daemon is pre-loaded with the first TRC.
- **Certificate request:** Similarly to TRCs, AS certificates can be requested from the SCION daemon by sending a request that specifies the target AS (i.e., its ISD and AS identifiers) and the version of the requested certificate (the field can be empty if the most recent certificate that the SCION daemon has is requested). To serve the request, the SCION daemon first searches for the certificate in its local cache; if absent, it contacts the local certificate server.

Error Codes

In case of a failure of any of the above requests, the SCION daemon returns an appropriate error code. The most frequent error codes are as follows:

- the requested object does not exist,
- the SCION daemon encountered a problem while processing the request (this message is followed by a more detailed error description),
- the local recursive server (i.e., path or certificate server) encountered a problem while processing the request (this message can be followed by a more detailed error description).

9.3 Transmission Control Protocol (TCP/SCION)

To make TCP available in SCION, TCP was modified by adding (a) SCION addresses, (b) SCION forwarding paths, and (c) SCION packet extensions, and by extending the standard TCP checksum algorithm.

Regarding the first modification, the TCP stack was extended to handle SCION addresses (in addition to IPv4 or IPv6 addresses). A SCION address is a 3-tuple of the form `(ISD, AS, ADDR)`. SCION addresses introduce the notion of *special service addresses*, which are used for SCION service anycast communication (see details in Section 15.1.2). As implemented via TCP, the service addresses have to be handled by the TCP/SCION stack.

TCP/SCION sockets listen on the 5-tuple

```
(ISD, AS, ADDR, port, SVC)
```

where the first three fields specify the SCION address, the port field denotes the standard TCP port, and the SVC field specifies the service that is bound to that socket (e.g., the path service, when a path server listens on the socket). For standard (non-SCION service) sockets, the SVC field is set to `None`.

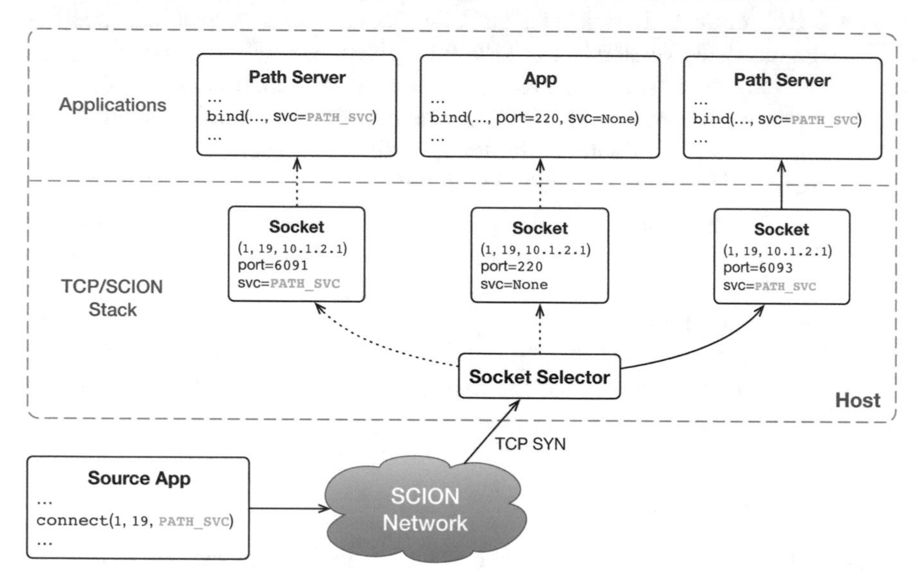

Figure 9.2: An example of SCION TCP service communication.

Example. An example of an initiation of a TCP/SCION service connection is presented in Figure 9.2. The source application from a remote AS (depicted at the bottom left) wishes to connect with a path service of another AS, for example $(1,19)$ in this case. The initial `SYN` packet that the source sends contains the SVC address, but has an unspecified port (i.e., set to `0`), as the source does not know the ports that are used by the path servers of the destination AS.

The ingress border router at the destination AS forwards the packet to a host that is designated to serve the path service (see details in Section 7.4). The packet then reaches the designated host, where two instances of the path server application are running. Both instances have sockets that are listening to the service address `SVC=PATH_SVC`. As SVC communication is anycast, the socket selector (which pairs the packets with the corresponding TCP states) selects one of the two listening service sockets at random and directs the packet there. (The socket selector ignores the port field of this packet.)

In the response, a TCP `SYN-ACK` packet is generated. In the presented example, this responding `SYN-ACK` packet will have set the source SCION address to `(1,19,10.1.2.1)` and the source port to `6093`. The socket selector at the source is modified to associate the responding `SYN-ACK` packet with the correct TCP state. The socket selector on the host is modified to handle the initiation of SVC connections. For all other cases, i.e., if a SYN packet has a destination address different from SVC, the socket selector associates it as in standard TCP/IP.

An established end-to-end TCP/SCION connection is identified by the 9-tuple

```
(src ISD, src AS, src ADDR, src port,
 dst ISD, dst AS, dst ADDR, dst port, protocol)
```

where the `protocol` field is set to the TCP protocol number (i.e., 6).

As another consequence of modifying the TCP stack with the SCION addresses, TCP's checksum algorithm needs to be extended. In TCP/SCION, the checksum algorithm accounts for different address sizes due to new SVC addresses and SCION's flexible destination addressing.

Our implementation of TCP/SCION offers flexibility with respect to path management: forwarding paths can be specified by an application or can be obtained by the stack itself. The former option allows path-aware applications to select optimal forwarding paths.

Another modified element of the TCP protocol is the calculation of TCP's maximum segment size (MSS). As an effective MSS depends on the path and on the extensions that are used, it must thus be calculated dynamically. The MSS computation algorithm is modified to incorporate the MTU of the currently used forwarding path and extensions used in a given packet. We also added an extension through which communicating parties can inform each other about their respective MTUs (see Section 15.1.4).

Currently, TCP/SCION is implemented using the userspace `lwIP` stack [74]. Our implementation exposes a high-level API for Python. An example of a client-server application is presented in the listing below. One difference from the standard (TCP/IP) API is that SCION uses triples as network addresses. Thus client and server have to bind and connect to instances of a special `SCIONAddr` class. Another difference is that a client, before calling `connect()`, has to obtain a path to initiate communication with a server. To this end, the `get_paths()` call is used, which communicates with the SCION daemon and returns a forwarding path with its metadata (e.g., the first-hop address, path MTU). Consequently, the `connect()` call is extended to pass a path and its metadata to the TCP/SCION stack.

```
from lib.packet.host_addr import haddr_parse
from lib.packet.scion_addr import ISD_AS, SCIONAddr
from lib.tcp.socket import SCIONTCPSocket
import lib.app.sciond as lib_sciond

# Set the server's address
srv_isd_as = ISD_AS("1-18")
srv_ip = haddr_parse("IPV4", "12.10.61.177")
srv_addr = SCIONAddr.from_values(srv_isd_as, srv_ip)
srv_port = 5000
```

```
# Set the client's address
cli_isd_as = ISD_AS("2-23")
cli_ip = haddr_parse("IPV4", "92.131.161.3")
cli_addr = SCIONAddr.from_values(cli_isd_as, cli_ip)

def server():
    # Bind to the server address and port
    sock = SCIONTCPSocket()
    sock.bind((srv_addr, srv_port))
    sock.listen()
    # Wait for connections
    while True:
        new_sock, addr, path = sock.accept()
        # Handle accepted connection.
        print("New connection accepted:", addr, path)
        new_sock.sendall(b"Hello from server!")
        new_sock.close()

def client():
    lib_sciond.init()
    # Get path(s) to AS (1, 18)
    reply = lib_sciond.get_paths(srv_isd_as, max_paths=5)[0]
    path_info = (reply.path().fwd_path(),
                 reply.first_hop().ipv4(),
                 reply.first_hop().p.port)
    # Create socket and connect
    sock = SCIONTCPSocket()
    sock.bind((cli_addr, 0))
    sock.connect(srv_addr, srv_port, *path_info)
    # Receive and print data
    print(sock.recv(1024))
    sock.close()
```

9.4 SCION Stream Protocol (SSP)

The SCION architecture provides path control and path transparency to allow end hosts and applications to select their communication paths. This facilitates multipath communication as end hosts know properties of the available paths and can even use several paths simultaneously.

In this section, we provide an overview of the SSP protocol, which is SCION's experimental multipath byte-stream transport protocol. It provides reliable communication using multiple SCION paths, and utilizes mechanisms for congestion control. In a future release of SCION, the SSP protocol may be deprecated — instead we plan to add native support for MPTCP [207] and a multipath variant of the QUIC protocol [105].

An SSP connection starts with the client sending an initial connect packet to the server using all available paths it has selected to use. The server acknowledges each incoming packet using the path it was received on. After this initial exchange, the client knows which paths are available and has an initial round-trip time (RTT) estimate for each available path. A subset (by default, two) of these paths will be used as *active paths*, while the rest are kept as *backup paths* in case one of the active paths fails.

For sending data, SSP uses one send buffer and one receive buffer per connection. These buffers contain all packets that were sent and received over all paths in a given connection. Internally, the send buffer includes both a list of freshly queued packets and a list of lost packets that need to be retransmitted. Retransmissions take priority over new packets in the scheduler. This behavior may change in the future when new scheduling strategies are implemented.

Under some conditions it is possible that a sender resends a packet even though the original packet was not lost, for instance if an acknowledgment was lost. In such cases of duplicate data, the receiver keeps the data that arrived first.

Due to the use of multiple paths with varying latencies, packet reordering will likely occur. SSP uses an in-order packet queue and a separate out-of-order queue. In-order packets are placed at the tail of the in-order queue while out-of-order packets are inserted in order into the out-of-order queue. When a new in-order packet arrives, the out-of-order queue is additionally scanned to determine whether the new packet has filled a hole in the data space.

A single scheduler thread is used to send a packet. The current default policy is to send the first available packet on the first available path. In the future, more sophisticated scheduling strategies will be available (e.g., to minimize latency).

Each path performs independent congestion control. The schemes currently implemented are Constant Bit Rate, PCC [71], TCP Reno, and TCP Cubic. Future work will include an integrated congestion control scheme such as that used by MPTCP [206, 207] to ensure fairness for flows that traverse the same physical link.

Figure 9.3 shows SSP's packet header format. All SSP packets contain a common header, then an optional acknowledgment header, and an optional path header.

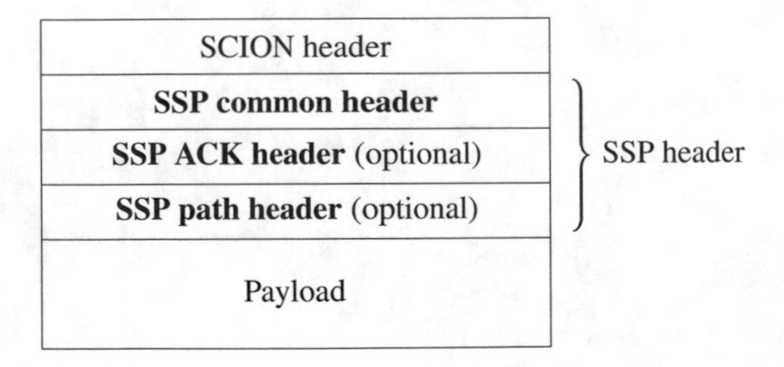

Figure 9.3: High-level layout of an SSP packet.

The mandatory *SSP common header* contains the *flow ID* field, which identifies the connection. Any packet on any path is identified with its connection using this ID. Because the identifier is independent of either endpoint's address, each host can use not only multiple paths but also multiple addresses with multiple ISD/AS associations. The header contains also the *port* field (which is identical to port fields used in TCP and UDP), and information about the data in the packet and connection state (e.g., *sequence number*, *flag fields*).

Acknowledgment packets contain the *SSP ACK header*, which informs the other communicating party about the received data. Another optional header is the *SSP path header*. For each connection, the first packet sent on each path will carry (within the path header) the list of hops that are traversed on the path. Each hop is represented by a concatenation of ISD, AS, and interface identifiers.

The SCION host network stack provides the fully implemented SSP protocol, and the implementation exposes high-level APIs for Python and C. SSP is a connection-oriented protocol; it thus provides calls such as `connect()`, `listen()`, and `accept()`. In SSP an application does not need to explicitly fetch forwarding path(s), as the SSP stack transparently takes care of path selection.

10 Deployment and Operation

YIH-CHUN HU, TOBIAS KLAUSMANN, ADRIAN PERRIG,
RAPHAEL M. REISCHUK, STEPHEN SHIRLEY,
PAWEL SZALACHOWSKI, ERCAN UCAN

How can the deployment of SCION be initiated? Which ISDs exist in the beginning? How can an ISP or end domain start using SCION and what benefits are obtained? This chapter discusses deployment and operation aspects of SCION and provides answers to these questions.

More precisely, this chapter makes two important points: (a) contrary to what might be expected of a future Internet architecture, the deployment of the SCION infrastructure requires surprisingly little effort for ISPs and domains, as few servers and routers need to be installed since SCION re-uses the existing intra-domain communication network, and (b) even early adopters can benefit from SCION's features.

Chapter Contents

10.1 ISP Deployment

From a global perspective, we envision three stages for deployment of SCION: early, intermediate, and full.

A. Perrig et al., *SCION: A Secure Internet Architecture*, Information Security and Cryptography, https://doi.org/10.1007/978-3-319-67080-5_10

In the ideal, *full* deployment case, each ISP and each domain in the world would deploy SCION. Before that happens, however, achieving an *early* deployment ranging from the current 20 domains to about 100 domains with dozens of deploying ISPs will be an important stage in maturing the technology and infrastructure, and creating a critical mass of users. The early deployment is likely to be characterized by highly specialized communication patterns, such as those involving corporations that desire high availability for safety-critical communications. During its early deployment, SCION will largely operate as an overlay network, making extensive use of the current Internet to connect deploying entities. When SCION operates as an overlay, the BGP routing system is used. The security guarantees offered by such an overlay deployment are discussed in Section 13.9.

Between early and full deployment lies a long stretch of *intermediate* deployment, which will be characterized by an increase in the diversity of applications that benefit from SCION's features. Teleconferencing and bitcoin mining pools may be among the applications to benefit from SCION's high availability [12]. Section 10.1.2 provides further details and discussion of this intermediate deployment phase, which we also refer to as incremental deployment.

We envision that high-availability applications will drive the initial phase of deployment. We discuss the details of incentives for deployment that will drive this process in Section 2.5. In the long term, we anticipate that deployments will diminish in their requirement for high availability over time, with an increase in demand for other features (e.g., SD-WAN).

In this context, the goal for SCION is to achieve a widely dispersed global deployment by domains and ISPs, providing the advantage of high availability due to path diversity [8]. A dispersed deployment will also provide kernels around which SCION islands can grow, steadily strengthening the properties that can be achieved locally. In the following sections, we discuss different options for ISPs to deploy SCION.

10.1.1 ISP Deployment Scenarios

An ISP that deploys SCION needs to first find other SCION ISPs to connect to. The connection can happen over an existing network link or ideally through a dedicated link, as we explain further in this section. The ISP then needs to obtain a certificate for their AS for each ISD it participates in. The details of this operation are explained in Section 10.6.

A deploying ISP needs to set up SCION border routers and services. Figure 10.1 contrasts the minimal, intermediate, and ideal deployment scenarios. In the minimal deployment case, it can have a single SCION border router and services all deployed at a single location, perhaps even on a single physical host. The minimal deployment places more reliance on the existing network infrastructure and therefore fewer guarantees are achieved, as an adversary might overload the legacy network with legacy traffic, clogging the same links that SCION traffic would use — we discuss such attacks in more detail in Section 13.9. In the intermediate deployment case, the ISP would deploy several SCION border routers adjacent to existing legacy border routers, along with multiple servers distributed over their network to achieve tolerance to failures. In the ideal deployment case, in addition to having multiple servers distributed over the network, the SCION border routers would be directly connected to the neighboring ISPs' SCION border routers.

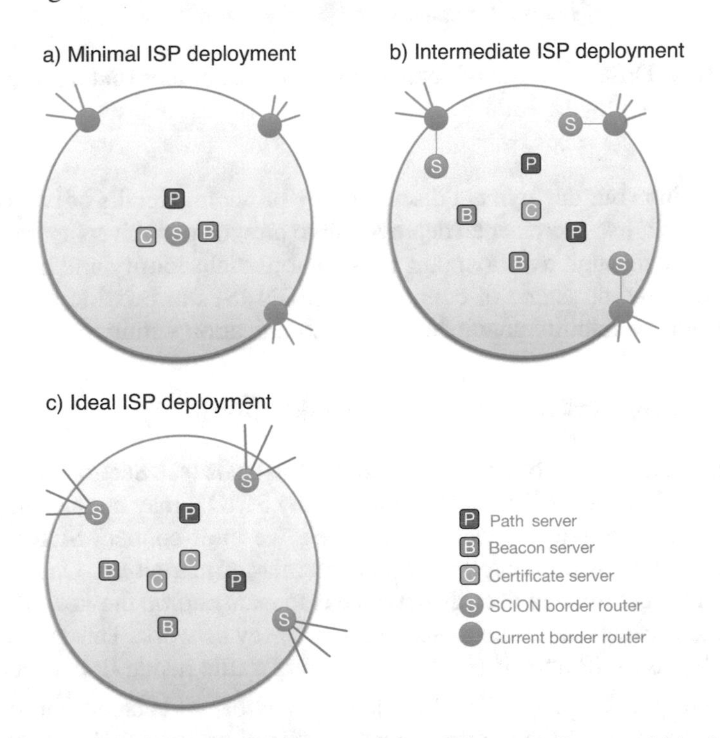

Figure 10.1: ISP Deployment: minimal, intermediate, and ideal.

The deploying ISPs can install their border routers in two different ways. Figure 10.2 shows the possible types of connections that can be established between border routers: a) depicts a simple deployment model with a "router-on-a-stick" configuration, where the SCION border router is attached to an existing legacy border router and communicates with the neighboring SCION border router over a short overlay connection. The advantage of this deployment is that the SCION border router does not interfere with legacy IP traffic, but the disadvantage is that legacy IP traffic may interfere with SCION traffic in case the legacy link is congested. b) shows the ideal type of SCION router deployment, where the two SCION border routers are directly connected via a cross-connect between the deploying ISPs, achieving the strongest security and availability properties.

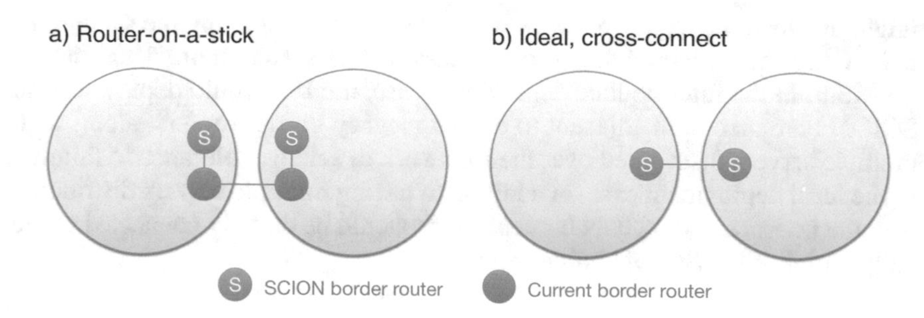

Figure 10.2: Different types of connections between border routers of different deploying ISPs.

We continue our deployment discussion by providing details of two complementary scenarios: incremental deployment to provide path diversity and global connectivity through overlay paths (ensuring partial security and availability properties), and an island of connected SCION ISPs to provide full SCION security and availability guarantees for communication within the island.

10.1.2 Early and Intermediate Deployment

During the early stages of deployment, SCION islands (see Section 10.1.3) may not be directly connected, i.e., ASes that deploy SCION may not be contiguous. To enable connectivity between such ASes, we inter-connect SCION ASes via overlay tunnels utilizing the current Internet infrastructure. One common approach for creating an overlay network is to *encapsulate* the new network's traffic into packets that can be routed over the legacy network. This is commonly done today using IP tunnels (i.e., encapsulating traffic inside IP packets).

When relying on properties of the legacy network, it is clear that the availability guarantees of the overlay network depend on the reliability of the IP tunnels. That is, IP tunnels should be resilient to failure, but should also be robust if exposed to any attack that degrades availability. In particular, the tunnels themselves should be robust when facing IP-prefix-hijacking attacks. To achieve this, tunnel IP addresses are ideally announced using /24 prefix blocks via BGP, which are very specific prefixes and thus cannot be hijacked as part of a more general, less specific announcement. Through a series of large-scale simulations, we show in Section 13.9 that IP tunnels are more resilient to hijacking attacks than end-to-end paths on the current Internet. Peter et al. [199] propose a similar high-availability strategy by using IP tunnels, albeit designed for the current Internet.

For our deployment strategy, we define the following tunnel types:

(a) Access tunnel: A tunnel with which a deploying ISP D_1 can offer high-availability services to customers whose ISPs do not support SCION.

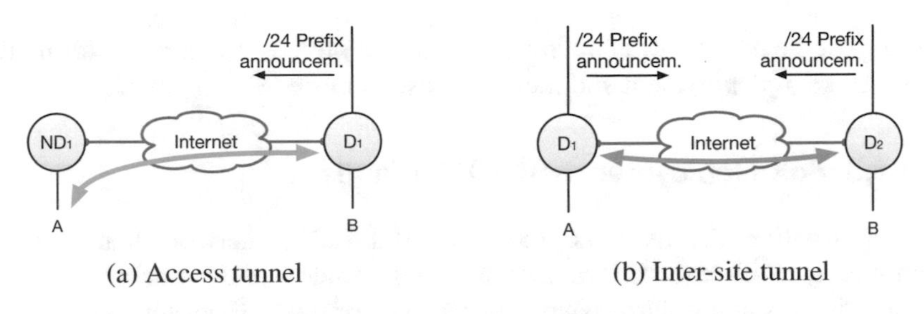

(a) Access tunnel (b) Inter-site tunnel

Figure 10.3: Tunnels used for incrementally deploying SCION. In each subfigure, circles marked D_i represent ASes that have deployed SCION, and the circle marked ND_1 represents a non-deploying AS. Colors are used to differentiate the different tunnel types.

As shown in Figure 10.3a, to protect the tunnel between A and D_1 from hijacking, the deploying AS D_1 announces a /24 prefix via BGP that contains the IP address used for that tunnel. This announcement reduces the probability of an adversary successfully hijacking this tunnel if the adversary is further away from A than D_1.

(b) Inter-site tunnel: A tunnel that connects two non-adjacent deploying ASes D_1 and D_2 over the Internet (see Figure 10.3b). The ASes need to protect the tunnels that link deploying sites from prefix-hijacking attacks. As with the access tunnels above, each of the two ASes would ideally announce a /24 prefix block that contains the IP address used as its tunnel end-point address.

We note that the /24 prefix announcements cannot always prevent all hijacking attacks, including hijacking attacks against the tunnels. As the length (measured in the number of AS-level hops) of the tunnel increases, the resilience to hijacking decreases. This is because BGP's path selection algorithm gives preference to shorter paths. From the attacker's perspective, hijacking a path requires the announcement of a shorter or more specific path than the advertised tunnel paths. Short tunnels thus reduce the possible locations from where an attacker can launch a successful hijacking attack. In Section 13.9, we verify this observation through BGP simulations.

One might argue that end domains could also announce /24 prefixes in BGP to achieve higher resilience to prefix hijacking. However, if every end domain were to announce a /24 prefix for higher availability, the BGP routing tables would become too large. Instead, in our approach, since only a few initially deploying ASes announce /24 prefixes to protect their tunnels, the overhead on BGP routing tables is small. Furthermore, beyond initial deployment, our approach benefits from a natural scalability property; as more ISPs deploy SCION, fewer /24 announcements are needed, since two contiguous ISPs that deploy SCION can communicate directly through a static route. Consequently, we expect the number of announced /24 tunnel prefixes to increase during the early stage of deployment and then decrease as more ISPs deploy.

10.1.3 Full Deployment: SCION Islands

A full or native SCION deployment means a SCION network that does not rely on any BGP-based information to provide end-to-end connectivity. Thus, connectivity is assured irrespective of the state of the BGP system.

In a full deployment, the security and availability properties that SCION offers are ideal, as has been discussed in earlier chapters. Therefore, a full SCION deployment at one or several directly connected ISPs will result in strong communication properties amongst their respective customers. We refer to such contiguous deployments as *SCION islands*.

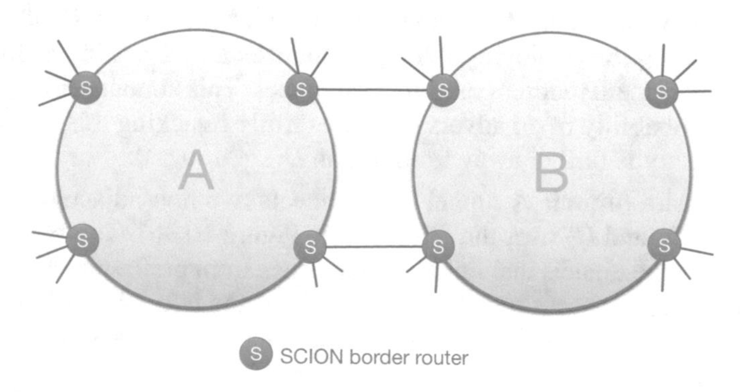

Figure 10.4: A SCION island consisting of two directly connected ASes.

In our early deployment phase, we already have one such SCION island in Switzerland, with the ISPs *Swisscom* and *SWITCH*. In this setting, several corporations interested in highly available and secure communication have started test deployments using this local infrastructure. We describe the current deployment status in the next section.

10.1.4 Current SCION Deployment

SCION is being tested by several ISPs, and is being evaluated by several corporations. The largest contiguously deployed infrastructure is currently in Switzerland. Before describing the deployment path we took in Switzerland, we would like to mention that there also is a growing SCION infrastructure on other continents, mainly in North America and Asia. In addition to the physical infrastructure, several ASes are running on the Amazon Web Services (AWS) EC2 platform in the US, Japan, Ireland, Australia, and Brazil.

In order to display the installed machines across the world, SCION provides a Google Maps API to show the various locations and data centers where SCION machines have been installed. Figure 10.5 shows the March 2017 map centered on Zurich city. The map is located at `http://www.scion-architecture.net/status/`.

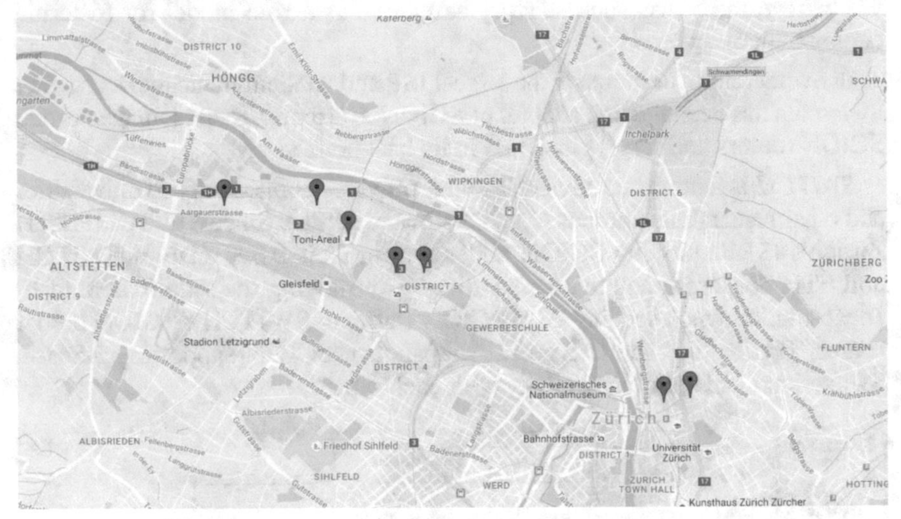

Figure 10.5: Map of deployed SCION routers, March 2017, Zurich, Switzerland.

Equipment

SCION services and routers run on all PC platforms supporting Ubuntu Linux 16.04. The high-speed version of the border router requires Intel DPDK and the hardware required for DPDK. For our initial deployment, we used a system based on an HPE ProLiant DL20 Gen9 Server. The DL20 is a commodity 1U rack-mounted server. It is powered by an Intel E3-1220 v5 processor. The server has 16 GB of RAM, and two 1 Gb/s network interfaces. Where necessary (depending on inter-connections to other locations), we additionally equipped the DL20s with dual 10 Gb/s Intel NICs (X520 series). These NICs, as well as the on-board NICs (Broadcom BCM5720), support DPDK.

Our codebase is designed for Ubuntu 16.04. Most code is written in Python, but the performance-critical parts are written in C, C++, and Go.

The infrastructure services run on servers that are distributed throughout an AS, to achieve fault tolerance to local (internal to the AS) network outages. Services can also run in a virtualized environment, e.g., in an internal cloud environment where available.

Deployment in Switzerland

We next discuss the details of the deployment path we have taken in Switzerland. We believe the Swiss success story can serve as a role model for fostering deployment elsewhere.

In Switzerland, *Swisscom* is the largest ISP and telecommunications provider. Swisscom has been the first adopter of SCION. Currently, Swisscom has several SCION routers and servers installed in its backbone network.

SWITCH is a non-profit Swiss ISP that provides connectivity to universities and other research institutions (e.g., CERN and EPFL). Since SWITCH is ETH Zurich's ISP, deploying SCION nodes within their ASes enables SWITCH and ETH border routers to connect directly, removing the requirement of IP encapsulation when sending ETH SCION traffic to SWITCH's SCION border router. Figure 10.6 shows a recent snapshot of SCION-supporting ASes in Switzerland and their connectivity with each other.

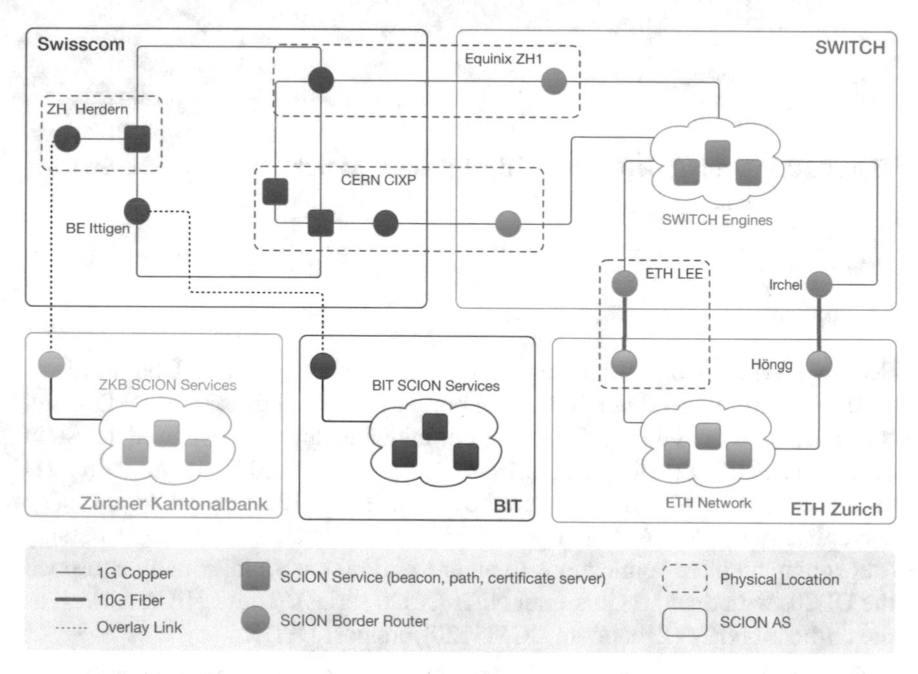

Figure 10.6: Native Deployment ISD Topology, March 2017, Switzerland.

SWITCH also peers with Swisscom and other ISPs, allowing us to incrementally add SCION partners at entities that peer with SWITCH. For this deployment, all the dedicated routers we deployed within SWITCH were provisioned as border routers, while beacon, path, and certificate servers were deployed in a virtualized infrastructure running on SWITCH's cloud-based compute and storage service called *SWITCHEngines*.

As shown in Figure 10.6, the SCION network in Switzerland features several cross-connects (direct physical connections) between the ASes. Hence, no BGP routing is required for communication within this infrastructure. The Swisscom SCION network peers directly with the SWITCH SCION network at two different locations: Zurich and Geneva. The SWITCH infrastructure also peers with the ETH infrastructure in two different locations in Zurich, offering genuine multipath communication even though ETH only has a single provider. We are continuously growing this infrastructure to provide strong communication properties to an increasing number of entities.

10.2 End-Domain Deployment

This section discusses the strategy, results, and lessons learned from deploying SCION-capable routers and servers at domains and ISPs. While certain aspects of the deployment (e.g., evangelism, coordination, troubleshooting) are beyond the scope of this book, we hope that the details provided herein will help the reader recognize the simplicity of deploying SCION.

10.2.1 End-Host Operation with Native SCION Support

Native SCION support for end hosts is available in the Ubuntu 16.04 operating system. Chapter 9 describes the host environment in detail where SCION communication is enabled via SCION's built-in UDP, TCP, and SSP protocols.

10.2.2 End-Host Operation Without Native SCION Support

SCION offers two operation methods that do not require end hosts to upgrade. The first method is based on HTTP(S) proxies, the second on VPN tunneling. A more general approach is called the SCION-IP gateway, which we present in Section 10.3.

SCION HTTP(S) Forward and Reverse Proxy

The SCION HTTP(S) Proxy, as depicted in Figure 10.7, is an approach that enables a legacy host to browse the web over the SCION infrastructure. The SCION proxy consists of two parts: Forward (Bridge) Proxy and Reverse Proxy. The forwarding proxy, which runs on an end host, receives the incoming HTTP(S) requests from a standard web browser, such as Chrome or Firefox, and communicates these requests through a SCION multipath socket to the reverse proxy across the SCION network. The reverse proxy fetches the HTTP(S) requests from the desired web site on the standard Internet, or from a SCION web server running on the same machine.

Figure 10.7: Deployment via SCION proxy.

The SCION proxy can communicate with a Chrome browser extension running on the legacy host, which is developed as a command and control center and a visualization tool for the proxy and the multipath socket. First, the extension is used to visualize the traffic statistics of the multipath socket. It records the data for every HTTP(S) session performed on the proxy, and shows the multiple paths used for each session on a graphical AS topology or on a Google Maps API with the locations of the relevant ASes and ISDs. Second, the extension can also be used for route control. The extension offers an easy-to-use control panel for ISD white- and black-listing, which allows selection of ISDs to be used for the communication between the forward and reverse proxies.

VPN-Based Deployment

SCION offers a gateway appliance to route VPN traffic between two VPN servers over the SCION network. This VPN gateway, which was developed in collaboration with Swisscom, encapsulates a VPN's UDP packets and sends them over a SCION multipath-UDP (MPUDP) connection, and at the other end decapsulates the packets back into UDP packets. The advantage is that the VPN server does not need to be changed in any way, yet the VPN connection can benefit from SCION's higher availability and the dynamic route optimization of the multipath socket.

Corporations with remote offices benefit from connecting their networks to unify IT services and general communications. One way to connect remote sites is through a leased line, which is a point-to-point connection between two networks. Leased lines are expensive because, unlike traditional Internet connections, their use is not shared among several customers. Instead, the connection is always on, forwarding traffic exclusively for the leaseholder.

An alternative to leased lines is to set up a virtual private network between the remote locations over the SCION network (see Figure 10.8) using a pair of SCION VPN gateways. A set of persistent connections can be made between sites, carrying encrypted traffic between virtual private network (VPN) end-points. The benefits of this deployment are cost and management; it is cheaper

Figure 10.8: Site-to-site SCION VPN.

to deploy than a leased line, and it requires no coordination or interaction with a third party (i.e., the leased-line provider).

While discussing SCION deployment with various organizations, we have found that site-to-site VPNs are pervasive across organizations of all types. Companies using site-to-site VPNs can begin using the SCION network transparently by adding the SCION VPN gateways in front of their VPN endpoints as depicted in Figure 10.8. In this deployment scenario both endpoints are assumed to be known, so the SCION VPN gateways are configured to forward the VPN traffic between themselves. With SCION, resilience to DDoS attacks, high availability thanks to multipath communication, and path control can all be achieved without paying the high cost of a leased line.

10.3 The SCION-IP Gateway (SIG)

Successfully deploying a new Internet architecture requires being able to interoperate with the existing Internet. This section describes the SCION-IP Gateway and illustrates how it enables SCION to interoperate with the legacy IP world. In particular, our mechanism enables legacy IP end hosts to benefit from a SCION deployment by transparently obtaining improved security and availability properties.

We first describe the challenges we faced when solving the problems that arise when designing any new Internet architecture: ensuring interoperability with the current Internet through minimally invasive changes, enabling transparent operation, and preventing downgrade attacks to the legacy Internet should the more secure SCION Internet be available for a given destination. We then introduce the SCION-IP gateway (SIG) and explain our mechanisms for addressing the challenges presented. We show hands-on examples on a case-by-case basis to demonstrate how interoperability is achieved. Finally, we discuss some rare cases that are not covered by the current design, simply because the design goal was to enable efficient operation for the common communication cases.

10.3.1 Overview of the Problem Space

We first describe the requirements of the SCION-IP gateway and provide an overview of the problems we intended to address with our design.

IP-in-SCION Encapsulation

Transporting legacy IP traffic over a SCION network requires encapsulating the IP traffic in SCION packets. The encapsulation protocol should be specifically *non-reliable*, to avoid problems with stacking retransmission timers.[1]

Recall that the maximum payload size of a SCION packet varies depending on the path length and MTU, and other factors (e.g., extensions used). As IP-path MTU discovery only allows an MTU to be decreased (it has no mechanism to increase an MTU again), the IP MTU for an encapsulated connection will decrease over time as paths change, which results in wasted bandwidth. Thus, the encapsulation protocol should insulate the IP traffic from the underlying SCION maximum payload size.

Routing and Connectivity

Providing proper interoperability requires that legacy IP connectivity should be transparently supported (i.e., communicating legacy hosts should not be aware that SCION is involved, nor should their connectivity be impacted by SCION's involvement). This means that traffic routing must be fully supported between two legacy IP hosts — one in a legacy (i.e., non-SCION) AS and one in a SCION AS. The same applies to traffic exchanged between two legacy IP hosts that both reside in SCION ASes.

As a consequence, the same routeability rules apply regarding public and private (RFC 1918 [210]) IP ranges. Hosts in SCION ASes that wish to be reachable by legacy hosts in other ASes must have public IP addresses.

Addressing

As legacy hosts (and clients) will not have support for SCION's name resolution service (RAINS), they will still rely on the legacy name resolution service (DNS). The latter does not provide any specific routing information to the legacy host, as the SCION AS is not mentioned in DNS; nor would a legacy host know how to route to a SCION AS in any case. As a consequence, interoperability requires that bare IP addresses are sufficient for legacy addressing of hosts in SCION ASes.

[1] Tunneling a reliable protocol over another reliable protocol can cause retransmission storms in the event of packet loss.

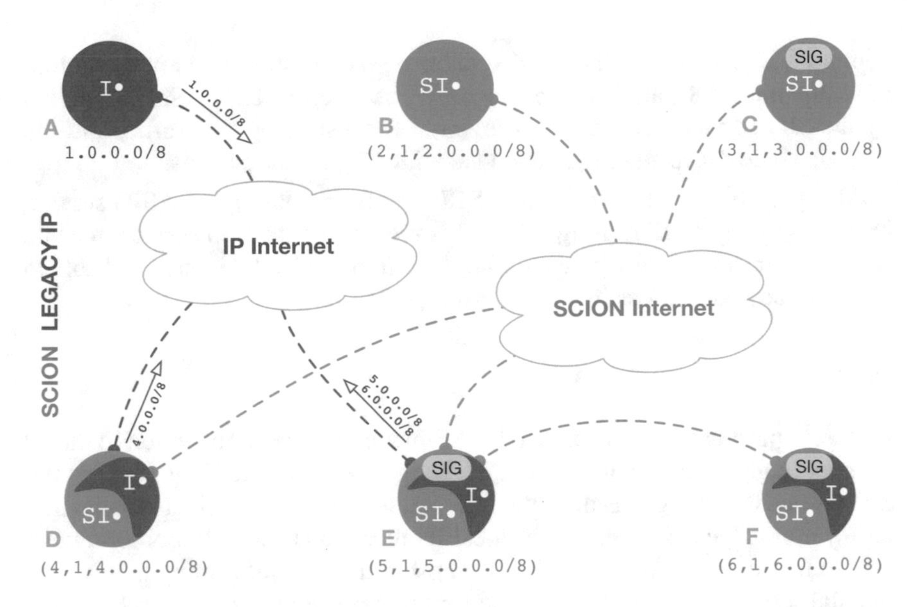

Figure 10.9: Types of networks we consider, differing in whether they connect to the SCION Internet, and whether they deploy a SCION-IP Gateway (SIG) service. The SCION components appear in blue, and the legacy IP components appear in red.

Support for Layer-4 Protocols

The legacy Internet heavily uses TCP and UDP, but it also uses many other layer-4 protocols (such as SCTP, L2TPv3, IPIP, ICMP[2], etc.). Any interoperability solution for SCION must be layer-4 agnostic, i.e., it must work for any layer-4 protocol that is in use by legacy traffic.

Support for SCION-only ASes

Some SCION ASes (e.g., AS E in Figure 10.9) may decide to be directly connected to both the legacy IP Internet and the SCION Internet. Other networks (e.g., AS F) may decide they do not want (or need) a direct connection to the IP Internet. Both of these cases should be fully supported.

10.3.2 Interoperability Between SCION and IP

The SCION-IP gateway (SIG) service is responsible for providing interoperability between SCION and the legacy IP world. Every SCION AS that wants to enable legacy IP connectivity between its legacy hosts and those in other ASes

[2]ICMP is counted as layer-4 in this context.

deploys a SIG service. The service takes care of routing and encapsulation of legacy inter-AS traffic. All legacy traffic between SCION ASes is handled by the SIG service, with the sending side encapsulating the traffic, and the receiving side decapsulating it again back into regular IP packets.

All legacy traffic into (or out of) a SCION AS goes through the SIG service, by means of legacy IP routing rules. This means that the SIG service must be fast, in order to keep up with the traffic flow. It must also be robust, and able to deal with any packet loss in the encapsulated traffic.

Routing

In the simplest case, that of a SCION AS having a direct connection to the IP Internet (e.g., AS E in Figure 10.9), all outgoing legacy traffic is sent via the SIG service by setting it as the default IP gateway inside the AS.[3] For incoming legacy traffic, the AS advertises its local IP allocations via its IP border routers (see solid red arrows in Figure 10.9). Those routers' local routing tables have the SIG service set up as the next hop for the local IP allocations, and forward the incoming traffic there.

In the case of a SCION AS without direct connection to the legacy Internet (e.g., AS F in Figure 10.9), connectivity is achieved by having another SCION AS (e.g., AS E) offer the use of its SIG service and legacy connection. Outgoing legacy traffic in F is routed to the local SIG service, which encapsulates the legacy traffic and sends it to the SIG in E, which decapsulates it into regular IP packets and sends it over its connection to the legacy IP Internet. Incoming legacy traffic is handled in an analogous way; AS E advertises F's IP allocations on the legacy Internet, and forwards incoming legacy traffic to F via E's SIG service.

Mapping Legacy IP Addresses to SCION ASes

When the SIG service receives a legacy IP packet, it needs to determine the SCION AS to which the destination IP belongs, if any. This mapping from public IP to SCION AS needs to be verifiable, to prevent an AS from claiming IP space it does not own (either maliciously, or through misconfiguration). Such verifiability is achieved by each SCION AS exporting an *IP allocation config* (IAC) via its certificate service. The IAC contains a list of IP allocations the AS owns, together with the public key of the AS. Figure 10.10 shows the IACs of three different ASes.

Each list of IP prefixes is signed by the corresponding RPKI authority, i.e., by the IANA, by a regional Internet registry (RIR), or by a local Internet registry (LIR). (The IAC additionally contains the SCION ISD-AS identifier

[3] Briefly as background, the default gateway in IP is the default router to which a host sends an IP packet if the destination host is outside the local network.

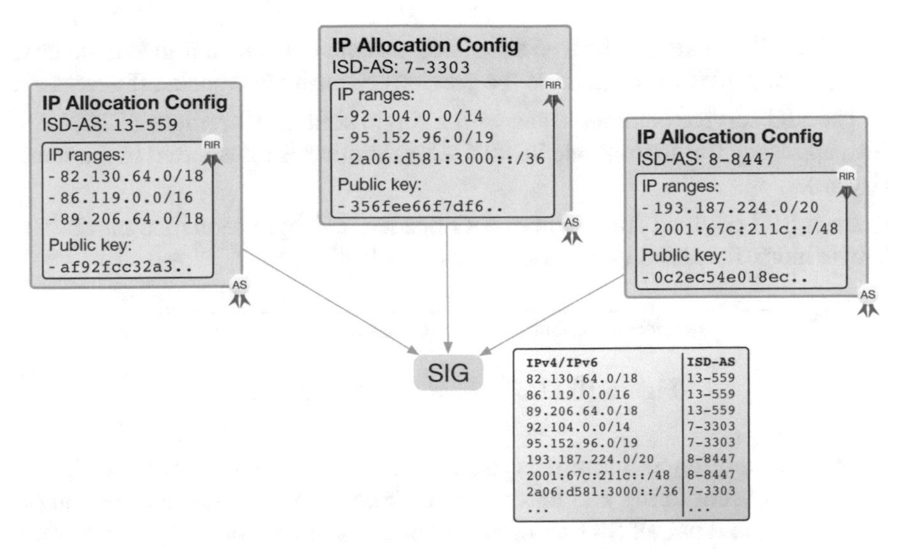

Figure 10.10: IACs of three ASes collected by the SIG service, and compiled into a mapping from IP address ranges to SCION ISD-AS numbers.

and timestamps, which we omit here for brevity's sake.) Each IAC is finally signed with the AS's private key, which is the private key that corresponds to the public key contained in the signed IP prefix list.

The SIG service periodically fetches an updated IAC from every remote SCION AS, by sending a request to the SIG service address for each AS.[4] The SIG also offers its IAC to the remote SIG, announcing its presence to quickly enable operation after a SIG is set up. Each SIG then verifies the signatures on the IAC. Based on all the received and valid IACs, the SIG service constructs a local mapping of IP ranges to SCION ASes. The SIG service makes this mapping available for query by local SCION clients, in case they want to look up the SCION AS for a given IP address.

If a legacy IP packet arrives with a destination address that is *not* covered by the map, it is assumed to be in a legacy (i.e., non-SCION) AS, and is routed to the legacy Internet.

Encapsulation

The SIG encapsulation protocol is built on top of UDP/SCION. It converts legacy IP packets into a byte stream to the remote SIG service (see Figure 10.12). The stream contains the original layer-3 (i.e., IP) and above contents of the en-

[4] To assemble the necessary list of all ASes, the coordination protocol among ISDs (as described in Chapter 5) is used to provide a list of all ISDs, and the core certificate servers of the ISDs provide information about the ASes inside the ISDs.

capsulated IP packet(s). Each encapsulated packet starts on an 8-byte boundary, with padding after the payload of the previous encapsulated packet if necessary.

The SIG service communicates with a single stream per remote SCION AS (i.e., all legacy traffic from one SCION AS to another is transported in the same stream).

Each SIG payload starts with a SIG header: a 4-byte sequence number, a 2-byte index field, and two unused bytes.

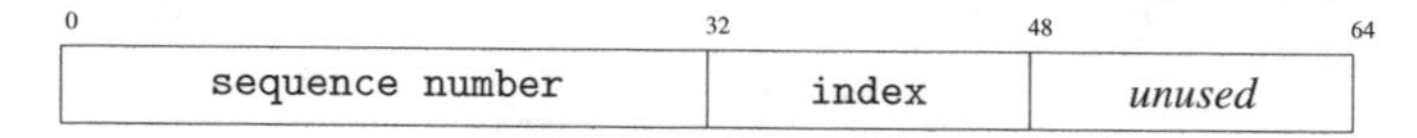

Figure 10.11: Format of the SIG header.

- The `sequence number` is used by a receiving SIG service to detect packet reordering and loss. It starts from zero for a given direction of traffic and pair of SIG services, and increases monotonically by one with every SIG packet. It resets whenever the sending SIG service restarts or the value reaches 2^{32}. Figure 10.12 shows increasing sequence numbers for the encapsulation of five IP packets in four UDP/SCION packets.
- The `index` field is used to allow the receiver to be resynchronized in the event of packet loss. It points to the next start of an encapsulated packet in the SIG payload, if any. The index is multiplied by eight to get the byte offset from the start of the SIG payload. If no encapsulated packets start in this payload, the value is zero (e.g., the packet with sequence number 2 in Figure 10.12). An index value of one indicates that the encapsulated packet starts at the beginning of the payload.

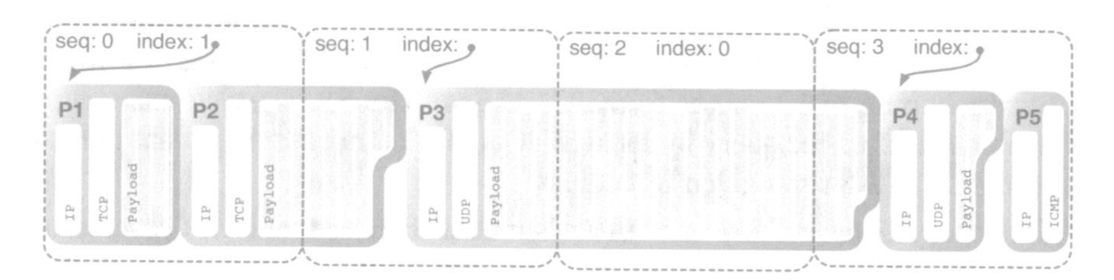

Figure 10.12: Encapsulation of five IP packets.

With the combination of sequence number and index, packet loss can be detected and the receiving SIG service can resume operation at the start of the next encapsulated packet in the stream, ensuring efficient recovery.

SIG Negotiation

When a SIG service wants to send encapsulated traffic to another AS, it needs to determine to which address and port to send the traffic. While it can contact the remote SIG service by sending a message to the SIG SVC address for that remote AS, this is not the intended procedure for high-volume traffic.[5]

Instead, the sending SIG service first sends a query to the remote SIG SVC address, and the remote instance responds with its own address, SCION control port, and dedicated encapsulation port. The latter is used because encapsulation traffic cannot be distinguished from SCION control traffic.

This query is sent every 500 ms, for as long as the sending SIG service desires to send traffic to the remote AS. If no response is received to two consecutive queries, the sending SIG service will use the remote SIG SVC address again, and start sending packets to the new address it receives in response. This allows failover in case the remote SIG instance becomes unreachable.

If there is no SIG service in the remote AS (or there are no instances running), the remote border router will send an SCMP `Unknown Host` error in response.

Client Protocol Negotiation

When faced with a name to resolve or a bare IP address, client end hosts supporting SCION need to decide what type of connection to try first, possibly necessitating additional lookups. In general, we assume a legacy host does not have access to RAINS data and cannot extract information from SCION addresses. Below, we therefore only consider SCION-enabled clients and their behavior.

Recall that where we mention SCION hosts, it is implied that the host is dual-stacked, i.e., the host supports both IP and SCION networking. Also note that for local traffic within an AS, the SIG is not involved.

Without DNS/RAINS. A SCION client presented with a SCION address first tries to make a SCION connection. If this fails, it tries a legacy IP connection. When presented with an IP address, a client queries the IAC of the SIG for the corresponding ISD-AS. If a mapping is found, the client tries to make a SCION connection. If this fails, or if there is no mapping, the client attempts to connect using legacy IP.

With DNS/RAINS. In the case where the destination is identified by a host name (i.e., not by a SCION or IP address), the SCION client performs both RAINS and DNS lookups on the host name. The client always prefers the RAINS answer over the DNS answer for positive entries. That is, if both RAINS and DNS have an entry for a name, the client uses the RAINS answer. If there is no entry for the name in RAINS, the client *can* fall back to using the DNS answer, if any. If the RAINS answer contains a bare IP (i.e., not a SCION

[5] Border routers perform heavier processing on packets with local SVC destinations, in order to select a specific service instance.

address), the client treats it as a legacy host. If the answer is a SCION address or if it only gets a positive answer from DNS, the client then treats this case as if it had just been given the address directly.

Protocol mismatch. In some cases, a client tries to connect with a protocol the destination does not support. In these cases, errors must be handled unambiguously and the client must be informed appropriately.

- **IP client → SCION service:** The destination host generates an error depending on the protocol used (for example ICMP `port unreachable` for UDP). The reply is routed back to the source (if the destination is remote, its SIG service is used), and the client is informed by its operating system.
- **SCION client → IP service on IP host:** The destination host will generate an ICMP `protocol unreachable` reply, which is routed to the client in the same way as in the previous case. The ICMP error does not contain enough information to match it to a specific application or socket on the client host. However, the error applies to all SCION connections to the destination host. Thus, the operating system or network stack delivers it to all applications that have outstanding connection attempts to the legacy IP host.
- **SCION client → IP service on SCION host:** The destination host generates an SCMP `port unreachable` error, which is routed back to the client host as normal SCION traffic, and the client is informed by its operating system.

10.3.3 Scenarios (Life of a Packet)

We next discuss the most frequent scenarios that occur in the translation between SCION and IP. We refer to Figure 10.9 throughout the description of the scenarios. To enhance the clarity of the discussion, the SCION components appear in blue, whereas the legacy IP components appear in red.

1) IP Host in SCION AS → IP Host in SCION AS

We begin with the case of two IP hosts in two different SCION ASes communicating with each other. In Figure 10.9, this could be `E.I` → `F.I` (or vice versa). We do not consider connections from/to AS `D` in this scenario since `D` does not deploy a SIG.

We assume that client host `E.I` connects to a web server, say `F.I`, and that `E.I` knows the corresponding destination IP address, say `6.6.6.6`, (possibly through a DNS lookup beforehand). The web server listens on port 80 and uses TCP as transport protocol. `E.I` uses `E`'s SIG service as its default gateway.

Once the first packet from E.I arrives at the source SIG service (in AS E), the address and port of the destination SIG service (in AS F) is needed to send the encapsulated traffic. To request that address and port, E's SIG service queries its aggregated IAC table (see Figure 10.10) to look up the SCION ISD-AS identifier that corresponds to F.I's IP address, and sends a query to the remote SIG SVC address. The remote SIG service instance responds with its own address, say (6,1,6.0.0.1), a SCION control port P, and an encapsulation port Q. E's SIG service will then encapsulate E.I's IP packets as shown in Figure 10.12 and send them to (6,1,6.0.0.1) on the encapsulation port Q.

If there is no SIG service in the remote AS (as for example in AS D), the remote border router will send an SCMP `Unknown Host` error in response. In this case, the packet will be sent via the legacy Internet.

The remote SIG service instance decapsulates the traffic and forwards the response to the web server F.I on port 80. The web server replies using the source address of client E.I and the specified client port. In particular, the legacy web server does not notice any encapsulation steps that have occurred on the way from the client. It simply sets the local SIG service as its default IP gateway.

2) IP Host in SCION AS w/ IP Connectivity → IP Host in IP AS

In this case, a connection from an IP host in a SCION AS (with direct IP connectivity, such as AS E) is established to an IP host in a non-SCION AS (such as AS A). The SIG service in source AS E acting as default gateway for the IP hosts in AS E tries to map the destination IP address, say 1.1.1.1, to a SCION ISD-AS identifier. It fails to do so since AS A is not aware of SCION.

The SIG service in AS E thus needs to fall back to IP routing: since AS A has announced its IP prefixes through its BGP-speaking border routers, the SIG service in AS E uses the legacy inter-domain IP link to send (non-encapsulated) IP packets to AS A.

3) IP Host in IP AS → IP Host in SCION AS w/ IP Connectivity

In this case, we model the reverse direction of the connection setup described in the previous case: a connection from an IP host in a non-SCION AS (such as AS A) is established to an IP host in a SCION AS with direct IP connectivity (such as AS E). As above, since AS E has announced its IP prefixes using its BGP-speaking IP border routers, AS A sends (non-encapsulated) packets to AS E.

The incoming packets at AS E are routed through E's SIG service. As the destination of the packets is within AS E, the SIG service will not modify or encapsulate the packets, but forward them inside AS E to the specified recipient.

4) IP Host in IP AS → IP Host in SCION AS w/o IP Connectivity

We next extend Case 3 above and assume that AS A connects to a SCION AS that has no direct IP connectivity, say AS F. In this case, AS F relies on some other SCION AS with direct IP connectivity, say AS E, to advertise IP prefixes on F's behalf to other legacy ASes.

Neither the client in AS A nor AS A itself notices the proxy and thus they send the IP packets destined for a client in AS F, say host F.I with IP address 6.6.6.6, to AS E. The border router at AS E forwards such proxy traffic to a SIG service instance, which then encapsulates the legacy traffic and forwards it to a SIG service instance in AS F. The negotiation protocol is similar to the one described in Case 1.

The return traffic is covered in the description of Case 5.

5) IP Host in SCION AS w/o IP Connectivity → IP Host in IP AS

This case models the return traffic of Case 4. The destination address of A.I, say 1.1.1.1, does not appear in the default gateway's IAC table, which lets the SIG encapsulate the return traffic and send it to its proxy AS E. The SIG service in AS E uses its IP connectivity to forward the (decapsulated) return traffic to legacy IP AS A.

10.3.4 Cases not Covered

Some possible combinations of IP and SCION networking are not covered by the functionality of the SIG service. Typically, these are rare cases or ones where there is only minimal gain in supporting them.

SCION Hosts Without IP Stack

A SCION host without an IP stack is a special case that adds a lot of complexity when communicating with IP hosts. The SIG service described in this chapter is designed for SCION hosts with IP stacks, which we expect to be the default case. Hence, SCION hosts without IP stacks are currently not supported.

Transparent Layer-3 Translation (AS Level)

The SIG service only offers transparent layer-3 translation under the following conditions:

- The connection to the remote side needs a fixed maximum payload size (meaning either fixed path, or a small payload size). See the above Encapsulation requirement section for more details.

- There is an L4 protocol the involved SIG services understand and can translate (so the SIG service(s) can update checksums, for example).
- All application traffic/logic above layer-3 is completely L3-independent (i.e., not something like IPSEC in AH mode, nor something which embeds layer-3 addresses).

The SIG service is unable to support these conditions for general traffic, and the last one in particular is unknowable, so this case is not covered by the SIG design.

Transparent Layer-3 Translation (Service Level)

In order to support connections between hosts in the IP Internet and hosts in SCION ASes, the translation would need to be done on the SCION host. However, doing the translation at that point provides no benefits over simply connecting via IP.

10.4 How to Try Out SCION

This section describes the different ways in which users can download the SCION source code for inspection, run it on a local system, and contribute to the project.

Retrieving the Code

SCION runs on Ubuntu 16.04 LTS operating systems. The SCION codebase is hosted on GitHub (`https://github.com/netsec-ethz/scion`). The repository is open-source and users are encouraged to clone a copy of the repository, test the software, deploy it, and submit pull requests with enhancements.

The main repository contains a detailed `README.md` file (in the root directory), which provides all the necessary instructions for installation and system setup. Users are encouraged to use the most up-to-date version, as the code constantly evolves and therefore parts of the system may deviate from the information in this book.

At the time of writing, the directory structure of the source code is laid out as follows:

- `docker/` provides a self-contained image for testing SCION builds
- `endhost/` the parts of the code used on end hosts (e.g., SCION daemon)
- `go/` contains parts of the system written in GoLang (e.g., border router, RAINS server, discovery service)
- `infrastructure/` contains the code of the SCION infrastructure implemented in Python (e.g., beacon server, path server, certificate server, and border router)

- `lib/` shared libraries used across SCION components
- `proto/` Cap'n Proto [218] definitions for SCION packets and messages
- `sphinx-doc/` definitions for building a Sphinx documentation from the SCION codebase
- `sub/` the submodules used by SCION
- `supervisor/` code for the supervisor engine, which helps run and manage SCION processes
- `test/` unit tests and integration tests
- `tools/` Wireshark plugin to parse SCION packets
- `topology/` scripts to generate the SCION configuration and topology files; also includes scripts for running SCION on Mininet

Installing Required Dependencies

The `README.md` file in the root directory includes instructions for downloading and installing all needed external dependencies. In general, this involves adding necessary locations to the user's `PATH`, and then running the provided script to retrieve and install dependencies.

Topology Generation

For development and testing purposes, we have created several complete SCION topologies to run locally on a commodity machine. Such an infrastructure provides end-to-end communication across many autonomous systems inside several isolation domains. The default topology is shown in Figure 10.13. It includes core ASes, provider-customer links, and peering relationships, both intra- and inter-ISD.

Note that running all the services and routers of the default topology on a single host requires at least 4 GB of RAM. Users running SCION with less RAM may try our smaller alternative topology (`topology/Tiny.topo`). These topologies suffice to use basic SCION features, but will not be representative of Internet-scale topologies.

Before the SCION infrastructure is run locally (see below), the chosen topology is loaded from disk (`topology/Default.topo`). From this layout, configuration files are built for all ASes in the topology. These configuration files include IP addresses, port numbers, certificates, and process identifiers through which the process supervisor system will manage the overall status of SCION processes.

Running SCION Locally

It is possible to start up a complete SCION infrastructure by running all elements of a selected topology on the local system. This is the fastest and easiest way

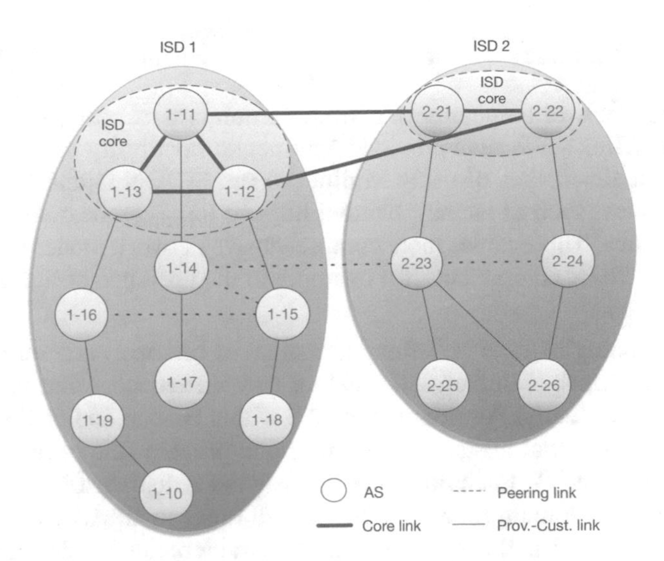

Figure 10.13: Default topology used in SCION test and development. ASes inside gray areas form the ISD core in each of the ISDs. Bold solid lines represent links between core ASes, solid lines represent customer-provider links, and dashed lines are peering links.

to get started; however, users should remember that real SCION networks will deploy separate physical systems for each of the infrastructure elements. Our codebase is designed such that each service is bound to an IP address and port, which allows us to run all services on a single machine (i.e., binding all services to the loopback interface), or connect physically remote systems provided there is IP connectivity between them.

The local infrastructure can be started via the `scion.sh run` command. This launches one process per infrastructure element, binds each process to an IP and port, and begins beaconing, path exploration, and path registration. Once the local infrastructure is up and running, applications may begin using the SCION daemon for retrieving paths and establishing sockets with servers. When run on a single machine, all the traffic will be forwarded through several processes, each of which corresponds to a SCION infrastructure element, before arriving at its final destination. After testing, the local infrastructure can be stopped by executing the `scion.sh stop` command.

The SCION local infrastructure works well for testing different topologies and ensuring that changes to the codebase work. Due to its loopback-interface-bound architecture, however, all services will appear to be directly connected to all other services with no packet loss, very high throughput, and near-zero latency.

SCION on Mininet

Mininet is a network emulator that allows emulation of hosts, switches, controllers, and links. We have provided Mininet compatibility for SCION. One of the main advantages of using Mininet to run SCION is that Mininet link characteristics (such as latency, bandwidth, and loss) can be configured with specific values. This enables more comprehensive network modeling and local testing of SCION components, for example in the presence of link failures or network attacks.

When running SCION on Mininet, instead of binding each service to the loopback interface, Mininet will create a new virtual host that will run the service independently. For example, an AS with a beacon and path server will be run by two virtual hosts, each running one process (the beacon and path server, respectively). Each host will be assigned a distinct IP address in the same broadcast domain (one per AS), and all hosts in an AS will connect to each other via a virtual switch. The link characteristics for any link can be specified via a configuration file (see `topology/mininet/links.conf` for an example).

One of the caveats of running SCION on Mininet is that due to the additional emulation layer, more resources are required to run large topologies. The default topology requires around 8 GB of RAM to run on Mininet on a single host.

Contributing to SCION

The SCION codebase is hosted on GitHub as an open-source repository and employs the standard GitHub workflow for development. In order to contribute to the project, a collaborator follows the procedure below:

- Create a fork of the repository in your GitHub account using the GitHub web interface at `https://github.com/netsec-ethz/scion`.
- Clone the forked SCION repository into your local development environment (see also "Retrieving the Code" earlier in this section).
- Develop a new feature, enhancement, or bug fix in your local environment on a Git branch. When done, push the changes to your fork on GitHub.
- Create a pull request from your fork against the master branch of the SCION repository using the GitHub web interface.

The pull request with the suggested modifications will then be reviewed by the SCION team. Once approved, the pull request will be merged into the main repository.

10.5 SCION AS Management Framework

SCION offers an intuitive and easy-to-use web interface for setting up and managing ASes. The SCION AS Management Framework also enables an AS to establish native and overlay connections to other ASes that support SCION.

The architecture of the framework consists of two main components: a local component per authority managing an AS (the Local Management Service) and a coordination component (the SCIONLab Coordination Service), which mediates between ASes. These components are presented in the following sections.

Figure 10.14 shows a general overview of the framework. Each AS is managed through a Local Management Service. The Local Management Service of each AS communicates with other ASes' management services via the SCIONLab Coordination Service, for operations such as joining an ISD, sending/approving connection requests, etc. Section 10.5.1 describes the SCIONLab Coordination Service, and Section 10.5.2 describes the Local Management Service. The information and screen shot presented in this section reflect the current state of the system, which is likely to be extended in the future as the implementation continues to evolve.

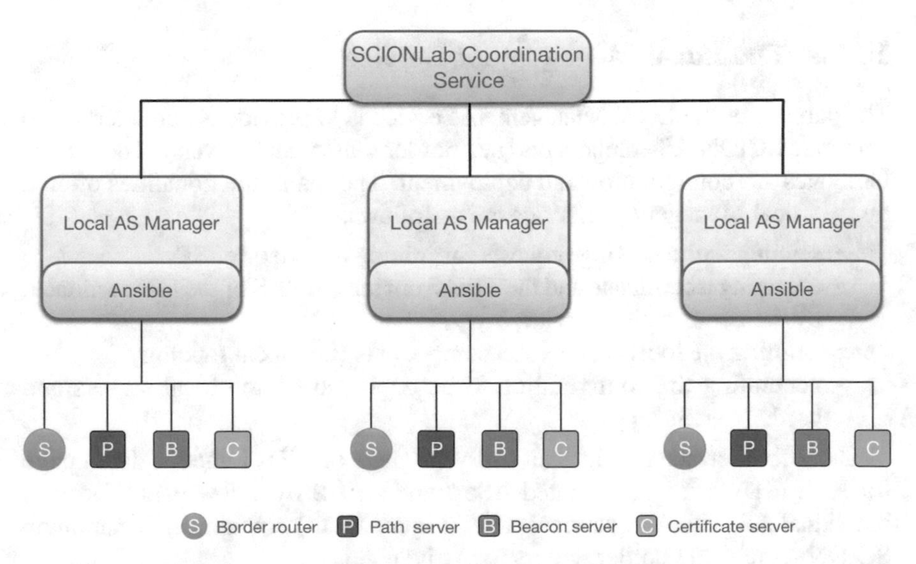

Figure 10.14: Deployment architecture overview.

10.5.1 The SCIONLab Coordination Service

The main role of the SCIONLab Coordination service (also referred to as the Coordination Service throughout this section) is to enable communication between instances of Local Management Services that belong to different ASes. It provides a public interface to obtain an overview of the existing ISDs and ASes, to facilitate creation of new ASes, and to mediate establishing connections between ASes. In addition, the coordination service also provides user account creation facilities for parties who want to register, manage, and deploy SCION.

The SCIONLab Coordination Service also assists with the operation of SCIONLab, which enables researchers to experiment with the SCION infrastructure. Section 10.7 describes SCIONLab in more detail.

The coordination service is not a requirement for running SCION, but merely facilitates operation during the early stages of deployment. With increasing deployment when more ISPs adopt SCION, the connections between neighboring ISDs and ASes will be established directly between the administrating parties of each entity. Therefore, at a later stage we expect that there will be no need any more for a centralized coordination mechanism, except for the purpose of SCIONLab.

10.5.2 The Local Management Service

The purpose of the Local Management Service is to provide a web interface to configure SCION AS components (i.e., border routers and servers). The service facilitates AS configuration and deployment. The main functionalities offered by the Local Management Service are as follows:

- sending and receiving requests for joining an existing ISD,
- obtaining a certificate and the TRC from the core AS of the ISD facilitated by the coordination service,
- defining the local SCION AS components (i.e., local topology),
- generating the configuration to be deployed on the local servers and border routers.

The web interface provides an overview page of ISDs joined, a detail page for each ISD with its associated ASes, and an AS overview panel for each individual AS. The AS overview panel (Figure 10.15) contains information on SCION servers and border routers for a given AS.

The AS overview panel also allows for easy navigation between the ASes managed by an administrator by providing a hyperlink to the AS to which a border router connects. Useful visualization options include the ability to view the AS topology as a graph, with clickable nodes to navigate between ASes. Further details on the operation of the local management service for deploying an AS are discussed in Section 10.6.

Figure 10.15: AS overview panel.

10.5.3 Communication between Local Management Service and SCIONLab Coordination Service

The SCIONLab Coordination Service mediates communication between ASes, for the purpose of registering a new AS in an ISD and establishing inter-AS connections. When an AS administrator creates an account using the SCIONLab Coordination Service, a set of credentials will be created connected to this account. These credentials will be used by the Local Management Service to communicate securely with the coordination service. The main types of interaction between the Local Management Service and the SCIONLab Coordination Service consist of the following:

- sending requests to join existing ISDs,
- sending connection requests to other ASes,
- sending status updates about the services running in the AS (e.g., path server, beacon server, certificate server, SIBRA server, etc.),
- receiving status updates about previous requests (pending/rejected/accepted),
- receiving notifications about new SCION versions.

10.5.4 Communication Between Local Management Service and Ansible

Ansible [118] is a platform for configuration management, application deployment, and task automation. SCION offers a suite of scripts, which are called playbooks and roles in Ansible's terminology. These playbooks are responsible for the deployment of SCION ASes, which requires configuring the involved hosts with the necessary Ubuntu packages, copying the configuration for each component of the AS into relevant locations on the designated hosts, cloning the SCION source code from the GitHub repository, compiling, and finally starting the processes.

The Local Management Service provides a simple front-end for users to input their AS configuration. Based on this information, the Local Management Service generates the necessary AS configuration (e.g., topology, process configuration files) for each server and border router, and finally packages them together with the already obtained certificate and TRC for the AS. The end result consists of deployable folder structures and hostfiles, which are used to perform the deployment. The hostfiles are used by Ansible to deploy each component onto the appropriate machine by relating the generated folders to the hosts to deploy.

Once the AS configuration is generated via the Local Management Service, Ansible deployment can be invoked via the command-line console, from the same host running the Local Management Service.

10.6 Deploying a New AS

This section describes the steps involved in deploying a new AS.

10.6.1 Obtaining an AS Identifier, Certificate, and TRC

AS certificates are generated by core ASes and are verified with a TRC. Certificate creation and distribution is facilitated via the SCIONLab coordination service as follows. The administrator of the AS to be registered sends a request to join an ISD using its Local Management Service. The request is sent to the SCIONLab Coordination Service, which then relays it to the core ASes of the relevant ISD. When a core AS approves the join request, it assigns an AS Identifier (e.g., AS1-3) to the new AS, creates the certificate, and uploads it to the Coordination Service, together with the TRC of the ISD, as part of the response. Upon receiving the response, the new AS obtains its certificate and the TRC, and saves it in its local database, ready to be copied into the relevant folders for Ansible to deploy the AS.

10.6.2 Creating the Discovery Service Configuration

The discovery service (described in Section 7.4.6) needs a configuration (more details in Section 16.3). The following fields are added for the creation of the discovery service configuration:

- the *MTU* (maximum transmission unit) for links within the AS, and
- the *Overlay Type* (e.g., IP, UDP/IPv4, UDP/IPv6) to be used in the AS.

In addition, the AS components need to be defined. For a functional deployment, at least one entry for each of the following components is necessary: beacon server, certificate server, path server, SIBRA server, and a border router. Each of the entries should be declared with the following information:

- a *server name*, which uniquely identifies the server through a combination of type prefix, ISD identifier, AS identifier, and instance number;
- a *public server address* and *public server port*; and
- a *private server address* and *private server port*, in case the component is running on a host behind a NAT.

The border routers need to be defined with additional information about their interfaces as follows:

- the ISD-AS identifier of the AS to which they connect,
- the interface identifier,
- the type of the link (e.g., PARENT, CHILD, PEER, CORE) to the neighbor AS,
- the link bandwidth (in kbit/s),
- the link MTU (which might be different from the AS MTU),
- the address and port of the remote border router,
- the address and port it exposes to the remote border router.

When creating the discovery service configuration of an AS, the administrator is also given the option to enter additional related information. The Local Management Service lists the addresses of the AS components and allows the user to specify a hostname for each address, as well as which cloud engine configuration Ansible should use to configure the machine, in case the host is a virtual machine in the cloud.

10.6.3 Establishing a Connection to an Existing AS

When an administrator provisions and configures a new border router with the Local Management Service, this router becomes available to create a new connection request to the neighboring AS. The connection request contains a free-text form to provide a motivation for the connection request as well the relevant parameters of the border router and link through which the interconnection is to happen, such as the following:

- the address and port of the border router,
- the overlay type,
- the bandwidth, and
- the maximum transmission unit (MTU).

The connection request is then sent to the SCIONLab Coordination Service, which relays the request to the remote AS. The request can have pending, accepted, or rejected status. When the connection request is approved by the remote AS, the initiating AS receives the border router parameters for the remote side of the connection. The option to update the configuration appears accordingly. The user can then confirm these settings and deploy the AS connection by invoking Ansible.

10.6.4 Initiating the Deployment

The deployment is performed via a series of Ansible playbooks corresponding to the deployment of the current AS. As Ansible playbooks are idempotent for the same input, no adverse effect arises when an AS is mistakenly deployed multiple times. After a successful deployment, all the AS components are in the running state and their status can be observed via the SCION monitoring application based on Prometheus [159], which is a monitoring service and timeseries database. When deployed, each AS component exports metrics to the Prometheus system, whereby the status of each component can be observed.

10.7 The SCIONLab Experimentation Environment

SCIONLab is an ongoing project that enables researchers to quickly and easily interface with the SCION network and perform experiments. The main idea of SCIONLab is that participants join the SCION network environment with their own computation resources and set up their own ASes, which get connected to the actual SCION network. The new ASes will actively participate in routing inside the SCION network. Consequently, SCIONLab enables realistic experimentation with the unique properties of SCION.

Figure 10.16 presents an example use case of SCIONLab. A researcher becomes a SCIONLab user by creating an account via the SCIONLab Coordination Service, creates ASes (in this example two) in her research institution, either on dedicated hosts or inside virtual machines, and connects these ASes to SCIONLab ASes, which are a subset of SCION ASes with the capability to accept (or auto-accept) connections from SCIONLab users. The user can then start sending and receiving packets through the SCION network.

The operation of SCIONLab leverages the existing mechanisms and the framework developed for SCION's deployment, such as the Local Management Service, the SCIONLab Coordination Service, and Ansible. Moreover,

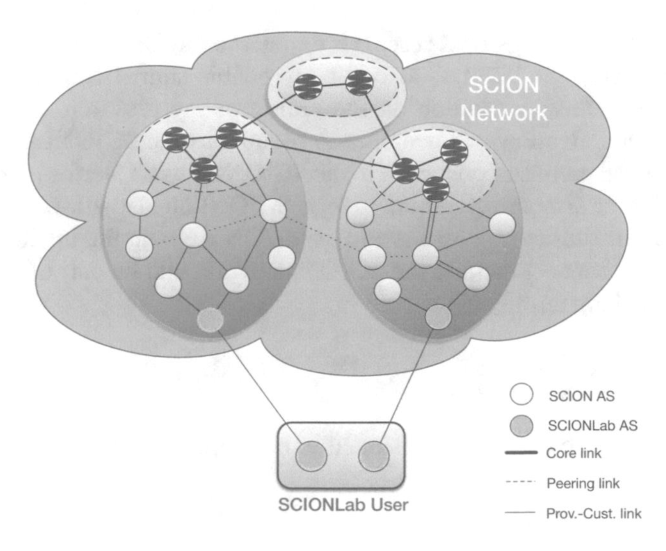

Figure 10.16: SCIONLab's vision is to form a unique testbed environment for researchers.

SCIONLab also aims at hosting test versions of SCION, with new and experimental features, and also to enable SCIONLab users to connect to one another.

10.7.1 SCIONLab Goals

SCIONLab targets the following goals:

- **Novel experiments:** SCIONLab enables researchers to actively participate in the SCION network and perform secure, fine-grained inter-domain route control, a property which cannot be achieved today by existing testbed platforms such as PlanetLab [58]. This property is available as a built-in feature of SCION. The ASes created by the users of SCIONLab are first-class citizens, which participate actively in the SCION routing infrastructure.
- **Allow organic growth:** SCIONLab aims to enable organic growth of SCION's deployment by opening up the existing SCION infrastructure to researchers.
- **Low management overhead:** SCIONLab is designed to require low administration overhead. The aim is to enable research personnel in universities and institutions that adopt SCION to easily join and experiment with SCIONLab with a minimal amount of human intervention, through an easy-to-use web interface.

- **Short- and long-lived research experiments:** In addition to providing easy setup, SCIONLab also aims to enable short-lived and long-lived experiments, depending on the nature of the research. For instance, SCIONLab allows a researcher to set up his own SCIONLab AS(es) on her local premises, connect it to the SCION network, perform experiments and later disconnect. Alternatively, SCIONLab also allows a researcher at an institute to set up a group of SCION ASes in the institution's data center, connect them to the SCION network, and be part of the network in the long term.

10.7.2 SCIONLab Use Cases

The following use cases are envisioned to be supported by SCIONLab.

Education

SCION's path transparency and control can help to teach networking concepts to students. The unpredictability of current-generation routing architectures (e.g., due to load balancing or traffic engineering), prevents students from clearly visualizing network paths. Network-monitoring and troubleshooting tools such as Traceroute do not always provide the correct paths [15], and often require hacks to get around the infrastructure and to manipulate packets at lower layers.

With SCION's source-selected paths, teachers and students gain explicit control over the ASes traversed by a packet. The routing information embedded in the packet helps users to explore and utilize the network topology.

Moreover, many existing routing protocols do not allow test domains to actively participate in the control plane. With SCION's isolation and scalability properties, the nodes in SCIONLab can also fully participate in the routing. This creates an experimentation ground for participation in the SCION inter-domain routing protocol, which is not possible in today's testbed platforms such as PlanetLab.

DDoS Defense Research

One of SCION's extensions is its bandwidth reservation architecture SIBRA (see Chapter 11). SIBRA allows SCIONLab users to obtain guaranteed quality of service on specified paths, which in turn enables reproducible network experiments. With SCIONLab, researchers can also leverage SIBRA to test DDoS attacks and defenses.

Multipath Communication Research

The current Internet does not natively support multipath communication at the network layer. The main problem today is that there is no deployed architecture that provides a meaningful multitude of path choices, on the order of a dozen diverse end-to-end paths. Although recent research has enabled multipath at the transport layer (MPTCP [207]), its use requires endpoints to enable this modified transport in their network stacks. To make matters more complex, MPTCP may not work as expected if middleboxes on the network path interfere with TCP headers. Currently, multipath researchers design and test features in datacenters [206] or virtualized environments, which precludes experimenting with multipath communication in a real-world context.

The SCION streaming protocol (SSP) socket supports multipath by default (Section 9.4), and enables a wide range of multipath experiments. With the use of SCIONLab, users can design experiments involving multipath communication.

Building SCION Extensions

As SCION is designed to be modular, new extensions can be written to extend its functionality. This design enables SCIONLab users to perform rapid prototyping of new network protocols that can leverage SCION's features. They can achieve this by modifying the open-source SCION codebase, and deploying this version on their hosts.

10.8 Example: Life of a SCION Data Packet

We describe the complete life cycle of a SCION packet: crafted at its source host, passing through a number of routers and middleboxes, and finally reaching its destination host. To this end, we assume that both source and destination are native SCION hosts (i.e., they both run a native SCION network stack). We note that the following description is also valid in the case where a SCION-IP gateway connects non-SCION hosts (as described in Section 10.3) and sends SCION packets on behalf of non-SCION hosts.

We start off with an intra-ISD case, i.e., all communication happens *within* a single ISD. We later extend this simplified example to the inter-ISD case.

10.8.1 Intra-ISD Case

Considering the topology depicted in Figure 10.17, we follow a SCION packet sent from `B` to `H` and we observe how it will be processed by each router on the path. We show simplified snapshots of the packet header after each such processing step. The packet header figures below show the most relevant information of the header, i.e., the SCION path, and IP encapsulation for local communication.

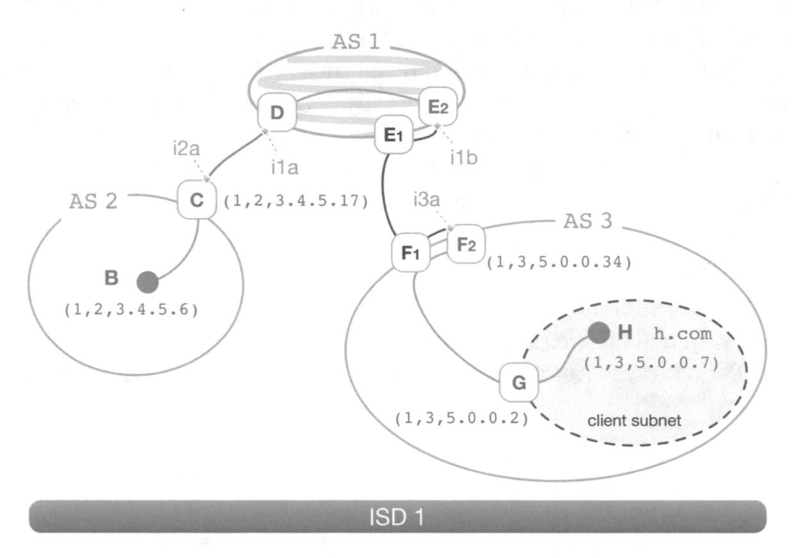

Figure 10.17: Sample topology to illustrate the intra-ISD life cycle of a SCION packet. AS 1 is a core AS of ISD 1, and AS 2 and AS 3 are non-core ASes. The red part of the path indicates a traditional IP connection between AS 1 and AS 3.

End host B first queries its local RAINS service (see Chapter 6) for the SCION address of h.com, which B obtains as (1,3,5.0.0.7). Next, B queries its local path server for a down-segment to AS 3, in which destination host H is located. The local path server (possibly after connecting to a core path server) returns up to k down-segments from the ISD core down to AS 3 (where the default value of $k = 5$). Figure 10.17 shows only a single path. Moreover, the local path server returns up to k up-segments from AS 2 to the ISD core. A path segment consists of the interfaces that each AS uses internally to refer to its inter-AS links. The interfaces have no significance outside the AS.

End host B selects and combines one up-segment with one down-segment, namely (•,i2a)(i1a,•) up and (•,i1b)(i3a,•) down. These two segments are combined to obtain an end-to-end forwarding path from B's AS to the destination AS. In our example, the resulting SCION forwarding path is IF1(•,i2a)(i1a,•) IF2(•,i1b)(i3a,•). It consists of two *info fields*, IF1 and IF2, and a series of *hop fields* that carry the ingress and egress interfaces of each AS, as described in Section 15.1.3.

1) B → C

TCP	P_S=35417, P_D=80
SC	SRC=B@(1,2,3.4.5.6) DST=H@(1,3,5.0.0.7) PATH=IF1(•,i2a)(i1a,•) IF2(•,i1b)(i3a,•)
IP	SRC=B@3.4.5.6 DST=C@3.4.5.17
Eth	SRC=B, DST=C

2) C → D

TCP	P_S=35417, P_D=80
SC	SRC=B@(1,2,3.4.5.6) DST=H@(1,3,5.0.0.7) PATH=IF1(•,i2a)(i1a,•) IF2(•,i1b)(i3a,•)
Eth	SRC=C, DST=D

3) D → E2

TCP	P_S=35417, P_D=80
SC	SRC=B@(1,2,3.4.5.6) DST=H@(1,3,5.0.0.7) PATH=IF1(•,i2a)(i1a,•) IF2(•,i1b)(i3a,•)
IP	SRC=D DST=E2
Eth	SRC=D, DST=E2

4) E2 → E1

TCP	P_S=35417, P_D=80
SC	SRC=B@(1,2,3.4.5.6) DST=H@(1,3,5.0.0.7) PATH=IF1(•,i2a)(i1a,•) IF2(•,i1b)(i3a,•)
IP	SRC=E2 DST=F2@5.0.0.34
Eth	SRC=E2, DST=E1

5) E1 → F1

TCP	P_S=35417, P_D=80
SC	SRC=B@(1,2,3.4.5.6) DST=H@(1,3,5.0.0.7) PATH=IF1(•,i2a)(i1a,•) IF2(•,i1b)(i3a,•)
IP	SRC=E2 DST=F2@5.0.0.34
Eth	SRC=E1, DST=F1

6) F1 → F2

TCP	P_S=35417, P_D=80
SC	SRC=B@(1,2,3.4.5.6) DST=H@(1,3,5.0.0.7) PATH=IF1(•,i2a)(i1a,•) IF2(•,i1b)(i3a,•)
IP	SRC=E2 DST=F2@5.0.0.34
Eth	SRC=F1, DST=F2

7) F2 → F1

TCP	P_S=35417, P_D=80
SC	SRC=B@(1,2,3.4.5.6) DST=H@(1,3,5.0.0.7) PATH=IF1(•,i2a)(i1a,•) IF2(•,i1b)(i3a,•)
IP	SRC=F2@5.0.0.34 DST=H@5.0.0.7
Eth	SRC=F2, DST=F1

8) F1 → G

TCP	P_S=35417, P_D=80
SC	SRC=B@(1,2,3.4.5.6) DST=H@(1,3,5.0.0.7) PATH=IF1(•,i2a)(i1a,•) IF2(•,i1b)(i3a,•)
IP	SRC=F2@5.0.0.34 DST=H@5.0.0.7
Eth	SRC=F1, DST=G

9) G → H

TCP	P_S=35417, P_D=80
SC	SRC=B@(1,2,3.4.5.6) DST=H@(1,3,5.0.0.7) PATH=IF1(•,i2a)(i1a,•) IF2(•,i1b)(i3a,•)
IP	SRC=F2@5.0.0.34 DST=H@5.0.0.7
Eth	SRC=G, DST=H

Step-by-Step Explanations

We next explain the packet header modifications at each router, by using the table above. Regarding the notation used in the table, each SRC and DST entry should be read as router (or host) followed by its address, separated by the @ symbol.

1. B → C SCION-enabled end host B creates a new SCION packet destined for H with payload *P*. B learns from the SCION discovery service (see Section 7.4.6) the mapping from interface fields to IP addresses of the corresponding border routers. For example, the interface i2a (as contained in the combined forwarding path) is mapped to border router C's IP address 3.4.5.17. Based on this information, B knows that it needs to send its packets (for the chosen forwarding path) to border router C, which will then consider the SCION path that B has added to the packet's SCION header. B adds a temporary IP header for the local delivery to C, utilizing AS 2's internal routing protocol.

   The info field pointer in the SCION header is set to IF1, which is indicated in the packet header figures above by a line below it. The pointer to the current hop field is indicated by a wave below it. Once the information in the path is consumed, the pointers are moved forward.

2. C → D Router C inspects the SCION header and considers the info field of the specified SCION path that is pointed at by the current info field pointer. In this case, it is the first info field IF1 with its first hop field, which instructs the router to forward the packet on its interface i2a. After reading the current hop field, C moves the pointer forward by one position.

   Note that, at this point, no IP header is necessary, since the routers C and D are directly connected.

3. D → E2 When receiving the packet, router D checks whether the packet has been received through the ingress interface i1a as specified by the current hop field. Otherwise, the packet is dropped by D. The router

notices that it has consumed the last hop field of the current path segment, and hence moves the pointer of the current info field to the next info field. There, it starts processing the first hop field, which instructs the router to perform intra-domain routing to transport the packet to the specified egress interface `i1b`, which is on router `E2`.

4. `E2 → E1` The dual setup with `E1`/`E2` and `F1`/`F2` models a *router-on-a-stick* configuration, in which `AS 1` and `AS 3` are connected through BGP-speaking IP routers `E1` and `F1`. This rather conservative setup guarantees that legacy IP traffic is not affected if one of the SCION routers (`E2` or `F2`) fails.

   `E2` inspects the current hop field in the SCION header, uses interface `i1b` to forward the packet to `F2` (which is part of a static configuration between `AS 1` and `AS 3`), and moves the current hop-field pointer forward. It adds an IP header to reach `F2`.

5. `E1 → F1` Router `E1` forwards the IP packet to the IP border router of `AS 3`, according to its IP forwarding table.

6. `F1 → F2` Router `F1` performs intra-domain forwarding of the IP packet to `F2`.

7. `F2 → F1` SCION router `F2` detects the SCION header and realizes that the packet has reached the last hop in its SCION path. Therefore, instead of stepping up the pointers to the current info or hop field, `F2` inspects the SCION destination address and extracts the end-host address `5.0.0.7`. It creates a fresh IP header with this address as destination and with `F2` as source. The intra-domain forwarding will first send the IP packet to router `F1`.

8. `F1 → G` Router `F1` continues the intra-domain forwarding and sends the packet to the next router on the path to `H`, which in this case is `G`.

9. `G → H` Router `G` delivers the packet to end host `H`.

When `H` sends an answer to the sender, it will flip the source and destination addresses in the SCION header, reverse the SCION path, and set the pointers to the info and hop fields to the beginning. `H` sends a response via border router `F2`, which has been used for the inbound direction. The address of `F2` can be learned either from the source IP address of the inbound packet or from the SCION discovery service.

10.8.2 Intra-ISD Case with Private Addresses Behind a NAT

We next consider a slightly modified topology as depicted in Figure 10.18, where we assume that SCION host `A` sends a packet to `H`. The packet header will be processed by each router on the path, as before, but here additionally with a network address translation (NAT) step at router B.

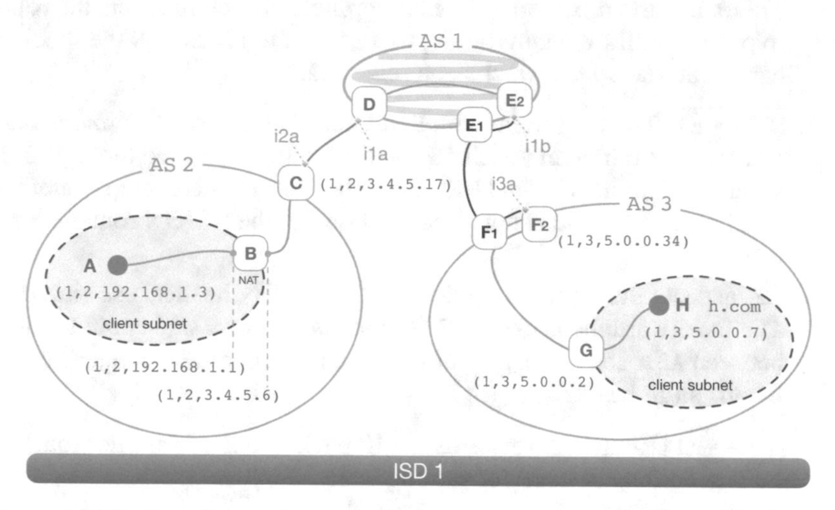

Figure 10.18: Sample topology (similar to that of Figure 10.17) with the difference that the source host resides in a subnet, which uses network address translation (NAT) to connect to AS 2.

As in the case before, end host A first queries its local RAINS service for the SCION address of h.com, which A obtains as (1,3,5.0.0.7). In the following, we will only show the differences to the case in Section 10.8.1 without NAT. These two differences occur in the first two hops.

1) A → B

TCP	P_S=22741, P_D=80
SC	SRC=A@(1,2,192.168.1.3) DST=H@(1,3,5.0.0.7) PATH=IF1(•,i2a)(i1a,•) IF2(•,i1b)(i3a,•)
IP	SRC=A@192.168.1.3 DST=C@3.4.5.17
Eth	SRC=A, DST=B

2) B → C

TCP	P_S=35417, P_D=80
SC	SRC=B@(1,2,3.4.5.6) DST=H@(1,3,5.0.0.7) PATH=IF1(•,i2a)(i1a,•) IF2(•,i1b)(i3a,•)
IP	SRC=B@3.4.5.6 DST=C@3.4.5.17
Eth	SRC=B, DST=C

Step-by-Step Explanations

1. A → B SCION-enabled end host A creates a new SCION packet destined for H with payload *P*. A learns from the SCION discovery service that it needs to contact the border router C that will consider the SCION path that A added to the SCION header of the packet. The packet's IP

header thus points to router C. To reach border router C, the packet is first sent to router B, which is the default gateway for host A. The intra-domain forwarding inside AS 2 will then deliver the packet to router C (as explained in Step 2).

The SCION header has its pointer to the current info field set to IF1 (which is again indicated in the packet header figures above with a line, and the pointer to the current hop field is again indicated with a wave). Once the information in the path is consumed, the pointers are moved forward.

2. B → C Router B performs network address translation (NAT) since end host A resides in a sub-network with a private address space. More precisely, router B replaces A's private IP address with B's public IP address, and assigns a fresh transport-layer port number. Router B maintains a NAT table as follows:

NAT table at B

dst address	dst port	temp. src port	src port	src address
(1,3,5.0.0.7)	80	35417	22741	(1,2,192.168.1.3)
⋮	⋮	⋮	⋮	⋮

It is important that the destination address of the NAT entry is the SCION address of H; the IP address of H or the IP address of C are not sufficient as they might not uniquely identify answers from H: The IP address of H may not have significance outside AS 3, and the IP address of C may be useless if the reverse path(s) from H to A do not go through router C.

When H sends an answer to the sender, it will flip the source and destination addresses in the SCION header, reverse the SCION path, and set the pointers to the info and hop fields to the beginning. H sends a response via F2, which has been used for the inbound direction. Router B will translate the addresses accordingly and deliver the packet to A.

10.8.3 Inter-ISD Case

We next discuss the case in which a SCION packet travels from one ISD to another, as depicted in Figure 10.19.

This inter-ISD case is slightly more complex than the previous case inside an ISD. The increased complexity is not due to the forwarding process (it works exactly as before with a longer path descriptor), but it comes with a slightly more complex path resolution process, which we will explain next.

The source end host A requests a path to the destination AS, AS 4 in ISD 2, from its local path server. The local path server may have a cached path to the destination AS, or requests one from the core path server located in its ISD core, AS 1. In this case, the core path server returns up to k down-segments. As the destination AS resides within a different ISD, the core path server requests

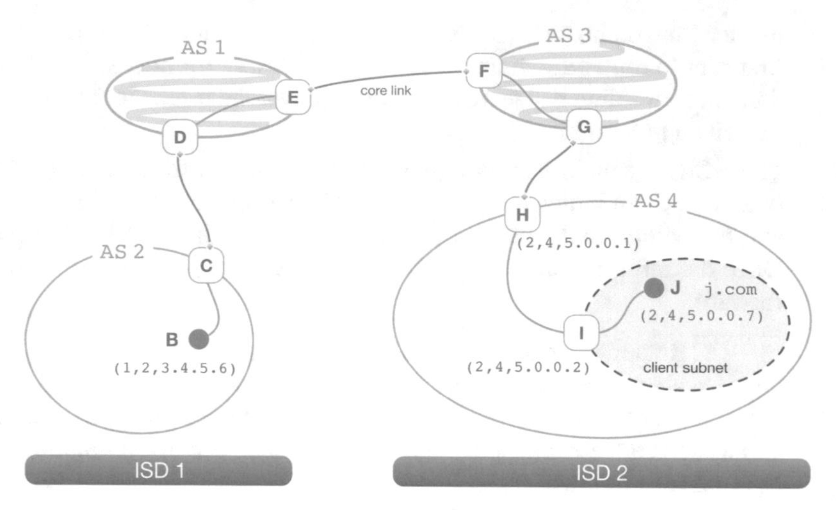

Figure 10.19: Sample topology to illustrate the *inter-ISD* life cycle of a SCION packet.

the down-segments from the remote (destination) ISD's core path server and returns these segments to the local path server, together with core-segments connecting ISD 1 to ISD 2.

10.9 SCION Path Policy

SCION can support a rich set of path policies, providing ISPs with fine-grained control over permissible paths. This is an important property, as ISPs need mechanisms that implement their traffic flow policies to match their business model. In today's Internet, ISPs define their routing policies through BGP; so in this section, we will highlight the differences between BGP routing policies and SCION path policies.

Before we discuss the differences in more detail, we would like to make an observation. BGP is sometimes described as the gold standard for routing policies. However, BGP is mainly able to express destination-based policies, but no general source-based policy. It might be concluded from this that current Internet routing policies have evolved towards what is expressible by BGP, and probably not that BGP has evolved to accommodate the most-desired routing policies. In fact, we anticipate that familiarity with SCION path policies may raise awareness of BGP's inadequate expressiveness.

Three fundamental points complicate the definition of SCION path policies. First, because SCION's path exploration is fundamentally different from BGP, policy construction differs from that in today's Internet. Second, announcing multiple paths creates a challenge (compared to the current situation where

ISPs only need to approve and provide a single path to each destination). To our knowledge, there are currently no multipath policy definitions available. Third, client-based path selection can disrupt an ISP's business model, as a client may select a more expensive path, incurring a higher cost for the ISP.

In the remainder of this section, we will first explain the fundamental differences between BGP routing policy and SCION path policy in terms of expressiveness. We will then describe the SCION path policy framework, illustrate with specific examples how BGP policies can be translated, and what policies SCION can naturally express that BGP cannot. Finally, we discuss how client-based path selection interacts with ISP-based traffic engineering.

10.9.1 Differences Between SCION Path Policy and BGP Routing Policy

Path exploration in SCION starts from core ASes and extends paths towards the leaf ASes — whereas paths in BGP are constructed from leaf ASes towards all other ASes. This difference suggests that SCION can express a different set of path policies from BGP (but as we describe below, beacon extensions and hop-field encryption can enable a full set of policies).

Example. To illustrate the differences between BGP routing policies and SCION path policies, we present an example in Figure 10.20. In BGP, the routing updates originate at the destination and are flooded through the network. For instance, AS E sends a BGP update message to its provider D, which further disseminates it upstream to A and C. Similarly, C further disseminates the update to A and B, and B sends it on to A. At this point, A can decide how E will be reached. Traffic follows the reverse path of the updates, so traffic destined for E can traverse path A-D-E, or path A-C-D-E, or path A-B-C-D-E. This example demonstrates the limitation that an AS can only control over which downstream path traffic is sent, but has no upstream control. Specifically, traffic destined for E cannot be controlled by E; it will flow over the link that A selects. Moreover, BGP cannot express source-based policies (except the next hop to which an update is sent) as we are going to illustrate later in this section.

Due to the bidirectional nature of SCION paths, simply expressible policies are neither source nor destination based. Instead, SCION policies relate to the paths towards the ISD core. Consider that AS A in Figure 10.20 is part of the ISD core and initiates PCBs that it sends to B, C, and D. B and C continue to send the PCB to D. D can now decide which beacon to send on to E based on its path policy. Since SCION is a multipath architecture, D would most likely forward the three PCBs, but to illustrate the policy options we assume it only sends a single PCB. This PCB represents both the path towards the destination (used as a down-segment), as well as the path from the source (used as an

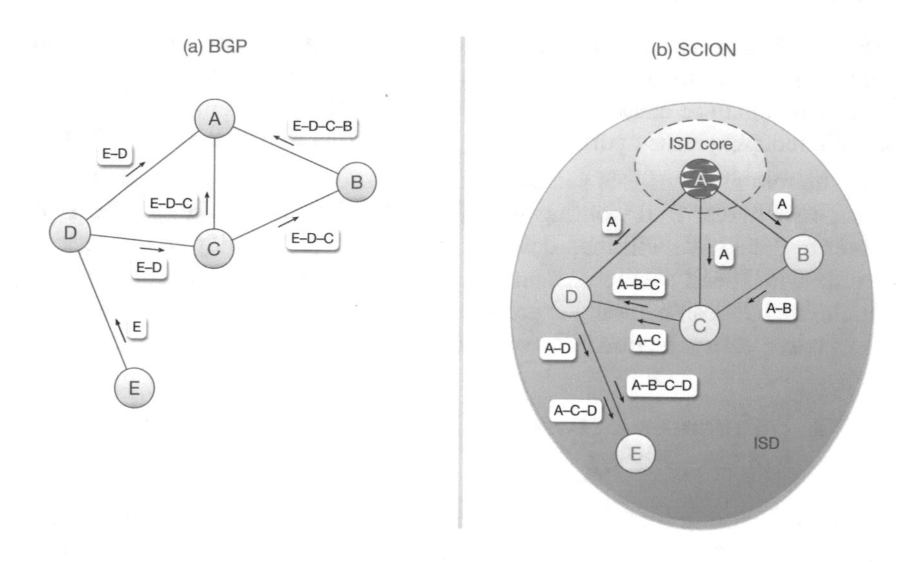

Figure 10.20: Simple network topology and two path exploration approaches: BGP (destination emits BGP update) and SCION (ISD core emits beacons).

up-segment). Thus, it is challenging to compare the policy expressiveness of SCION directly with that of BGP.

Another difference is that BGP routing policies can be based on IP prefixes, but since in the SCION data plane neither the source nor destination address influences inter-domain forwarding, the end host addresses in the SCION header can be selected from private address space (e.g., RFC 1918 [210]). Consequently, SCION path policies are purely based on ASes and ISDs.

10.9.2 Approaches to Implementing Path Policies

SCION offers three approaches to implementing path policies:

- **Beaconing control:** An AS decides which beacons to send on and which peering links to add to the beacon. This enables implementation of a first basic level of path policy. An AS can decide which upstream ASes should be avoided when propagating beacons to downstream ASes, or which path properties are preferred (see Section 7.1.4).
- **Explicit path policy transmitted as beacon extension:** An AS adds information to the PCB to explicitly indicate which paths are permissible. This can include a list of downstream ASes which are (or are not) permitted to use the PCB's path. The granularity of path policy can be fine grained to the level of per-link policies, so even peering links can be

annotated with a policy in the PCB (see Section 15.3.4). Downstream ASes that violate usage policy are accountable for their actions as they sign the PCB and a policy violation is detectable when the path is registered. When the path is only used against the explicit path policy but not registered, detection is more challenging. To detect such misuse, an AS can monitor hop fields (HFs) used in traffic and in the case of HFs that were not registered by any of the downstream ASes, it can verify whether the source or destination AS is allowed to use the path. Furthermore, violation by an intermediate AS can be detected by tracing the intermediate ASes in a sequence of HFs and verifying compliance with the explicit path policy. Although detection requires operational effort, it is likely to be a sufficient deterrent for misbehavior.

- **Hop field encryption with explicit path activation:** An AS that intends to encode more sophisticated policies can encrypt the hop field in the PCB to make it unavailable unless it is activated by the end domain. Activation requires sending a special packet through the network with the entire end-to-end path, so that on-path ASes can inspect and activate the path by decrypting the hop field if they permit the path. A unique policy identifier can be added to the PCB to enable end domains to optimize which paths are attempted to be activated.

At the time of printing, the current implementation of SCION supports basic beaconing control and explicit policies, but not yet explicit path activation, which will be included in a future release.

We observe that if an ISP only announces encrypted hop fields that require explicit path activation, end-to-end path setup is slowed down and in the worst case requires several attempts to find a working path. We thus require that each AS must make at least one upstream path available, called the *default path*, that supports arbitrary end-to-end paths. This ensures quick establishment of an end-to-end path that is supported by all upstream ASes. To achieve high availability and rapid failover, two disjoint paths that support arbitrary end-to-end paths should be made available. The multipath system will then continue to seek additional paths as the connection progresses, finding new paths that optimize latency, bandwidth, loss rate, ASes traversed, etc.

10.9.3 Sample Path Policies

To demonstrate how one can express path policies in SCION, we will consider some popular routing policies that are used in BGP, and present some policies that cannot be expressed in BGP. For the current BGP policies, we discuss the policies presented by Gill et al. [94]: the GR model, next-hop routing, consistent export, and most stable path. We also discuss hot-potato routing, an example of a complex BGP policy, and finally a source-based policy that BGP cannot express.

Gao-Rexford Model

The Gao-Rexford (GR) model [91] captures BGP routing policies that are believed to be realistic due to commercial relationships between ASes. A policy is compliant with the GR model if it implements the two following sub-policies:

- **GR preference:** The preference policy is based on the observation that ISPs want to maximize profits: when an ISP has a choice of where to send traffic to, then the preferred order is first towards a customer, second over a peering link, or finally to a transit provider. The reason is that traffic sent to a customer earns a profit from the customer, traffic sent over a peering link has zero marginal cost, but traffic sent to a transit provider incurs a cost. In BGP, the way these preferences are expressed is through the `LocalPref` setting, which is assigned a different value depending on whether the update was received from a customer, a peering link, or a provider. Gill et al. report that over 85% of ISPs utilize this policy [94].

  In SCION, paths are usually bidirectional, so once a path is announced, it can be used in both directions and it thus incurs costs if it is used for traffic towards a provider. Moreover, SCION traffic makes use of multipath communication, and thus a multitude of paths are simultaneously used. We discuss how an ISP can influence the flow of traffic to maximize its revenue in Section 10.9.5.
- **GR export:** The export policy is needed for route stability. Customer routes are announced everywhere, but routes learnt over peer and provider links are only announced to the customer. This policy ensures *valley-free* routing [91], where traffic never flows "down" the AS hierarchy to a customer and back "up" towards the destination. By following this simple policy, Gao and Rexford were able to show that BGP converges [91]. Gill et al. find that over 70% of ISPs enable this policy.

  Since SCION has no convergence problems, this policy is not required. However, domains can still ensure that they do not re-send PCBs to any of their providers and peers.

Next-Hop Routing Policy

Gill et al. conducted a study where they investigate the various BGP routing policies used in practice [94]. It was found that the majority of ISPs use simple routing policies that are easy to configure and maintain. For instance, around 60% of ISPs use a next-hop routing policy, which implies that the BGP `LocalPref` setting is solely based on the next hop (i.e., the incoming link of the BGP update) and the destination, and not on the intermediate path. This creates a predictable, simple routing policy, and enables implementation of the GR preference policy.

SCION enables control over which PCBs are sent to which next-hop ASes, allowing for policies that take the previous and next hop into account. As SCION paths are bidirectional, both source- and destination-based policies can be expressed, rather than only destination-based policies as in BGP.

Consistent Export Routing

Gill et al. also report that consistent export routing is a popular policy, with 65% of the ISPs deploying it [94]. In this policy, if a route with `LocalPref` $= \ell$ is exported, then routes with `LocalPref` $\geqslant \ell$ are also exported. This represents a monotonicity property, which again leads to predictable behavior.

The reason why policies in BGP need to be simple and predictable lies in the danger of connectivity loss in case a link fails or another ISP changes its policy and withdraws a route. With complex policies, outages and loss of connectivity are common [54, 103, 172, 243]. SCION does not need such a rule — given its path exploration mechanism and the default path, which always guarantees at least one working path.

Route Along the Most Stable Path

BGP supports a preference for more stable paths, which is expressed based on the age of a path. Such an approach is not critical in SCION as several paths can be active simultaneously, and one of them can be a path that has been stable over an extended time period.

Hot-Potato Routing

Hot-potato routing denotes the strategy of sending a packet off as quickly as possible to the next AS, limiting the amount of resources consumed within the AS. This strategy can lead to asymmetric paths, as packets traveling from A to B may thus take a different path than packets traveling from B to A. Since SCION defines the specific ingress and egress links, hot-potato routing cannot be achieved by standard SCION. However, if two neighboring ISPs do want to perform hot-potato routing, they can assign the same interface identifier to two different links, and send packets across the closer link. This approach would sacrifice opportunities for multipath communication, so we do not expect it to be used in practice.

Example of Complex BGP Policy

Consider the example depicted in Figure 10.21, where B is an educational network that only permits traffic with either a source or a destination that is also inside an educational network. Since F is an educational network, traffic destined for it can traverse B. In BGP, D learns the path originated by F (i.e., F-E-D), and the path originated by G (i.e., G-E-D). D has a choice of which links to propagate the updates to. Since F is an educational network, its update can be sent to both B and C, but G is a corporate network so its update can only be sent on to C. Traffic follows the reverse path of the updates, so traffic destined for F can come to D from B or C, but traffic destined for G cannot flow across link B-D.

Such a policy can be expressed in SCION, although the path exploration process is conducted in a top-down manner. Considering A is in the ISD core, then D receives PCBs with paths A-B and A-C, extends them (by adding information about itself), and propagates them downstream to E. The beacons sent contain a path policy extension, where D states that only F can use the path A-B-D. This policy is enforced at several points:

- E, on learning D's statement, does not send the beacon containing B to G.
- Core path servers refuse path registrations attempted by entities other than permitted ASes (i.e., other than F in our example).
- D can sample traffic to check whether the policy is violated (since SCION addresses contain AS identifiers).

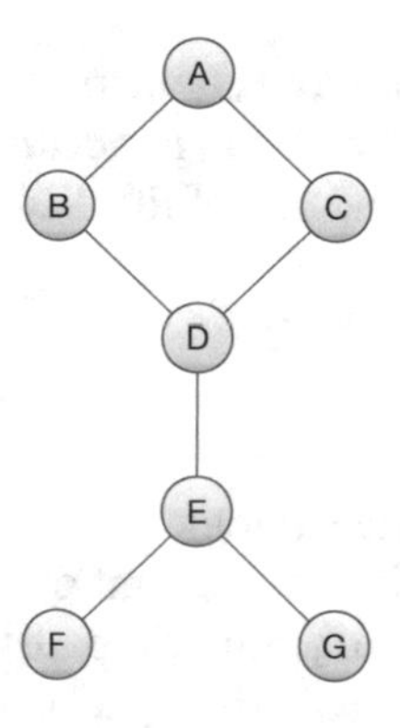

Figure 10.21: Network topology, where AS B represents an educational network (e.g., Internet2) that provides transit only to educational entities. AS F is an educational institution, which is allowed to receive traffic through B. All other entities are commercial ASes. In the case of SCION, we consider AS A to be a core AS.

Source-Based Downstream Path Policy

For a given destination, BGP can only announce a single update, preventing a diversified routing policy based on the source of the packet. In SCION, path exploration is realized top-down and an AS receives several beacons with diverse paths. Moreover, a provider has better control in SCION over paths learned by its customers.

For instance, consider the topology in Figure 10.21. Using BGP, *D* must select which path to *A* it will forward to *E*, either *D-B-A* or *D-C-A*. Even if *E* knew of both paths, all traffic destined for *A* would either traverse *B* or *C*. Thus, *E* has no choice but to forward the relevant BGP update for *A* on to *F* or *G* — it cannot have *F* use path *E-D-B-A* and *G* use path *E-D-C-A*.

With SCION, both *F* and *G* can use paths through *B* or *C*, as long as *D* has sent one beacon with the path *A-B-D* and another with *A-C-D* to *E*, and *E* in turn has extended the beacons and passed them to *F* and *G*. However, if it is desired, *D* can decide not to reveal the connection with *B* (if, for instance, it is a backup link), and can keep sending to *E* only beacons with the path *A-C-D*. Similarly, if *E* has beacons with the two paths, it can send both PCBs (i.e., traversing *B* and *C*) to *F*, and send only one PCB to *G*, the one traversing *C*.

10.9.4 Secrecy of Routing Policies

Routing policies are often sensitive information for an ISP's business. ISPs thus guard their policies, even though the actual routing decisions leak some information about their policy. An important question is whether SCION leaks more policy information than BGP. In the case of explicit path policies, the policy is directly published in the PCB. We anticipate that non-sensitive policies will be published this way. Standard beaconing also discloses policy information, and as a natural consequence of multipath path discovery, SCION discloses more information than BGP — as any multipath routing protocol would naturally disclose more information. If ISPs were to use encrypted hop fields to hide their policy (even though encrypted hop fields require a higher overhead for path setup), then extensive probing of paths with encrypted hop fields would reveal the policy. However, in that case an ISP can monitor how much probing is performed and which paths it intends to permit.

So in summary, despite SCION revealing more policy information than BGP, ISPs can monitor and control the amount of disclosed information. Much of the revealed information is a fundamental consequence of multipath communication. We believe that this is a worthwhile tradeoff to make, given the advantages offered by multipath communication.

10.9.5 Conflict Between End-Host Path Control and ISPs' Traffic Flow Policies

SCION end hosts have more control over paths than in today's Internet. This can represent a problem for ISPs, as clients may select communication paths that incur a higher cost than a default path.

In the same vein, multipath communication is fundamentally at odds with ISPs' single-path-routing-along-cheapest-link policy. If multiple paths are made available, then naturally some of them will be more expensive than the cheapest one. Consequently, ISPs may incur a higher cost to support multipath communication.

To resolve these issues, ISPs can enforce their traffic flow policies by making use of bandwidth allocation to guide flows towards less costly paths by providing higher available bandwidth. The multipath path exploration mechanism will continuously optimize the set of paths used for the communication, and thus the paths and traffic will naturally migrate towards paths with more available bandwidth.

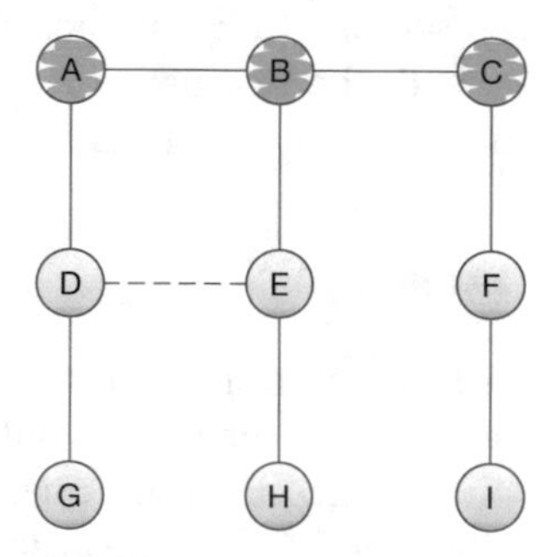

Figure 10.22: The vertical links indicate provider-customer relationships, and the horizontal links indicate peering relationships. ASes *A*, *B*, and *C* are part of the ISD core.

There is, however, a challenge: Consider Figure 10.22, in which a host *g* in AS *G* desires to communicate with host *h* in AS *H*. If host *g* makes use of the peering link *D*-*E*, then ASes *D* and *E* save money as the traffic does not flow through their respective providers *A* and *B*. SCION's path control, however, enables hosts *g* and *h* to select the path *G*-*D*-*A*-*B*-*E*-*H* for their communication, which would incur a cost for *D* and *E*. *D* and *E* cannot easily force the sender to use the peering link, as host *g* needs to be able to use the path *D*-*A* as a backup path in case the peering link fails, or also if it wants to communicate with host *i* in AS *I*.[6]

The bandwidth control we describe above only helps in a limited fashion, as *D* needs to provide ample bandwidth on link *D*-*A* for communication with the remainder of the network.

We see the following approaches to how *D* and *E* can achieve their desired path policy outcome:

[6]A similar example can be seen in Figure 10.21. Consider a host in AS *F* creating a path to a host in AS *G*, merging up- and down-segments. A non-malicious host would create path *F*-*E*-*G*, but a malicious host can create path *F*-*E*-*D*-*E*-*G* to harm *E* (or help *D*) by incurring additional cost for *E*.

- An ISP could inspect the source and destination and decide on the amount of bandwidth granted for each flow. Unfortunately, this approach incurs high overhead on the router and is thus not recommended.
- Hop-field encryption with explicit path activation can be used to deny paths that should instead traverse a peering link. Since this approach imposes additional overhead for path establishment, its use is not recommended for the majority of traffic. Perhaps a hybrid approach can be used, where best-effort traffic can obtain a small amount of bandwidth, and additional bandwidth is only available for activated paths or SIBRA-based paths described in Chapter 11.
- An ISP may charge more for SCION than for a traditional Internet connection. As domains obtain benefits, such as obtaining better service thanks to multipath communication and path control, they should be willing to pay a higher price.
- Per-path pricing can be used to accurately reflect the cost of each path. We are working on a pricing architecture for SCION, which will become available in a future version.

From our interactions with ISPs, we have made the following observations. ISPs' cost structures have changed over the past decade, so the cost is largely due to the fixed costs of maintaining the infrastructure, while the marginal cost of sending a packet is becoming negligible. Evidence for this observation is that transit costs have greatly declined over the past decade, and that utilization-based billing is being replaced with alternate pricing models. Another observation is that in modern networks, the path quality is inversely correlated with the price, so the best path is often also the cheapest. Therefore, rational senders will automatically pick the best paths and thus also reduce the ISPs' costs.

In conclusion, while there is a conflict between end-host path control and ISPs' traffic flow policies, SCION offers mechanisms to mitigate the conflict. However, based on the economics of modern ISP networks, the conflict seems to be vanishing. Consequently, we anticipate that clients will be able to benefit from path control and multipath communication without incurring higher cost.

Part III

Extensions

11 SIBRA

ADRIAN PERRIG, RAPHAEL M. REISCHUK,
STEPHEN SHIRLEY, PAWEL SZALACHOWSKI

This chapter presents SIBRA, the *Scalable Internet Bandwidth Reservation Architecture*, which enables global bandwidth resource allocation. End hosts can use resource allocations to obtain end-to-end bandwidth guarantees to defend against DDoS attacks, which continue to be a menace on today's Internet.

SIBRA provides scalable inter-domain resource allocations and *botnet-size independence* — two important properties that prior DDoS defense systems could not achieve. Intuitively, botnet-size independence enables two end hosts to set up communication regardless of the size of distributed botnets. SIBRA thus ends the arms race between DDoS attackers and defenders.

SIBRA can be implemented with per-flow stateless fastpath operations on transit routers for reservation renewal, flow monitoring, and policing, which results in highly efficient data-plane operation on core routers. SIBRA enables dynamic inter-domain leased lines (DILLs), which offer new business opportunities for ISPs.

The text in this chapter is based on the paper "SIBRA: Scalable Internet Bandwidth Reservation Architecture" by Cristina Basescu, Raphael M. Reischuk, Pawel Szalachowski, Adrian Perrig, Yao Zhang, Hsu-Chun Hsiao, Ayumu Kubota, and Jumpei Urakawa, which was published in the Proceedings of the Symposium on Network and Distributed System Security (NDSS), 2016 [22]. Some concepts have been revised or extended, such as bandwidth allocation among core ASes.

Chapter Contents

A. Perrig et al., *SCION: A Secure Internet Architecture*, Information Security and Cryptography, https://doi.org/10.1007/978-3-319-67080-5_11

11.1 Motivation and Introduction

A recent discussion among network administrators on the NANOG mailing list [187] pointedly reflects the current state of DDoS attacks and the trickiness of suitable defenses: defenses typically perform traffic scrubbing in upstream ASes or in the cloud, but attacks surpassing 20–40 Gbps can still overwhelm the upstream link bandwidth and cause congestion that traffic scrubbing cannot handle. Amplification attacks of up to 400 Gbps [202] and direct, non-amplified attacks from a large army of Internet-of-Things (IoT) devices of up to 620 Gbps [140] and 1,156 Gbps [188] have been witnessed recently, plaguing websites and critical infrastructures, without any viable solution on the horizon that can defend the network against such large-scale flooding attacks.

Quality of Service (QoS) architectures at different granularities, such as IntServ [254] and DiffServ [16], fail to provide end-to-end traffic guarantees at Internet scale: with billions of flows traversing the network core, routers cannot handle the per-flow state required by IntServ, whereas the behavior of DiffServ's traffic classification across different domains cannot guarantee consistent end-to-end connectivity.

Network capabilities [10, 150, 183, 258, 260] are not effective against attacks such as Coremelt [231] that build on *legitimate* low-bandwidth flows to swamp core network links. FLoc [150] in particular considers bot-contaminated domains, but it is ineffective in the case of dispersed botnets.

Fair resource reservation mechanisms (*per source* [181], *per flow* [65, 254, 258], *per destination* [260], *per computation* [196], and *per class* [16]) are necessary to resolve link-flooding attacks, but are not sufficient: none of them provides *botnet-size independence*, a critical property for viable DDoS defense.

Botnet-size independence is the property in which a legitimate flow's allocated bandwidth does not diminish below a guaranteed allocation when the number of bots in other ASes increases. Per-flow and per-computation resource allocation, for instance, will reduce their allocated bandwidth towards 0 when the number of bots that share the corresponding resources increases.

To illustrate the importance of botnet-size independence, we observe how previous systems suffer from the *tragedy of the network-link commons*,[1] which refers to the problem that the allocation of a shared resource will diminish toward an infinitesimally small allocation when many entities have the incentive to increase their "fair share". In particular, per-flow fair-sharing allocations (including per-class categorization of flows) suffer from this fate, as each source has an incentive to increase its share by simply creating more flows. However, even when the fair-sharing system is not abused, the resulting allocations can be too small to be useful.

To explain in more detail, denoting by N the number of end hosts in the Internet, per-source or per-destination schemes could ideally conduct fair sharing of $\mathcal{O}(1/N)$ based on all potential sources or destinations that traverse a given link. However, with increasing hop-count distance of the link from the source or to the destination, the number of potential sources or destinations that traverse that link increases exponentially. Per-flow reservation performs even more poorly, allocating a bandwidth slice of only $\mathcal{O}(1/M^2)$ in the case of a Coremelt attack [231] between M bots, and only $\mathcal{O}(1/(MP))$ during a Crossfire attack [129] with P destination servers that can be contacted. In the presence of *billions* of end hosts engaged in end-to-end communication, the allocated bandwidth becomes too small to be useful.

SIBRA's novel bandwidth allocation system operates at Internet scale and resolves the drawbacks of prior systems. In a nutshell, SIBRA provides interdomain bandwidth allocations (which enable the construction of *dynamic interdomain leased lines* (DILLs), and in turn enable new ISP business models). SIBRA's bandwidth reservations let an AS guarantee a minimal amount of bandwidth to its end hosts by limiting the possible paths in end-to-end communication. An important property of SIBRA is its per-flow stateless fastpath operation on transit routers for reservation renewal, monitoring, and policing, which results in scalable and efficient router operation.

11.2 Goals and Adversary Model

The goal of SIBRA is to defend against *link-flooding attacks*, in which distributed attackers collude by sending traffic to each other (*Coremelt* [231]) or to publicly accessible servers (*Crossfire* [129]) in order to exhaust the bandwidth of targeted servers and Internet backbone links. In the case of *Coremelt*, it may be impossible to limit the traffic volume (e.g., by TCP congestion control) since the participating hosts are under *adversarial* control and can thus run any protocol. In the case of *Crossfire*, distributed attackers collude by sending traffic

[1] We use this term following Garrett Hardin's *Tragedy of the Commons* [107], which according to the author has no technical solution, but instead "requires a fundamental extension in morality." As we should not expect attackers to show any of the latter, we believe in a technical solution — at least for the Internet!

to *legitimate* hosts in order to congest network links leading towards selected servers. We note that many other known attacks constitute a combination of the two cases above.

Adversary Model

We assume that ASes may be malicious and misbehave by sending large amounts of traffic (bandwidth requests and data packets). We furthermore assume any AS in the world can contain malicious end hosts (e.g., as parts of larger botnets). In particular, there is no constraint on the distribution of compromised end hosts. However, attacks launched by routers that intentionally modify, delay, or drop traffic cannot be handled by SIBRA.

Desired Properties

Under the defined adversary model, we postulate the following properties a link-flooding-resilient bandwidth reservation mechanism should satisfy:

- **Botnet-size independence:** The amount of guaranteed bandwidth per AS does not diminish (below a reserved cap) with an increasing number of bots in ASes other than source or destination.
- **Per-flow stateless operation:** The mechanism's overhead on routers should be small. In particular, border routers of transit ASes should not require per-flow, per-source, or per-destination state in the fastpath, which can lead to state exhaustion attacks. Our analysis of real packet traces on core links (see Section 11.9.2) supports the validity of this property.
- **Scalability:** The overhead of the system should scale to the size of the Internet, including management and setup, AS contracts, router and end-host computation and memory, as well as communication bandwidth.

To achieve the above properties, SIBRA directly uses SCION's concepts of isolation and path control, and performs a hierarchical *bandwidth decomposition* at the granularity of ASes: Figure 11.1 depicts an example of four ISDs, in which the two end hosts *S* and *D* in *different* ISDs are connected by stitching three types of path segments together: an *up-segment* from *S* to its ISD core, a *core-segment* within the Internet core (from source ISD to destination ISD), and a *down-segment* from *D*'s ISD core to end host *D*. Intuitively, SCION's isolation property applied to SIBRA enables ASes inside an ISD to establish paths with bandwidth guarantees: *SIBRA steady paths* inside the ISDs, and *SIBRA core paths* between ISD cores. The SIBRA steady paths are set up independently of bandwidth reservations in other ISDs. Finally, a third type, end-to-end bandwidth reservation, called *SIBRA ephemeral paths*, will then be based on the reservations *inside* and *between* the ISDs, but will be lower-bounded for each AS. In particular, malicious entities will not be able to reduce the guaranteed long-term allocation of other ASes.

SIBRA scales to the size of the Internet because the SCION network contains only a small number of ISDs, each with a small number of core ASes. Therefore, it is possible to perform resource allocation per neighboring core AS. Similarly, within an ISD, resources can be allocated based on customer-provider contracts.

11.3 Design Overview

This section describes the design of SIBRA, in particular bandwidth reservations and their enforcement. After a brief overview, we describe SIBRA's reservation types in detail.

A key insight of SIBRA is that hierarchical decomposition of the bandwidth allocation problem can make allocation management and configuration scale to the size of the Internet. Specifically, SIBRA uses three types of paths:

SIBRA Core Paths
between core ASes
across ISDs

SIBRA Steady Paths
between ASes
inside ISD

SIBRA Ephemeral Paths
between end hosts
for end-to-end communication

SIBRA core paths (the double continuous lines in Figure 11.1) can scalably be established between core ASes due to their relatively small number. Within each ISD, providers sell bandwidth to their customers, and customers can establish intermediate-term reservations for specific *intra*-ISD paths, which we call *SIBRA steady paths* (the dashed lines in Figure 11.1). Steady paths are used for connection setup traffic: core and steady paths in conjunction enable the creation of short-term end-to-end reservations *across* ISDs, which we call *SIBRA ephemeral paths* (the solid green lines in Figure 11.1). Ephemeral paths, in contrast to steady paths, are used for the transmission of high-throughput data traffic.

SIBRA paths are established over SCION links with the following allocation (see also Figure 11.2): 85% of the bandwidth of each SIBRA link is allocated for SIBRA traffic (i.e., steady and ephemeral traffic), and the remaining 15% for best-effort traffic. These proportions are flexible system parameters; we discuss the current choice in Section 11.9.1. Note that the proportion for steady and ephemeral traffic constitutes an upper bound: in case the steady and ephemeral bandwidth is not fully utilized, it is allocated to best-effort traffic (see Section 11.6).

An important feature of SIBRA is that steady paths, besides carrying the control traffic of links inside an ISD, also limit the bandwidth for ephemeral

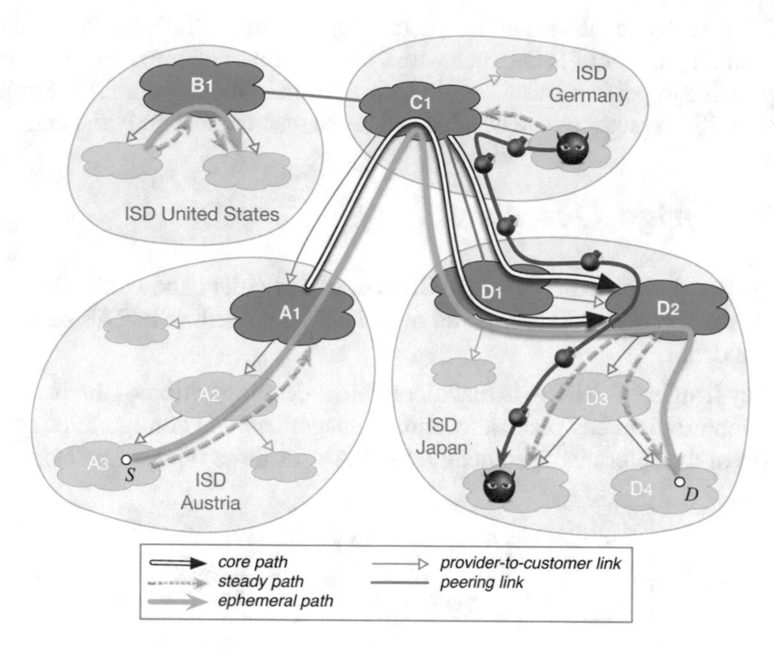

Figure 11.1: Sample topology with four ISDs and their ASes (the core ASes are drawn in dark blue, non-core ASes in light blue). The SIBRA ephemeral path (green) from end host *S* to *D* is created along a SIBRA steady up-path, a SIBRA core path, and a SIBRA steady down-path. The attack traffic (red) cannot diminish the reserved bandwidth on SIBRA ephemeral paths.

paths: an ephemeral path is created by launching a request through existing steady paths, whose amounts of bandwidth limit the bandwidth of the requested ephemeral paths. More precisely, an ephemeral path is created through the combination of (a) a SIBRA steady up-path in the source ISD, (b) a SIBRA core path, and (c) a SIBRA steady down-path in the destination ISD.[2] The ephemeral path request uses only steady and core paths, while the actual data traffic uses only the ephemeral path. The more bandwidth on steady (and core) paths is purchased locally within an ISD (and between ISDs), the larger the fraction of ephemeral bandwidth an end host can obtain to any other end host on the Internet.

If an AS is dissatisfied with its reservation, it can purchase more bandwidth for its SIBRA steady paths, as well as request its core AS to purchase a larger allocation for the SIBRA core path, which the AS would likely need to pay

[2]For instance, Figure 11.1 shows an ephemeral path from host *S* in AS A_3 to host *D* in AS D_4. If the source and destination are located in the same ISD, then the SIBRA core path may not be necessary.

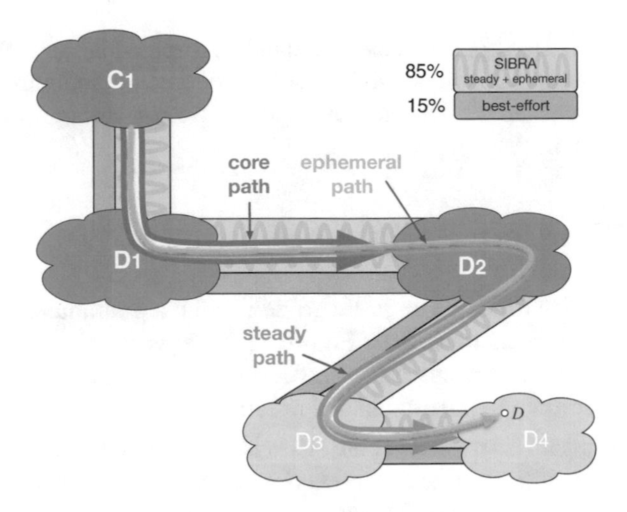

Figure 11.2: The anatomy of SIBRA links: 85% of the link bandwidth is used for SIBRA traffic (i.e., core/steady/ephemeral traffic), and 15% for best-effort traffic. In case the 85% SIBRA traffic is not fully used, the remaining bandwidth can be utilized by best-effort flows.

for. Alternatively, a different core path that provides higher bandwidth could be used instead.

Based on these ideas, it becomes intuitively clear how botnet-size independence is achieved and how the tragedy of the network-link commons is resolved: each pair of ASes can obtain a guaranteed bandwidth allocation, based on the respective SIBRA steady paths and based on the SIBRA core paths. A botnet cannot influence this guaranteed allocation, no matter what its size and distribution. A bot can only use up the bandwidth allocated to the AS it resides in, but not lower the guaranteed allocation of any other AS. It is thus the responsibility of an AS to manage its allocations, and thereby to prevent bots from exhausting the resources of other hosts within that AS.

To make SIBRA viable for practical applications, it is important to ensure that all aspects of the system are scalable and efficient, which holds in particular for frequent operations such as flow admission, reservation renewal, and monitoring and policing. For instance, all fastpath operations are per-flow stateless on transit border routers to avoid state-exhaustion attacks and to simplify the router architecture. The *SIBRA service* in each deploying AS relieves the SCION border routers from dealing with the setup of SIBRA steady/core paths; it processes steady/core reservation requests and updates accounting and policing tables at the border routers.

To protect the SIBRA control plane, we mandate that a requester authenticates its requests. A request contains a list of MACs, each corresponding to an on-path AS (the requester derives the corresponding keys through the DRKey protocol (Section 12.5).

11.4 SIBRA Core Paths

Core ASes establish *SIBRA core paths* to determine a guaranteed amount of bandwidth for ephemeral traffic (see Figure 11.2, between AS C_1 and AS D_2). If one of the core ASes sends more traffic than agreed on, the AS is held accountable, according to the established contracts.

A SIBRA core path between two core ASes is established by either of the two ASes and provides reciprocal reservations (i.e., in both directions with possibly differing bandwidth amounts). The approach relies on two main techniques, which are *aliasing* and *telescoping*.

Telescoping and Aliasing

Telescoping permits a SIBRA core path to be nested inside another core path. An AS can leverage telescoping to *extend* a given reservation by adding an AS at the reservation's end, e.g., a reservation on the path $(A_1 \rightarrow C_1)$ can be extended to $(A_1 \rightarrow C_1 \rightarrow D_1)$, and then to $(A_1 \rightarrow C_1 \rightarrow D_1 \rightarrow D_2)$, as well as to $(A_1 \rightarrow C_1 \rightarrow D_1 \rightarrow C_2)$, assuming sufficient bandwidth capacities (see below). To enable efficient policing, the telescopically extended core path is said to be *nested* within its base core path, e.g., $(A_1 \rightarrow C_1 \rightarrow D_1)$ is nested inside $(A_1 \rightarrow C_1)$.

The way the bandwidth accounting is implemented is through aliasing, so that different nested paths are accounted to the same reservation. Aliasing permits a requester to use multiple identifiers to refer to reserved bandwidth on a link, which naturally models the fact that multiple end-to-end paths share some of the links between the core ASes. For example, in Figure 11.3, the three paths $(A_1 \rightarrow C_1)$, $(A_1 \rightarrow C_1 \rightarrow D_1)$, and $(A_1 \rightarrow C_1 \rightarrow C_2)$ all share the link A_1-C_1, thus aliasing enables the reservations for paths $(A_1 \rightarrow C_1 \rightarrow D_1)$ and $(A_1 \rightarrow C_1 \rightarrow C_2)$ to utilize the same reservation as $(A_1 \rightarrow C_1)$. Aliasing of identifiers simplifies accounting and enables telescoping.

Reservations are usually reciprocal, that is, every request contains (a) a demand for outgoing bandwidth, and (b) suggestions for incoming bandwidth the requester accepts (which can be zero). This ensures atomic requests, i.e., requests are either accepted and immediately valid; or requests are denied, in which case hints are given that enable a quick follow-up request with modified bandwidth values. If accepted, reservations have a lifetime of a few minutes (3 minutes in the current design).

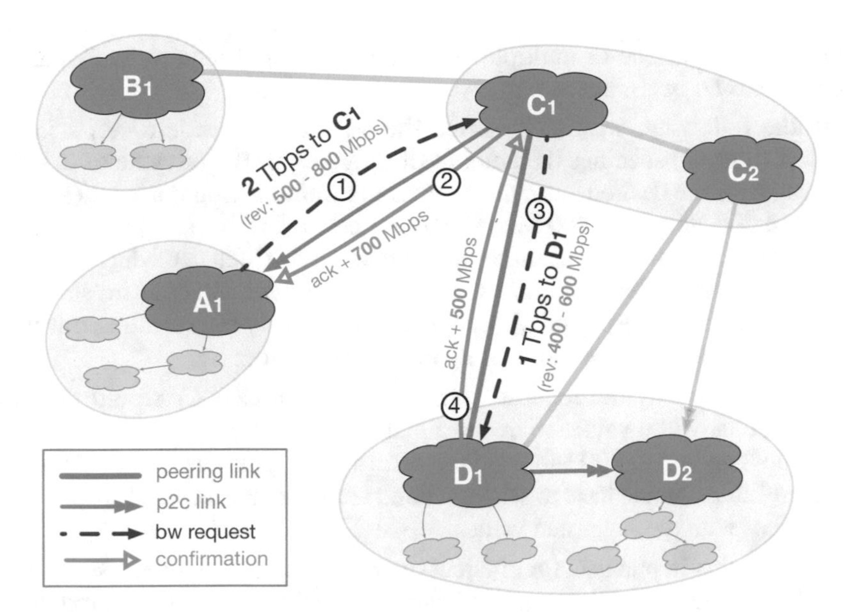

Figure 11.3: SIBRA core path established between core AS A_1 and AS D_2.

We differentiate between the *renewal* and the *extension* of a SIBRA core path. A renewal prolongs the lifetime of an existing core path (i.e., the reservation validity duration), but keeps its structure (i.e., the order of the involved core ASes); an extension extends the core path by one (or more) hops (i.e., the involved core ASes), but retains its validity time.

Setup of SIBRA Core Paths

The setup of SIBRA core paths is best explained by means of the example provided in Figure 11.3. The SIBRA core path from core AS A_1 to core AS D_2 is established by A_1 when requesting bandwidth to D_2 (using the corresponding core-segment between A_1 and D_2). More specifically, A_1 has two options for sending a *SIBRA core path request* along the path to D_2:

- initial: AS A_1 sends an *initial request* directly to AS D_2, which each core AS on the path acknowledges or denies (see below), or
- telescope: AS A_1 first sets up a SIBRA core path to AS C_1 (using an initial request as in Step ① in Figure 11.3) and later *extends* the established core path by sending a *telescope request* (Step ③) such that the extended SIBRA core path includes the next core AS, D_1 in this case. This extension step is repeated until the core path terminates at D_2.

We note that an initial request can span one or multiple hops (e.g., A_1's initial request can terminate at C_1, but also at C_1, D_1, or D_2). Likewise, a telescope request can span one or multiple hops (e.g., an initial request to C_1 can be extended to D_1 or C_2, but also to D_2).

In the following, we first describe the telescope procedure. A_1's initial request (Step ①) specifies the amount of outgoing traffic (a value of 2 Tbps) and suggests the desired amount of incoming traffic it would accept (a range between 500 Mbps and 800 Mbps). C_1 can now decide to accept the request and send a confirmation message ("ack") back to A_1 (Step ②), which includes a return request of, say 700 Mbps, that A_1 will accept and confirm since the offered bounds are respected. A_1 can then immediately start sending traffic of up to 2 Tbps, and receive traffic of up to 700 Mbps from C_1.

C_1 can also deny the request from A_1, in which case C_1 would propose different bandwidth values to be used instead, for instance 1.5 Tbps (instead of 2 Tbps) and up to 900 Mbps on the return path. This case is not shown in Figure 11.3. A_1 would have to send a second request that is adapted accordingly, if it agrees with the suggested values.

We note that in practice, for efficient operation, an AS does not specify arbitrary bandwidth values (or ranges); instead the AS chooses from a predefined set of bandwidth classes.

Telescope Extension

If AS A_1 wants to extend its SIBRA core path (so that it terminates at AS D_1), it sends a new SIBRA core path request, say of 1 Tbps, along the path to AS D_1 (Step ③), and declares that the new path should be an *alias* (on the link A_1-C_1) for the previously established core path to AS C_1. For the sake of illustration, we introduce core path identifiers such as $\langle A_1C_1 \rangle$ for the SIBRA core path from request ①, and $\langle A_1C_1D_1 \rangle$ for its first extension.[3]

In this context, $\langle A_1C_1D_1 \rangle$ is an *alias* for $\langle A_1C_1 \rangle$ on the link A_1-C_1, which means that from the perspective of the ingress border router at AS C_1, any data packet that contains either $\langle A_1C_1 \rangle$ or $\langle A_1C_1D_1 \rangle$ will be treated equally with respect to checking that the base reservation on the ingress link, in this case 2 Tbps, is not exceeded. This is irrespective of whether $\langle A_1C_1 \rangle$ or $\langle A_1C_1D_1 \rangle$ is used. We annotate base identifiers with a small rectangle ▪, and aliased identifiers with a star $*$ in front of the identifier:

Link A_1-C_1, at ingress C_1	
▪$\langle A_1C_1 \rangle$	2 Tbps
$*\langle A_1C_1D_1 \rangle$	/

However, to let A_1's border router correctly police the outgoing traffic, it considers the extended identifier $\langle A_1C_1D_1 \rangle$ *nested* inside the base identifier

[3] Our implementation uses fixed-length, non-guessable randomized identifiers, which we omit here for the sake of readability.

⟨A₁C₁⟩. This entails not only the constraint that 2 Tbps of its base reservation are met, but also that A_1's traffic to D_1 stays within the reserved 1 Tbps. We annotate nested identifiers with an arrow ↳ as follows:

Link A_1-C_1, at egress A_1	
■$\langle A_1C_1\rangle$	2 Tbps
↳$\langle A_1C_1D_1\rangle$	1 Tbps

C_1's egress border router on C_1-D_1, however, does not consider the two identifiers as aliases or nested when policing the outgoing traffic to D_1. It will forward only the traffic that contains the identifier $\langle A_1C_1D_1\rangle$ and only up to 1 Tbps.

Link C_1-D_1, at egress C_1	
■$\langle A_1C_1D_1\rangle$	1 Tbps

The SIBRA core path can be extended further (not shown in Figure 11.3): AS A_1 can send yet another extension request for a new path $\langle A_1C_1D_1D_2\rangle$ of 800 Mbps and declare $\langle A_1C_1D_1D_2\rangle$ an alias of $\langle A_1C_1D_1\rangle$. Consequently, on the link A_1-C_1, all three core paths are aliases of the same reserved 2 Tbps. A_1's border router considers the identifiers to be nested (left table). On the link C_1-D_1, there are only two aliased identifiers for the reserved bandwidth (right table).

Link A_1-C_1, at egress A_1	
■$\langle A_1C_1\rangle$	2 Tbps
↳$\langle A_1C_1D_1\rangle$	1 Tbps
↳$\langle A_1C_1D_1D_2\rangle$	800 Mbps

Link C_1-D_1, at egress C_1	
■$\langle A_1C_1D_1\rangle$	1 Tbps
*$\langle A_1C_1D_1D_2\rangle$	/

The rule of thumb here is that egress border routers of the source ASes use nested identifiers to ensure that outgoing traffic does not exceed the reserved bandwidth margins. The same applies to ingress border routers of the destination ASes. All transit border routers, however, enable more efficient operation in that aliases are used instead. We provide more examples of aliasing and nesting later in this section.

Initial Requests

In the case of an initial request that is sent directly to the destination AS, say D_2, every transit AS on the path accepts or denies, as above. In the denial case, modified bandwidth values are suggested by *each* transit AS and collected in the request packet (see also Case (d) in Figure 11.6 on Page 260). The destination will return the suggested bandwidth values to the requester, who can then issue a second request with bandwidth values adjusted accordingly.

Algorithm 8 Bandwidth determination at a transit core AS.

1: demandLinkAS: a map that stores demanded bandwidth per a (link, AS) pair.
2: demandTransit: a map that stores demanded bandwidth between two links.
3: reservedLinkAS: a map that stores reserved bandwidth per a (link, AS) pair.
4: limit: a map that for a ($link_1$, $link_2$) pair stores a bandwidth limit ($\leqslant min(link_1$.capacity,$link_2$.capacity)). This map can express a traffic matrix (derived from average traffic patterns) or can be configured by an AS operator.
5: $\delta \in (0,1)$: an AS's parameter.
6: **procedure** HANDLEREQUEST(*req*)
7:     *bwReq* = min(*req*.BW, limit[*req*.inLink, *req*.outLink])
8:     demandLinkAS[*req*.inLink, *req*.src] += *bwReq*
9:     demandLinkAS[*req*.outLink, *req*.src] += *bwReq*
10:     demandTransit[*req*.inLink, *req*.outLink] += *bwReq*
11:     *bwDet* = DetermineBW(*bwReq*, *req*.inLink, *req*.outLink)
12:     **if** *bwDet* == *req*.BW **then**
13:         Reserve the bandwidth
14:     **else**
15:         Send *bwDet* as a hint
16:     **end if**
17: **end procedure**
18: **procedure** DETERMINEBW(*bwReq*, *inLink*, *outLink*)
19:     *inAvailable* = *inLink*.capacity - sum(reservedLinkAS[*inLink*, ***any***])
20:     *propIn* = *bwReq* / sum(demandLinkAS[*inLink*, ***any***])
21:     *bwInIdeal* = *inLink*.capacity**propIn*
22:     *bwIn* = min(*bwReq*, *bwInIdeal*, *inAvailable* $*\ \delta$)
23:     *totalTransitDemand* = 0
24:     **for** *inLinkT*, val $\in$ demandTransit[***any***, *outLink*] **do**
25:         *totalTransitDemand* += min(val, *inLinkT*.capacity, *outLink*.capacity)
26:     **end for**
27:     *linkDemand* = min(demandTransit[*inLink*, *outLink*], limit[*inlink*, *outLink*]) /
28:                 *totalTransitDemand*
29:     *outAvailable* = *outLink*.capacity - sum(reservedLinkAS[***any***, *outLink*])
30:     *bwOutIdeal* = (*bwReq* / demandTransit[*inLink*, *outLink*])**linkDemand*
31:     *bwOut* = min(*bwOutIdeal*, *outAvailable* $*\ \delta$)
32:     return min(*bwIn*, *bwOut*)
33: **end procedure**

Algorithm 8 describes how a transit core AS could decide how much bandwidth a given request should be granted. To enable scaling, decisions are made per neighbor link. The first step of the algorithm is to normalize a request and to add the demand to the bandwidth demand maps.[4] Then, the bandwidth determination procedure is executed. First, bandwidth for an incoming link (*bwIn*) is computed. This is the minimum of the weighted available bandwidth of the incoming link[5] and the proportion between the demanded amount and

[4]The algorithm operates on 3-minute request windows, and bandwidth reserved and stored in the maps automatically expires at the end of its request window.

the sum of all demands on the incoming link. The bandwidth for the outgoing link (*bwOut*) is computed in a similar way, but the proportion is calculated between the demand for the pair of incoming and outgoing links and the sum of all demands that use the outgoing link. The offered bandwidth is the minimum of *bwIn* and *bwOut*. If that value equals the requested bandwidth, then the bandwidth can be reserved. Otherwise, the requester AS is informed about the offered bandwidth. The requested bandwidth for link pairs can be limited by an AS through the *limit* map.

In case all ASes have agreed, a successful request is delivered back to the requester (Case (c) in Figure 11.6). This message contains all necessary identifiers so that the request is considered active and the requester can instantaneously use the reserved bandwidth until its expiration.

Requests for the Reverse Direction

As previously mentioned, a SIBRA core path request contains not only the fixed bandwidth value that the sending core AS requests to the destination core AS but also a bandwidth range, which the destination core AS uses as a guide to choose a bandwidth value and then to set up a reservation in the reverse direction.

Figure 11.4 shows how bandwidth in the reverse direction is established: with the confirmation ("ack") to the initial request ("req"), a reciprocal request is issued that reserves bandwidth in the reverse direction. After both requests have been confirmed, the border routers are updated to perform policing and accounting for traffic carrying the request identifiers. In Figure 11.4, the border routers are updated twice: once after the first request, and once after the telescopic extension. The changes from the previous allocation table are highlighted in blue.

Example: Nesting and Aliasing at SCION Border Routers

In the following, we will explain each step shown in Figure 11.4. A_1 requests a core path to its neighbor C_1, according to the topology shown in Figure 11.3. A_1 thus issues a request of 2000 Mbps using a fresh identifier $\langle \mathrm{A_1C_1} \rangle$. A_1 specifies a range of 500 to 800 Mbps as a suggestion for traffic in the reverse direction. The bullet • indicates an empty value for *reference identifiers*, meaning that the request is an *initial request*: it neither extends an existing core path, nor does it create an alias.

C_1 accepts the request and issues a *reverse request* of 700 Mbps, for which it chooses a fresh identifier $\langle \mathrm{C_1A_1} \rangle$. As the identifier serves as a response to the first request, C_1 links its new identifier to the received identifier $\langle \mathrm{A_1C_1} \rangle$ (as

[5] The available bandwidth on incoming and outgoing links is weighted by the parameter δ to leave some amount of bandwidth for new requests.

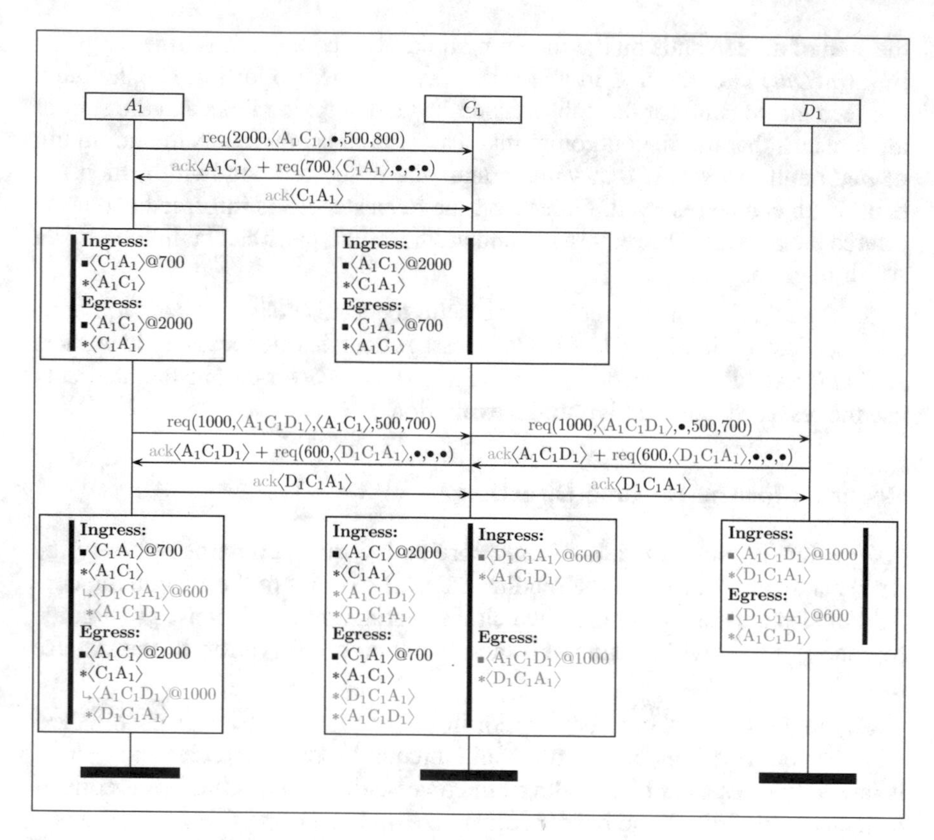

Figure 11.4: SIBRA core path request messages. Within each AS, the SIBRA service sends and processes requests.

specified in the acknowledgment). This link contains sufficient information for the SIBRA service to update the border routers that the requested identifier $\langle C_1A_1\rangle$ has the alias $\langle A_1C_1\rangle$. The reverse request does not contain a suggested bandwidth range since it has been set up with the first request.

A_1 accepts the reverse request and sends an acknowledgement back to C_1. Thereafter, both A_1 and C_1 update their border routers as specified in the boxes in Figure 11.4: overall, there are eight entries, but — thanks to the aliasing of identifiers — only four of them (the base reservations, marked with a rectangle ▪) are necessary for accounting and policing purposes. The aliases (marked with a star $*$) identify the reverse data traffic that uses the same identifier as the original traffic. For example, traffic initiated at C_1 and sent to A_1 will use the identifier $\langle C_1A_1\rangle$ and can send at up to 700 Mbps. When A_1 replies, it will keep the identifier $\langle C_1A_1\rangle$, but can send at a rate of 2000 Mbps.

A_1 next establishes a core path to D_1. As it extends its existing reservation to C_1, it will also specify (in addition to a fresh identifier $\langle A_1C_1D_1\rangle$ and the bandwidth of 1000 Mbps) the reference to the existing core path that it extends; in this case ⟨A1C1⟩.

When the request arrives at C_1, it will remove the reference to the extended core path ⟨A1C1⟩, since it terminates at C_1. In other words, the reference identifier will not be useful beyond C_1. After removing the reference, the request is forwarded to D_1. We assume that D_1 agrees and chooses 600 Mbps for the reverse reservation back to A_1, which confirms the reservation. All three ASes update their border routers as highlighted in blue in Figure 11.4.

Egress border routers use nested identifiers (a) when referenced identifiers are specified, and (b) when the border router is in the source AS (example: ⟨A1C1D1⟩ for A_1.`Egress` on the link to C_1). The ingress border router of the transit AS C_1 uses an alias instead (example: ⟨A1C1D1⟩ for C_1.`Ingress` on the link from A_1). Ingress border routers use nested identifiers when located in the destination AS (example: ⟨D1C1A1⟩ for A_1.`Ingress` on the link from C_1). All transit ASes use aliases or base identifiers.

Example: Telescoping After Two-Side Reservations

We next explain a more involved scenario, as depicted in Figure 11.5. The main difference to the example shown in Figure 11.4 is that both A_1 and D_1 establish reservations with C_1 before A_1 telescopically requests bandwidth to D_1.

A_1 sets up a reservation with C_1 as before. D_1 requests bandwidth to C_1 of 3000 Mbps. It creates a fresh identifier ⟨D1C1⟩ and accepts 1400 Mbps from C_1. This identifier is used in request ⑦ when D_1 sends its reverse request to A_1. In this case, D_1 knows it has a reservation with C_1 and thus creates a telescopic extension for it. To this end, it includes the base identifier as a reference. This reference will be removed by C_1 in the forwarded request ⑧ since ⟨D1C1⟩ terminates at C_1. Similarly, and as in the previous example, C_1 removes the reference identifier ⟨A1C1⟩ when forwarding request ⑤ to D_1.

As a consequence, the border routers are updated by the SIBRA service as shown in Figure 11.5.

Lifetime and Renewal

SIBRA core reservations have a lifetime of 3 minutes; this is a SIBRA system parameter that we chose to enable rapid adjustment to changing network conditions. Within the lifetime of a SIBRA core path, the core AS (that has been granted the corresponding reservation) has guaranteed access to the forward steady bandwidth as reserved. Client end hosts can obtain ephemeral bandwidth according to the steady bandwidth as described in Section 11.6 below.

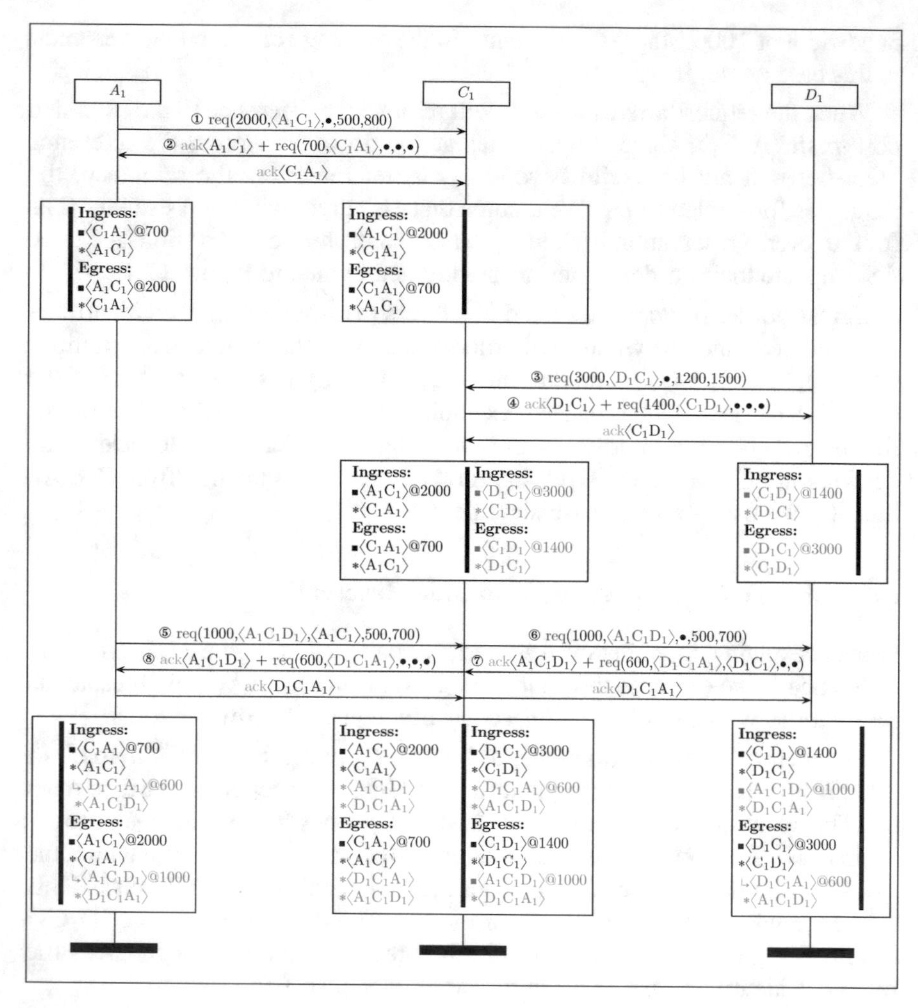

Figure 11.5: SIBRA core path request messages.

Authentication

SIBRA core path requests are authenticated using public-key signatures that state the requests' validity at AS-level granularity. The SIBRA service checks the signed requests using SCION's AS certificates.

As we will discuss later in the chapter, SIBRA relies on efficient message authentication codes (MACs) to enable stateless fastpath operation on transit border routers.

Payment

Core paths not only guarantee bandwidth between ISDs, they also regulate the traffic-related money flow between core ASes according to existing provider-to-customer (p2c) or peering (p2p) relationships (e.g., p2c between AS D_1 and AS D_2, and p2p between AS C_1 and AS D_1).

In line with today's state of affairs, we believe that market forces create a convergence of allocations and prices when ASes balance bandwidth with their peers and adjust contracts such that direct core AS neighbors are satisfied. The neighbors, in turn, recursively adapt their contracts to satisfy the bandwidth requirements of their customers. Paying customers thus indirectly indicate to core ASes the destination ISDs of core paths and their desired bandwidth.

Path Format

SIBRA paths are created using SCION paths and have a similar structure. In order to construct a SIBRA path, an AS creates cryptographically authenticated *reservation tokens* (RTs), which are similar to hop fields. A SIBRA path is a series of RTs prepended with an info field. An RT generated by AS_i is authenticated using a cryptographic key K_i known only to AS_i, by which AS_i can later verify if an RT embedded in the data packet is authentic. More specifically, the RT contains the authenticated ingress and egress interfaces of AS_i, and the reservation request information (such as reservation lifetime or amount of reserved bandwidth). In order to prevent an attacker from crafting a path from partial RT chunks, the RTs are onion-authenticated:

$$\begin{aligned} RT_{AS_i} = \; & ingress_{AS_i} \parallel egress_{AS_i} \parallel \\ & MAC_{K_i}(ingress_{AS_i} \parallel egress_{AS_i} \parallel Request \parallel RT_{AS_{i-1}}) \end{aligned} \tag{11.1}$$

where *Request* is defined as $BW_{req} \parallel ExpTime \parallel FlowID \parallel ResvID$.

The same path format is used by steady and ephemeral paths (described throughout the following sections).

11.5 SIBRA Steady Paths

Steady paths are intermediate-term reservations that are established by ASes *within* an ISD. For example, in Figure 11.1, AS A_3 sets up a steady path to A_1, and D_4 sets up a steady path to D_2. Steady paths expire, but their validity periods can be periodically extended (in our current implementation the default validity period is 3 minutes). An endpoint AS can voluntarily tear down its steady path before expiration and set up a new steady path. Steady paths are set up similarly to core paths, though not between core ASes *across* ISDs, but between ASes *inside* an ISD. SIBRA uses steady paths (in addition to core paths) as building blocks for ephemeral paths: to guarantee availability during connection setup and to perform weighted bandwidth reservations, as we explain in more detail below.

Reservation Request

SIBRA leverages SCION's PCBs (Sections 2.1 and 7.1.1), which disseminate top-down from the ISD core to the ASes. On their journey downstream, they collect AS-level path information as well as information about the current amount of available bandwidth for each link. When a leaf AS receives such a beacon with information about a path segment, the AS can decide to submit a reservation request for a steady path on that segment. In this case, the leaf AS (e.g., AS A_3 in Figure 11.1, or S_3 in Figure 11.6) computes a new flow ID, chooses the amount of bandwidth and the expiration time, and sends a *steady path reservation message* up the path to the core. The requested amount of bandwidth can be chosen from a number of predefined bandwidth classes, introduced for monitoring-optimization purposes (Section 11.7). The actual request messages for steady paths are established similarly to the way SIBRA core paths are established (Section 11.4).

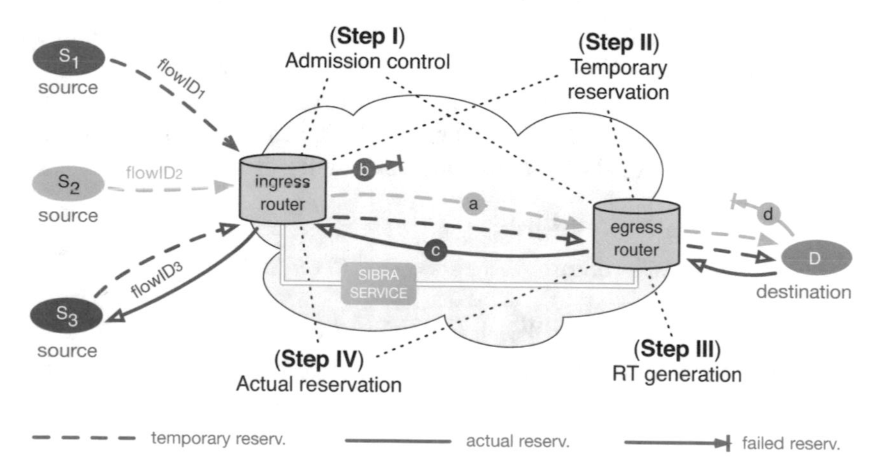

Figure 11.6: A transit AS processing reservation requests from sources S_1, S_2, S_3 to destination D.[6] Steady/core path requests are processed by the SIBRA service, while ephemeral path requests are efficiently processed by the border routers.

Each intermediate AS on the path to the core performs *admission control* by verifying the availability of steady bandwidth to its neighbors on the path (Step I in Figure 11.6). Given the fact that inbound traffic from multiple ingress routers may converge at a single egress router, admission control is performed at both ingress and egress routers. Specifically, the ingress router of AS_i checks the availability of steady bandwidth on the link $AS_{i-1} \rightarrow AS_i$, and the egress router of AS_i on the link $AS_i \rightarrow AS_{i+1}$. If enough bandwidth is available at both the ingress and the egress router (Case (a) in Figure 11.6), both routers *temporarily* reserve the requested bandwidth (Step II). Subsequently, the egress router of AS_i issues a cryptographically authenticated reservation token (RT) (see Equation 11.1) encoding the positive admission decision (Step III).

If at least one of the routers of AS_i cannot meet the request (Case (b) in Figure 11.6), it suggests an amount of bandwidth that could be offered instead, and adds this suggestion to the packet header. Although already failed, the request is still forwarded to the ISD core to collect suggested amounts of bandwidth from subsequent ASes. This information helps the source make an informed and direct decision in a potential bandwidth re-negotiation. At the same time, the denying AS immediately sends a *denial* response back to the requester to enable early notification.

As steady paths are only infrequently updated, scalability and efficiency of steady path updates are of secondary importance. However, AS_i can still perform an efficient admission decision by simply considering the current utilization of its directly adjacent AS neighbors. Such an efficient mechanism is necessary for reservation requests (and renewals) to be fastpath operations, avoiding accessing per-path state. In case of a positive admission decision, AS_i needs to account for the steady path individually per leaf AS from which the reservation originates. Only **slowpath*** operations, such as policing of misbehaving steady paths, need to access this per-path information about individual steady paths.

Confirmation and Usage

When the reservation request reaches the destination[6], the destination D replies to the requesting source (e.g., S_3) either with a *confirmation message* (Case (c) in Figure 11.6) containing the RTs accumulated in the request packet header, or with a *rejection message* (Case (d) in Figure 11.6) containing the suggested bandwidth information collected before. As the confirmation message travels back to the source, every ingress and egress router accepts the reservation request and switches the reservation status from *temporary* to *active* (Step IV).

11.6 SIBRA Ephemeral Paths

Ephemeral paths are short-lived end-to-end reservations for high-bandwidth communication between *end hosts* (unlike core/steady paths, which are established between ASes). In the spirit of fair allocation of joint resources, the

[6] We use the term *destination* in the following (and also in Figure 11.6) to stay as general as possible. For steady-path reservation requests, the destination is the ISD core; for ephemeral-path reservation requests, the destination will be another end host (Section 11.6).

lifetime of ephemeral paths is limited to 16 seconds in order to curtail the time of resource over-allocation; they thus require continuous renewals throughout the life of the connection. The source, the destination, and any on-path AS can rapidly re-negotiate the allocations. Figure 11.1 on Page 248 shows two ephemeral paths, one *inside* an ISD, and one *across* three ISDs — from end host S in AS A_3 to end host D in AS D_4.

Ephemeral Paths from Steady and Core Paths

An ephemeral path reservation is launched by an end host (as opposed to a steady path reservation, which is launched by an AS). The end host (e.g., host S in Figure 11.1) first obtains a steady up-path starting at its AS (e.g., A_3) to the ISD core, and a steady down-path starting at the destination ISD core (e.g., D_2) to the destination AS (e.g., D_4). Joining these steady paths with a SIBRA core path (e.g., from A_1 to D_2) results in an end-to-end path P, which is used to send the ephemeral path request from the source end host S to the destination end host D using the allocated steady bandwidth.

More specifically, S first generates a fresh, randomized flow ID of length 64 bits, say $\langle A_3 \mid\mid D_4 \rangle$ (we omit the randomness here for the sake of illustration), then chooses an amount of bandwidth from SIBRA's predefined ephemeral bandwidth classes, and sends the ephemeral path request along path P. The AS in which the source host resides (e.g., AS A_3) may decide to block the request in some cases, for instance if the client requires too much bandwidth. Each intermediate AS on path P performs admission control. Source and destination AS ensure through a weighted fair-sharing mechanism that the ephemeral bandwidth is split among all end hosts (who wish to obtain ephemeral reservations). The bandwidth reservation continues similarly to the steady path case, except that instead of the ASes' SIBRA service, the ASes' border routers process the reservation request for ephemeral paths (see Figure 11.6).

If bots infest source and destination ASes, these bots may try to exceed their fair share by requesting excessively large amounts of bandwidth. To thwart this attack, each AS is responsible for splitting its purchased bandwidth among its end hosts according to its local policy, and for subsequently monitoring the usage.

Efficient Weighted Bandwidth Fair Sharing

The intuition behind SIBRA's weighted fair-sharing approach for ephemeral bandwidth allocation is that purchasing steady bandwidth on a link enables ephemeral bandwidth to be requested on that link. We explain the details based on Figure 11.7: the bandwidth of the ephemeral path from end host S to D depends on the reserved bandwidth of the steady up-path from A_3 up to A_1, on the reserved bandwidth of the steady down-path from D_2 down to D_4, and also on the reserved bandwidth of the core path from A_1 to D_2. We explain the details of the three cases of intra-source-ISD links, core links, and intra-destination-ISD links in the following. In each case, we discuss how much ephemeral bandwidth can be given *to the entirety* of end hosts inside the source AS, that is, we do not differentiate between individual end hosts for admission control by on-path ASes.

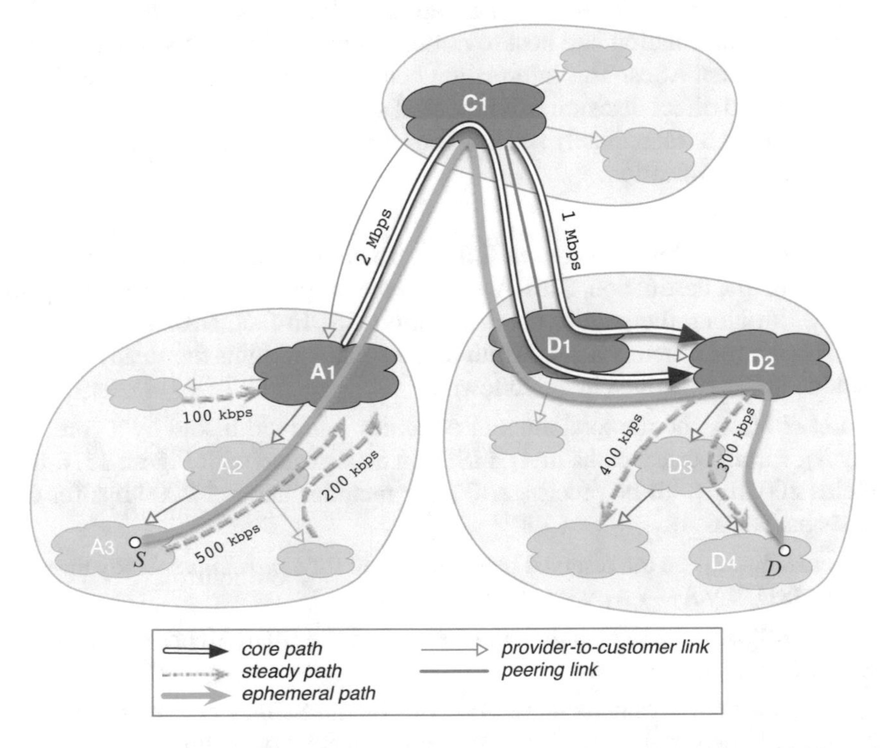

Figure 11.7: Ephemeral bandwidth amounts depend on steady bandwidth and the ratio with respect to other steady reservations.

Ephemeral bandwidth in the source ISD. Assuming A_3 has a SIBRA steady up-path of 500 kbps to its core A_1, then A_3 can offer its end hosts ephemeral bandwidth of up to 500 kbps on the link up to A_1 — regardless of the ephemeral path requests from any other AS to A_1:

$$eBW_{A_3 \to A_1} = sBW_{A_3 \to A_1} \tag{11.2}$$

The ephemeral bandwidth is shared among the end hosts of A_3, but not shared with the end hosts of other ASes. This sharing is at the discretion of A_3, that is, A_3 decides how much ephemeral bandwidth each end host should obtain (assuming that no more than 500 kbps is used in total). As every intermediate

AS on the path has to accept (or deny) the ephemeral request, a request can still fail, even if A_3 has accepted it. This, however, should only happen in case of failures or over-subscription.

In the case of a denied request, the procedure is the same as for steady path requests: if one of the routers cannot meet the request, it suggests an amount of bandwidth that could be offered instead, and adds this suggestion to the packet header (see Figure 11.6). Although already failed, the request is still forwarded towards the destination end host to collect suggested amounts of bandwidth from subsequent ASes. This information helps the requesting end host make an informed and direct decision in a potential second request. At the same time, the denying AS immediately sends a *denial* response back to the requester to enable early notification.

Ephemeral bandwidth on core links. To provide bandwidth guarantees on *every* link to a destination, SIBRA extends the influence of the steady up-path bandwidth along the path to the destination AS. In fact, SIBRA's weighted fair sharing for ephemeral bandwidth on core paths weights the steady up-path bandwidth, as explained in the following.

Let $sBW_{*\to A_1}$ be the total amount of steady bandwidth sold by a core AS, say A_1, for *all* steady paths in A_1's ISD. In accordance with Figure 11.7, this yields 800 kbps. To be precise, $sBW_{*\to A_1}$ includes another 200 kbps for the hosts inside AS A_1.

Let $sBW_{A_3\to A_1}$ be the reserved bandwidth sold for a *particular* steady up-path in this ISD, say $A_3 \to A_1$ with 500 kbps.

Let $cBW_{A_1\to D_2}$ be the bandwidth reserved on the SIBRA core path from A_1 to D_2, 2 Mbps in this case.

Then, the sum of ephemeral reservations on the SIBRA core path launched by the end hosts in A_3 (which are competing against other hosts whose ASes have steady reservations to the ISD core) can be up to

$$eBW_{A_1\to D_2} = \frac{sBW_{A_3\to A_1}}{sBW_{*\to A_1}} \cdot cBW_{A_1\to D_2} \tag{11.3}$$

which amounts to 1 Mbps for the example of Figure 11.7.

In other words, the ephemeral bandwidth on a SIBRA core path (reservable by the end hosts of a particular AS A_3) depends not only on the reserved bandwidth on the core path, but also on A_3's steady up-path bandwidth in relation to the total amount of steady up-path bandwidth purchased by other ASes in the ISD.

Thus, even if all end hosts in each AS of the source ISD, say A, establish reservations to hosts in one destination core AS, say D_2, then the total traffic will not exceed the reservation of 2 Mbps on the core path to D_2.

Ephemeral bandwidth in the destination ISD. In the destination ISD, the weighted fair sharing is slightly more complex, but follows the ideas of the previous cases: the weighting includes the ratio of steady bandwidth of all steady up-paths in the source ISD, the ratio of core bandwidth of the core paths ending in the core AS of the destination ISD, and the steady bandwidth down to the destination AS.

More precisely, the ephemeral path requests from A_3 to D_4 can obtain ephemeral bandwidth in the destination ISD of up to

$$eBW_{D_2 \to D_4} = \frac{sBW_{A_3 \to A_1}}{sBW_{* \to A_1}} \cdot \frac{cBW_{A_1 \to D_2}}{cBW_{* \to D_2}} \cdot sBW_{D_2 \to D_4} \tag{11.4}$$

where $cBW_{* \to D_2}$ is the total amount of bandwidth negotiated in the core paths between any core AS and D_2, and $sBW_{D_2 \to D_4}$ is the steady bandwidth reserved from D_2 down to D_4. In our example, we obtain $eBW_{D_2 \to D_4} = \frac{500 \text{ kbps}}{1000 \text{ kbps}} \cdot \frac{2 \text{ Mbps}}{3 \text{ Mbps}} \cdot 300 \text{ kbps} = 100 \text{ kbps}$.

Equation 11.4 looks similar to Equation 11.3, with an additional factor in the weighting that reflects the ratio of incoming traffic from other core ASes. Intuitively, this factor ensures that traffic from every other core AS obtains its share based on the bandwidth negotiated. This means that even if all hosts that have steady/core reservations to D_4 reserve ephemeral bandwidth to D_4, then: (a) no more traffic than the steady down-path reservation to D_4 will be received by D_4, and (b) every such host obtains its fair proportion according to the established reservations on steady and core paths.

There is one problem, though. The above equation takes into account reservations from hosts located in ASes *outside* D_2's ISD, i.e., ASes that have core paths set up to D_2. ASes *inside* D_2's ISD, however, (as they obviously have no core paths set up to D_2) are not considered. We thus extend the equation by adding a term to the denominator of the second fraction:

$$eBW_{D_2 \to D_4} = \frac{sBW_{A_3 \to A_1}}{sBW_{* \to A_1}} \cdot \frac{cBW_{A_1 \to D_2}}{cBW_{* \to D_2} + sBW_{* \to D_2}} \cdot sBW_{D_2 \to D_4} \tag{11.5}$$

where $sBW_{* \to D_2}$ represents the steady up-paths to core AS D_2. To be precise, this value does not include the steady up-paths from D_4, a detail that we omit here for brevity's sake. D_4 would not set up ephemeral reservations to itself through the ISD core.

In order to give D_2 more fine-grained control for balancing traffic, the calculated value $eBW_{D_2 \to D_4}$ can be weighted by a factor $\alpha \in (0,1)$.

Finally, the total bandwidth for all ephemeral paths between the end hosts in A_3 and D_4 can be up to

$$eBW_{A_3 \to D_4} = min(eBW_{A_3 \to A_1}, eBW_{A_1 \to D_2}, eBW_{D_2 \to D_4}) \tag{11.6}$$

Processing Ephemeral Path Requests

Upon receiving a SIBRA ephemeral path request, each border router on the path inspects the specified identifiers of the underlying steady/core paths. For example, assume end host S inside A_3 requests 50 kbps to end host D inside D_4, and uses the ephemeral path identifier $\langle A_3 \,||\, D_4 \,||\,$ `f09a...c` $\rangle$ to uniquely label the new ephemeral path. Figure 11.8 shows the steady/core paths, on which the ephemeral path request is based. We note that all identifiers are 64 bits and randomized, which is omitted from the figure.

The egress border router of A_3 compares the requested 50 kbps with the available bandwidth on the steady up-path. The router performs a table lookup for the identifier of the underlying steady path, $\langle A_3 \,||\, A_1 \rangle$ in this case, and finds that 150 kbps of ephemeral BW are in use.

Path ID	Reserved steady BW	Used ephemeral BW
$\langle A_3 \,\|\, A_1 \rangle$	500 kbps	150 kbps

As more than 50 kbps is available, the router grants the request and adds 50 kbps to the used ephemeral bandwidth. It also remembers to release the 50 kbps at the time when the ephemeral path expires. The expiration time is expressed in *SIBRA ticks* and one SIBRA tick is 4 seconds. Assuming that the requested ephemeral path has a lifetime of four SIBRA *ticks* (i.e., 16 seconds), then the router performs the following operations:

```
usedEphBW += 50
releaseBW[cur+4] += 50
```

The array `release` is a cyclic data structure that stores the amounts of bandwidth to be released in the future (time is expressed in SIBRA ticks). After each epoch, the pointer `cur` is stepped by one position, the bandwidth is released, and the data structure is reset:

```
cur += 1
usedEphBW -= releaseBW[cur]
releaseBW[cur] = 0
```

All operations referring to epochs are computed modulo the number of epochs.

The request is forwarded along the path and reaches the ingress border router of A_2, which will perform the same operations as A_3.

If a border router further along the path denies the request, the denying router will send a deny message back to all previous routers. The egress router of A_3

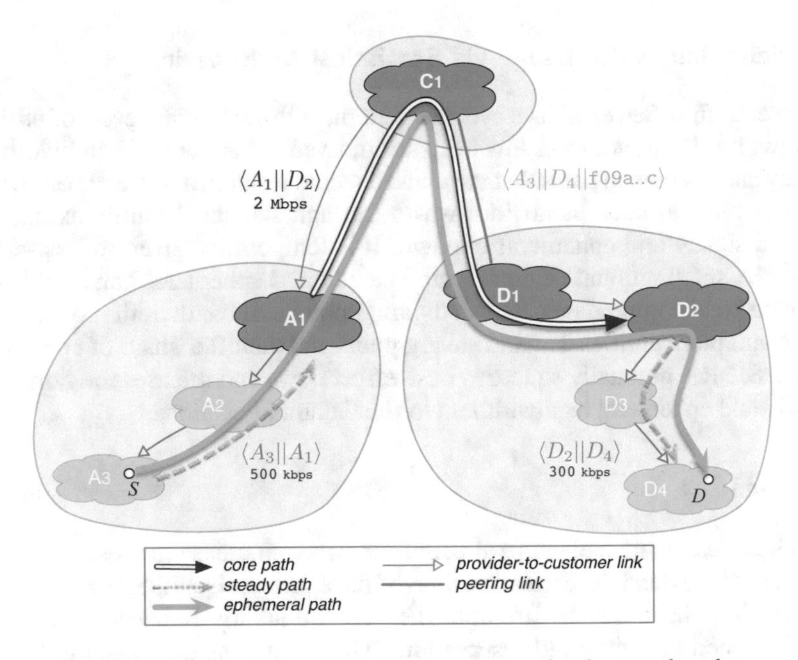

Figure 11.8: Identifiers for steady, core, and ephemeral paths.

will then subtract the 50 kbps from its used bandwidth and will also remove the future release value.

For the sake of simplicity, we have omitted the bandwidth values for the reverse direction. Requests and table entries also carry the reverse bandwidth values.

Fair Sharing of Steady Paths

A challenging question is whether a fair-sharing mechanism is necessary for steady bandwidth. A steady up-path is used solely by the AS that requested it, and its use is monitored by the AS, which splits the steady up-path bandwidth between its end hosts. In contrast, steady down-paths need to be revealed to several potential source ASes, either as *private* steady down-paths (e.g., for a company's internal services), or as *public* steady down-paths (e.g., for public services). To prevent a botnet residing in malicious source ASes from flooding steady down-paths, SIBRA uses a weighted fair-sharing scheme similar to ephemeral paths: each AS using a steady down-path obtains a fair share proportional to its steady up-path, and its ISD's core path.

Efficient Bandwidth Usage via Statistical Multiplexing

Internet traffic often exhibits a cyclical pattern, with alternating levels of utilized bandwidth. In situations of low utilization, fixed allocations of bandwidth for steady and ephemeral paths that are unused would result in a waste of bandwidth. SIBRA reduces such bandwidth waste through statistical multiplexing, i.e., unused steady and ephemeral bandwidth is temporarily given to best-effort flows. A small amount of unallocated steady and ephemeral bandwidth still remains to accommodate new steady and ephemeral bandwidth requests. As more and more entities demand steady paths and their fair share of ephemeral paths, SIBRA gradually squeezes best-effort flows and releases the borrowed steady and ephemeral bandwidth up to the default allocations.

Renewal

End hosts can launch ephemeral path renewals to increase the reserved bandwidth and/or extend the expiration time of the ephemeral path. Since ephemeral reservations have a short lifetime, they are frequently renewed. Renewals are launched using the old reservation, which contains the bandwidth class of the reservation; therefore routers can rapidly decide on the fastpath how much bandwidth they should allocate for the renewal, for instance if the bandwidth increased, decreased, or remained the same. Reservations are given a *reservation index*, incremented for each renewal of a specific ephemeral path. Reservations can be renewed anytime before they expire, and the end host is allowed to switch to the newer reservation at any time. However, the end host is not allowed to use both the old and the renewed reservation at the same time; Section 11.7.1 shows a mechanism to detect such misbehavior.

11.7 Priority Traffic Monitoring and Policing

Flows that exceed their reservations may undermine the guarantees of other legitimate flows. An ideal monitoring algorithm should immediately catch *every* such malicious flow. This, however, would be too expensive for line-rate traffic in the Internet core. Instead, as the first line of defense, SIBRA relies on edge ASes to perform fine-grained traffic monitoring. Edge ASes rely on flow IDs to check each flow's bandwidth usage and compare it against the reserved bandwidth for that flow ID, which is stored by each AS locally during the reservation request. Previous research has shown that per-flow slowpath operations are feasible at the edge of the network [230].

Monitoring on transit ASes, however, needs to be processed on the fastpath. To detect misbehaving ASes that purposely fail to regulate their own traffic, SIBRA deploys a lightweight monitoring mechanism in transit ASes. First, each AS monitors the bandwidth usage of incoming traffic from each neighbor

AS and compares it against the total bandwidth reserved for that neighbor. Such coarse-grained monitoring promptly detects a misbehaving neighbor that has failed to correctly police its traffic.

Why Per-Neighbor Monitoring Is Insufficient

There are cases, though, when per-neighbor monitoring in transit ASes is insufficient. Figure 11.9 depicts two flows originating in AS_0, each having reserved 5 Mbps. Flow 1 is malicious and sends traffic with 8 Mbps, while flow 2 underuses its reservation. AS_0 hence does not properly monitor its flows. When AS_1 performs per-neighbor monitoring, it can only notice that, in the aggregate, it receives 10 Mbps from AS_0 and sends 10 Mbps to AS_2. However, when the two flows diverge, AS_2 detects flow 1 as malicious and holds AS_1 responsible, although AS_1 properly performed per-neighbor monitoring.

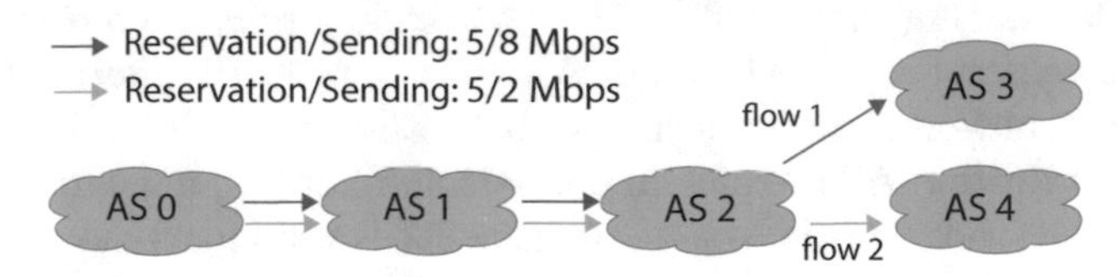

Figure 11.9: Per-neighbor monitoring may label benign AS_1 malicious.

For this reason, SIBRA additionally utilizes *fine-grained* probabilistic monitoring of individual flows at the transit ASes, using a recently proposed technique [255]. Each transit AS monitors, per given time interval, all the flows in a number of randomly chosen bandwidth classes. Recall that the bandwidth class of a flow is authenticated by the RTs in the packet header. If the average bandwidth utilization of a flow during that time interval exceeds the flow's bandwidth class, the flow is classified as malicious and added to a blacklist, preventing its renewal.

To localize the origin AS of the malicious flow, an AS informs the previous AS of the misbehaving flow. In response, the previous AS can simply monitor that specific flow explicitly. If the violation persists, the suspicious previous-hop neighbor is likely to be malicious. Then, the AS can punish it, for instance, by terminating their contract.

11.7.1 Flow Renewal Monitoring and Policing

A successful ephemeral path renewal replaces the old reservation, therefore the renewal receives the same flow ID as the old reservation. However, SIBRA paths allow for RTs with overlapping validity periods. Therefore, if multiple renewals were to occur before the ephemeral path expires, the source would be in possession of multiple sets of valid RTs: some corresponding to the ephemeral path with the previous bandwidth class and old expiration time, and the others corresponding to the new values for bandwidth class and expiration time, along the same path. Since all sets of RTs are associated with the same flow ID, routers would overwrite their per-flow entries with the new bandwidth class.

A malicious end host could thus exploit renewals by using both sets of RTs, old and new, during the overlapping validity time of the RTs, thus using more bandwidth than the reserved amount. To prevent such misuse, end hosts are not allowed to use old RTs after having used the renewed RTs. When renewals use the same bandwidth class as the old reservation, simultaneous use of old and new RTs is detected by the per-class monitoring mechanism (as described above) since the usage is jointly accounted under the same flow ID.

We now consider the case when the renewed bandwidth class is different from the old one. The edge AS performs per-flow stateful inspection and is supposed to filter out traffic that violates the sending rule. Therefore, the edge AS can be held accountable by other ASes for improperly filtering traffic. In transit ASes, however, we propose a probabilistic approach for detecting this type of misbehavior. ASes maintain one Bloom filter [36] per currently active expiration time and bandwidth class. Since an RT is valid for at most 16 seconds and the time granularity is 4 seconds (i.e., SIBRA tick), four Bloom filters are needed per bandwidth class to record flow IDs that use the bandwidth class within that time period. The details about time discretization are discussed in the SIBRA paper [22]. For an incoming packet with a reservation in a monitored class C, ASes simply store the tuple $\langle$*flow ID, reservation index*$\rangle$ in the Bloom filter of C. By checking these Bloom filters, each AS can notice whether a flow ID uses two different bandwidth classes during a time period.

The monitoring algorithm is further optimized as follows. SIBRA selects a small number of classes to monitor at a given moment in time, therefore ASes store Bloom filters only for the few monitored traffic classes. In addition, SIBRA does not investigate all Bloom filters: we observe that, when the renewed bandwidth is much higher or much lower than the previous bandwidth, using both the old and new reservations would incur an insignificant bandwidth overuse. Therefore, if a certain reservation index is used in class C, SIBRA investigates only the Bloom filters of the classes whose bandwidth values are comparable to C's bandwidth (the comparability of classes is discussed in the SIBRA paper [22]). SIBRA investigates whether in these Bloom filters an index $\textit{reservation index} + i$ is present, where $i \in \{0, 1, \ldots, 15\}$ chosen randomly ($i = 0$ detects whether the end host maliciously reuses the same reservation index). If found, ASes increment a violation counter for the source of that flow ID. The violation counter allows for Bloom filter false positives. When the violation counter exceeds a threshold, an alarm is raised for that sender. Therefore, the more packets an attacker sends, the higher the probability of detection.

11.7.2 Dealing with Failures

While bandwidth guarantees along fixed network paths allow for a scalable design, link failures can still disrupt these paths and thus render the reservations futile. In fact, leaf ASes and end hosts are interested in obtaining a bandwidth guarantee rather than obtaining a specific network path for their traffic.

SIBRA deals with link failures using two mechanisms: (a) a failure *detection* technique to remove reservations along faulty paths, and (b) a failure *tolerance* technique to provide guarantees in the presence of failures. For (a), SIBRA uses short expiration times for reservations and keep-alive mechanisms. Steady paths expire within 3 minutes of creation, but leaf ASes can extend the steady paths' lifetime using keep-alive messages. Ephemeral paths have a default lifetime of 16 seconds, which can be extended by source end hosts through renewals. Unless keep-alive messages or renewals are used, reservations are removed from the system within their default expiration time. By construction, a new reservation cannot be created on top of faulty paths. For (b), SIBRA allows leaf ASes to register multiple disjoint steady paths. We also envision source end hosts being able to use multiple disjoint ephemeral paths to the same destination.

11.7.3 Dynamic Inter-domain Leased Lines (DILLs)

Businesses use *leased lines* to achieve highly reliable communication links. ISPs implement leased lines virtually through reserved resources on existing networks, or physically separated through dedicated network links. Leased lines are very costly, can take weeks to set up, and are challenging to establish across several ISPs.

A natural desire is to achieve properties similar to traditional leased lines, but more efficiently. GEANT offers a service called "Bandwidth on Demand" (BoD), which is implemented through the Inter-Domain Controller Protocol [67] to perform resource allocations across the participating providers [92]. Although BoD is a promising step, the allocations are still heavyweight and require per-flow state.

With SIBRA's properties, ISPs can offer lightweight dynamic inter-domain leased lines (DILLs). A DILL can be composed of two longer-lived steady paths, connected through a core path, or dynamically set up with an ephemeral path that is constantly renewed. Thanks to the lightweight operation of SIBRA, DILLs can be set up within a packet round-trip (source-destination-source) setup message and are immediately usable. Our discussions with operators of availability-critical services have revealed that the DILL model has sparked their interest.

To enable long-term DILLs, valid on the order of weeks, the concept of ephemeral paths in SIBRA could be reframed: long-term DILLs could use the same techniques for monitoring and policing as ephemeral paths, but they would also introduce new challenges. To enable long-term DILLs, ISPs need to ensure bandwidth availability even when DILLs are not actively used, as opposed to ephemeral bandwidth, which can be temporarily used by best-effort flows. For this purpose, ISPs could allocate a percentage of their link bandwidth for DILLs, besides steady, ephemeral, and best-effort paths. Additionally, for availability in the face of link failures, ISPs would need to consider active failover mechanisms. For instance, in architectures that provide path choice, ISPs could leverage disjoint multipath reservations concentrated in a highly available DILL. We leave a detailed design to our future work.

11.8 Use Cases

With the flexible lifetime of DILLs, ranging from tens of seconds to weeks on-demand, SIBRA brings immediate benefits to applications where guaranteed availability matters. These applications comprise critical infrastructures, such as financial services and smart electric grids, as well as business applications, such as videoconferencing and reliable data sharing in health care. Setting up leased lines in many cases may take several weeks and may be prohibitively expensive: it is costly to install leased lines between each pair of domains, and also to connect each domain through a leased line to a central location in order to build up a star topology.

Critical Infrastructures

Financial services, for instance *transaction processing from payment terminals*, would become more reliable when using SIBRA DILLs: since DILLs guarantee availability even in the presence of adversarial traffic, payment requests and their confirmations would *always* obtain a guaranteed minimum bandwidth. Other use cases are connections to ATMs, which would be too costly to realize with leased lines. DILLs could also be used for *remote monitoring of power grids*: a guaranteed minimum bandwidth would be suitable to deliver the monitored parameters, independent of malicious hosts exchanging traffic. *Telemedicine* is another use case of practical relevance: the technology uses telecommunication to provide remote health care — often in critical cases or emergency situations where interruptions could have severe consequences.

Business-Critical Applications

Videoconferencing between the remote sites of a company gains importance as a convenient way to foster collaboration while reducing travel costs. Short-lived and easily installable DILLs provide the necessary guaranteed on-demand bandwidth for reliably exchanging video traffic. Another application is *reliable on-demand sharing of biomedical data* for big-data processing, complementing the efforts to improve health care quality and cost in initiatives such as *Big Data to Knowledge* launched by the US National Institutes of Health (NIH) [164].

11.9 Discussion

11.9.1 The Choice of Bandwidth Ratios on SIBRA Links

Recall that in Section 11.3, for the setting where a SIBRA connection is the default one, we assigned 85% of a link's capacity to SIBRA, and left 15% to best-effort traffic. The reason for this choice is that the majority of traffic constitutes persistent high-bandwidth connections: for example in Australia, Netflix's video connections contribute to more than 50% of the entire Internet traffic [104]. Given an additional amount of traffic from other large video providers such as YouTube and Facebook, we estimate ephemeral paths to require roughly 70–90% of a link's bandwidth.

Best-effort, however, is still important for some types of low-bandwidth connections: email, news, and part of the SSH traffic could continue as best-effort traffic, totaling 3.69% of the Internet traffic [146], as could DNS traffic totaling 0.17% of the Internet traffic [146]. In addition, very short-lived flows (that is flows with a lifetime less than 256 ms) with very few packets (the median flow contains 37 packets [241]) are unlikely to establish SIBRA reservations, simply to avoid the additional setup latency. Such flows amount to 5.6% of the Internet traffic [241] and can thus also be categorized under best-effort.

Since it is hard to specify the actual bandwidth proportions precisely, we use 85% and 15% as *initial* values and note that these values can be re-adjusted at any point in the future.

We recall from Section 11.6 that, in addition to the parameter choice, SIBRA's statistical multiplexing between the traffic classes helps to dynamically balance the traffic. We expect that in particular the long-lived reservations are not always fully utilized, in which case best-effort traffic can be transmitted instead. Consequently, an end host can by default use SIBRA (with constant bit rate) and multipath best-effort (with congestion control) to use up the remaining bandwidth.

11.9.2 Per-Flow Stateless Operations Are Necessary

To understand the amount of per-flow storage state required on the fastpath, we investigate the number of active flows per second as seen by a core router in today's Internet. We used anonymized one-hour Internet traces from CAIDA, collected in July 2014. The traces contain all the packets that traversed a

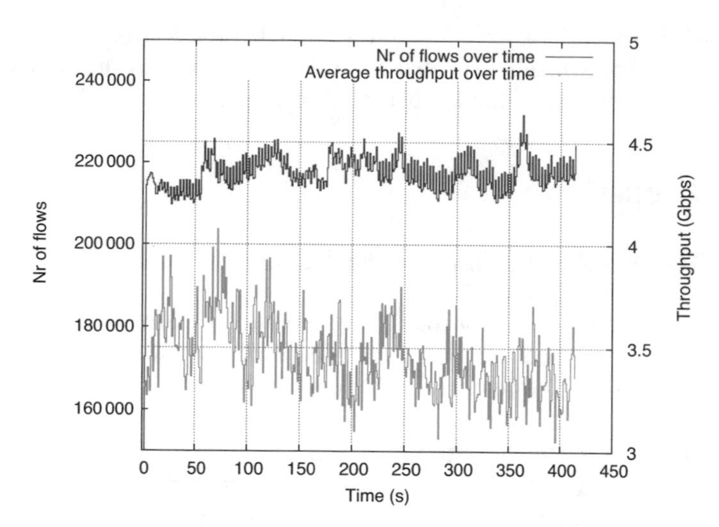

Figure 11.10: The number of active flows every second and their throughput, observed on a 10 Gbps Internet core link.

10 Gbps Internet core link of a tier-1 ISP in the United States, between San Jose and Los Angeles.

Figure 11.10 depicts our findings as the number of active flows on the core link at a granularity of 1 second, for a total duration of 412 seconds. We observe that the number of flows varies around 220,000, with a boundary effect at the beginning of the data set. These flows amount to a throughput between 3 and 4 Gbps — a link load of 30% to 40%. A large core router switching 1 Tbps (with 100 such 10 Gbps links) would thus observe 22×10^6 flows per second *in the normal case*, considering a link load of only 40%. *In an attack case*, adversaries could greatly inflate the number of flows by launching connections between bots, as in Coremelt [231]. Schuchard et al. already analyzed attacks that can exhaust the router memory [219]. All these results suggest that storing per-flow state in the fastpath can be vulnerable to resource exhaustion attacks.

11.9.3 Case Study: Ephemeral Bandwidth on Core Links

A central point of SIBRA is to guarantee a sufficient amount of bandwidth using today's infrastructure, even for reservations that span multiple ISDs. A central question is how much bandwidth an end domain could minimally obtain if globally all domains attempt to obtain their maximum fair share. To investigate this point, we considered a scenario with Australia as destination, and all non-Australian leaf ASes in the world reserving ephemeral bandwidth to Australia. We picked Australia because with its 24 million inhabitants, it represents a major economy, and it has already experienced infrastructure

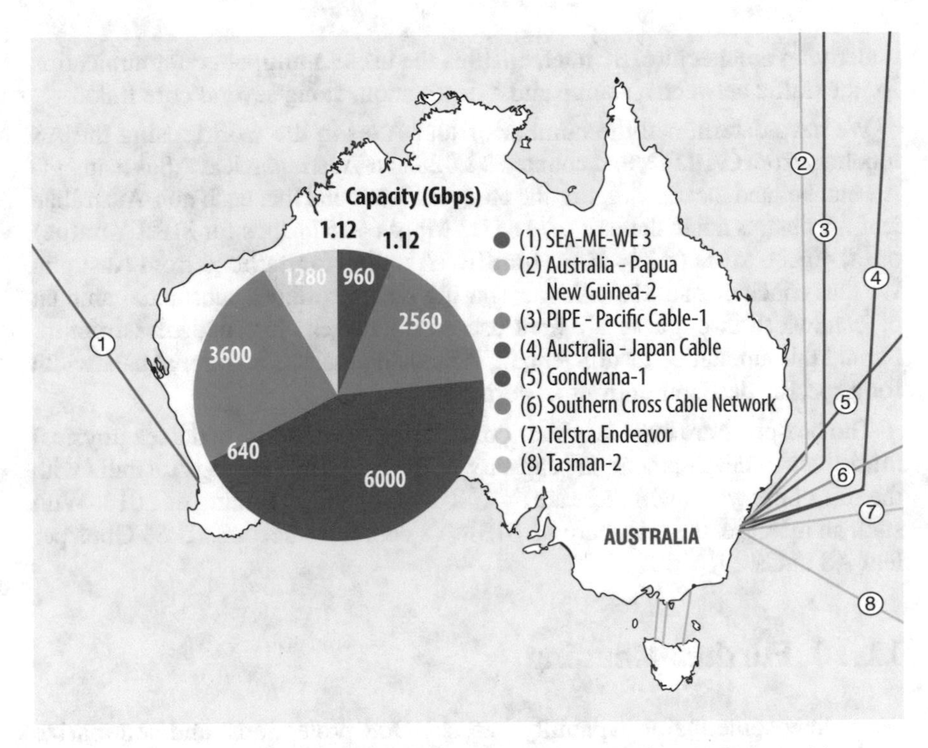

Figure 11.11: Australia submarine link map, including link capacities.

congestion in today's Internet [104]. While its geographical location hinders laying new cables, Australia is well suited for our study aiming to determine a lower bound on the amount of bandwidth SIBRA core links can expect. Other countries, especially those situated on larger continents, typically feature higher-bandwidth connectivity, as laying cables on land is easier than in the ocean.

Figure 11.11 illustrates the current submarine link map of Australia, including the name and capacity of the links.[7] The entire traffic traverses these links. For simplicity, we assume guaranteed bandwidth is split equally between leaf ASes. In practice, however, the bandwidth is proportional to the size of the steady paths of the leaf ASes (Section 11.3). We considered two cases: (a) the worst case, i.e., when all reservations are squeezed over the same link — in our case, we chose the highest-bandwidth cable, namely the Australia-Japan Cable (6 Tbps), and (b) the best case, i.e., when the reservations are distributed across all cables (totaling 15.04 Tbps). In contrast to other architectures, SIBRA's

[7] `http://www.submarinecablemap.com/` illustrates the submarine link map. The link capacities were obtained from various resources, e.g., the Australia-Japan Cable capacity from `http://www.ajcable.com/company-history/`.

underlying architecture, SCION, enables the use of multipath communication for the traffic between a source and a destination, along several core links.

We have determined the number of leaf ASes in the world, using the AS topology from CAIDA[8], and counted 32,428 non-Australian leaf ASes using the AS number and location[9]. After the analysis, we found that each non-Australian leaf AS obtains a fair share of (a) 185.02 Mbps ($\approx$157 Mbps for SIBRA traffic), or (b) 463.86 Mbps ($\approx$394 Mbps for SIBRA traffic) for traffic to/from Australia. We thus conclude that SIBRA's fair-sharing scheme offers a substantial amount of bandwidth through an efficient use of the current Internet infrastructure. Should this amount be insufficient, an AS could purchase additional bandwidth for a specific destination from its core AS.

The prospects are even brighter: considering the planned undersea physical infrastructure development, the capacity of the cables connecting Australia with the rest of the world will increase by 168 Tbps by the beginning of 2018. With such an increase, the fair share on SIBRA's core links becomes 5.64 Gbps per leaf AS in Case (b).

11.10 Further Reading

We review some major capability-based DDoS protections, and summarize resource allocation and reservation mechanisms.

Capability-Based Mechanisms

Capability-based mechanisms [10, 114, 150, 183, 196, 258, 260] aim at isolating legitimate flows from malicious DDoS attack traffic. Network capabilities are access tokens issued by on-path entities (e.g., routers and destination) to the source. Only packets carrying such network capabilities are allowed to use a privileged channel. Capability-based schemes, however, require additional defense mechanisms against denial-of-capability (DoC) attacks [14] and against attacks with colluding hosts or legitimate-looking bots [129, 231]. To address DoC attacks, *TVA* [260] tags each packet with a path identifier based on the ingress interface of the traversing ASes. The path identifier is used to perform fair queueing of the request packets at the routers. However, sources residing further away from the congested link will suffer a significant disadvantage. *Portcullis* [196] deploys computational puzzles to provide per-computation fair sharing of the request channel. Such proof-of-work schemes, however, are too expensive to protect every data packet. Moreover, Portcullis does not provide the property of botnet-size independence. *Floc* [150] fair-shares link bandwidth of individual flows and differentiates between legitimate and attack

[8] `http://www.caida.org/data/as-relationships/`
[9] `http://data.caida.org/datasets/as-organizations/`

flows for a given link. However, such coarse-grained per-AS fair sharing may not always be effective; in particular, low-rate attack flows can often not be precisely differentiated. *CoDef* [151] is a collaborative defense mechanism in which a congested AS asks the source ASes to limit their bandwidth to a specific upper bound and to use a specific path. Source ASes that continue sending flows that exceed their requested quota are classified as malicious. CoDef does not prevent congestion in the first place, but instead responsively handles one congested link at a time. Since congestion can still occur on links, sources cannot be given a guarantee of reaching a destination. *STRIDE* [114] is a capability-based DDoS protection architecture that builds on several concepts from SCION. Although STRIDE shares similarities with SIBRA (steady paths and ephemeral paths), STRIDE lacks intra-core and inter-ISD communication guarantees; STRIDE's intra-domain guarantees are built on the assumption of congestion-free core networks. Moreover, STRIDE lacks monitoring and policing mechanisms, as well as an implementation.

Resource Allocation

Several queuing protocols [193, 222, 230] have been proposed to approximate fair bandwidth allocation at routers. Their correctness, however, relies on the trustworthiness of the routers and flow identifiers. The Path Computation Element (PCE) architecture [84, 244] computes inter-AS routes and enables resource allocation across AS boundaries in Generalized Multi-Protocol Label Switching (GMPLS) Traffic Engineered networks. However, the discovery of inter-AS PCE path fragments discloses information about other cooperating ASes, such as the internal topology. Some ASes will be reluctant to share this information due to confidentiality concerns.

Resource Reservation

RSVP [264] is a signaling protocol for bandwidth reservation. Because RSVP is not designed with security in mind, the reservation may fail due to DDoS attacks. RSVP requires the sender (e.g., a host or an AS when RSVP aggregation is used as specified in RFC 3175 [17]) to make an end-to-end reservation to the receiver(s), causing a large number of control messages in the network and large state on intermediate routers.

12 OPT and DRKey

ADRIAN PERRIG, RAPHAEL M. REISCHUK,
DOMINIK ROOS, PAWEL SZALACHOWSKI

This chapter presents *Origin and Path Trace (OPT)* — lightweight, scalable, and secure protocols for shared key setup, source authentication, and path validation. In-network source authentication and path validation are fundamental primitives for constructing higher-level security mechanisms such as DDoS mitigation, path compliance, packet attribution, or protection against flow redirection.

We also describe an extension called *Retroactive Path Trace* that helps the destination to perform path validation with retroactive key setup and to detect coward attackers with small, constant overhead in the packet header, which enables implementation on software routers with minimal performance impact.

We further introduce the *Dynamically Recreatable Key (DRKey)* protocol, an important mechanism for enabling routers to efficiently derive secret keys "on the fly" for any destination, without requiring per-destination state.

The present chapter is based on the paper "Lightweight Source Authentication and Path Validation" by Tiffany Hyun-Jin Kim, Cristina Basescu, Limin Jia, Soo Bum Lee, Yih-Chun Hu, and Adrian Perrig, which was published in the Proceedings of ACM SIGCOMM 2014 [132]. A formal analysis of the OPT protocols discussed in the chapter was completed by Zhang et al. [263].

Chapter Contents

A. Perrig et al., *SCION: A Secure Internet Architecture*, Information Security and Cryptography, https://doi.org/10.1007/978-3-319-67080-5_12

12.1 Introduction

Source authentication and path validation are useful primitives to help mitigate various network-based attacks, such as DDoS, address spoofing, and flow redirection attacks [63]. *Path validation*, in particular, provides a way to enforce path compliance according to the policies of ISPs, enterprises, and data centers. End hosts and ISPs desire to validate service level agreement compliance regarding data delivery in the network: Did the packet truly originate from the claimed client? Did the client select a path that complies with the service provider's policy? Did the packet indeed travel along the path selected by the client?

Unfortunately, the current Internet provides almost no means for source authentication and path validation by routers or end hosts, so numerous attack surfaces are opened up. For example, a malicious ISP may forward a packet on an inferior path while claiming to its client that it forwarded the packet on the premium path. Alternatively, a malicious router may inject packets with a spoofed source address to incriminate a victim source node as being responsible for having sent an excessive number of packets. A malicious router may simply alter the contents of received packets as well. The inability to detect such attacks near the point of deviation wastes downstream resources.

End-to-end encryption and authentication mechanisms, such as TLS, do not solve any of the above issues, since they are agnostic regarding which path the packet takes. A stronger approach is needed that enables routers and destinations to perform source authentication and path validation. Existing solutions either require extensive overhead, or only partially address fundamental problems, affecting both feasibility and practicality in the existing network. For example, ICING [182] addresses both source authentication and path validation, but it requires each intermediate router on a path to store and look up keys shared with other routers; ICING requires 42 bytes per verifying router in the packet header. Furthermore, ICING requires each router to calculate a message authentication code (MAC) for *all other routers on the path*.

In contrast, the OPT protocol suite does not require any per-client state on routers; it requires only 16 bytes per AS hop, and a single MAC in the packet header. The computational overhead for an on-path AS is a PRF computation, *irrespective of the path length*. Moreover, the Retroactive-PathTrace instantiation (Section 12.4.2) preempts coward attacks [161], where an adversary only attacks when it is certain that the attack will not be detected. The OPT protocol suite, however, offers reduced security in the case of a malicious sender colluding with a malicious AS on the path. Since in the common case sender and receiver trust each other, the performance gain of $\mathcal{O}(1)$ MAC operation per AS instead of $\mathcal{O}(n)$ is worth the tradeoff.

12.2 OPT Problem Definition

Desired Security Properties

- **Source authentication and data authentication:** The destination and each intermediate AS should be able to determine whether the packet indeed originated from the claimed source and whether the packet content has not been altered en route. In this chapter, we let *source authentication* include *data authentication*.
- **Path validation:** The source, intermediate ASes, and the destination should be able to validate that the packet has indeed traversed the path selected by the source. Successful path validation ensures that the packet has traversed each honest AS on the path in the correct order.[1]

Elided Security Properties

- **No packet delivery guarantee:** As each router can decide at any point whether or not to forward packets, it is not the purpose of path validation to guarantee that packets will be delivered to the specified destination.
- **No detection of packet siphoning:** A misbehaving AS_m on the source-selected path can siphon packets and send them over a separate channel to a remote entity. Since AS_m can still forward the packet to AS_{m+1}, this attack is not detected. We consider AS_m to be obeying the protocol as long as it performs all protocol-compliant operations with the packet.
- **No locating of packet-altering and packet-dropping routers:** Locating routers that alter or drop packets is not the goal of authentication or path validation; it is the goal of *fault localization* — another challenging problem especially in inter-domain settings [21, 265]. Since path validation is a simpler problem than fault localization, the goal of this chapter is to present a more efficient protocol than heavier-weight fault-localization protocols.

[1] Unfortunately, stronger properties are challenging to achieve efficiently (e.g., without utilizing trustworthy computing hardware on routers). Thus, a *malicious* AS could misbehave by sending a packet through an additional set of ASes, then removing all traces of these additional ASes, and finally sending the packet on the regular path toward the destination. Besides increased latency or additional traffic sent on the affected network links, such misbehavior does not leave any traces that are detectable by the end hosts. Also, *colluding* ASes can misbehave: if malicious ASes AS_m and AS'_m exchange their secret keys with each other, then they can perform seemingly valid cryptographic operations on behalf of each other. Consequently, one malicious AS can perform the cryptographic computations on behalf of the other and thus claim to have forwarded the packet in the place of the other.

Adversary Model

We consider a computationally bounded network attacker that deviates from the protocol and violates the protocol's security goals by means of one (or more) of the following attacks.

- **Packet alteration:** A malicious AS can alter any part of the packet, such as source address, header information, or the payload data.
- **Packet injection:** A malicious AS can fabricate packets and send them towards destinations of its choice. A *packet replay attack* is a special case of packet injection.
- **Path deviation:** A malicious AS may cause packets to be forwarded along a path other than the path previously selected by the source. We subdivide this attack as follows:
  - **Path detour:** Malicious AS_m causes a packet to deviate from the intended forwarding path, but the packet later returns to the correct downstream AS_{m+1} to resume traversal of *all* ASes on the intended path.
  - **Router skipping:** A malicious AS redirects the packet and skips other AS(es) on the path. Thus, some ASes on the intended path do not forward the packet.
  - **Out-of-order traversal:** An adversary causes path deviations such that ASes on the intended path are not traversed in the right order.
- **Coward attack:** A *coward attack* [161] is an attack that is launched by an adversary only when the adversary believes that the attack cannot be detected. For example, an attacker diverts traffic only when the protocol is inactive.
- **Denial-of-service (DoS) attack:** As part of DoS attacks, we consider memory and computation exhaustion attacks on routers performing source authentication and path validation.
- **Collusion:** Protocol participants may collude to carry out any of the attacks listed above. For example, two or more intermediate ASes may collude to claim the use of an expensive path for monetary profit, or the source may collude with an intermediate AS to spoof authenticators for its downstream ASes if the destination prefers/trusts skipped ASes. Also, both the source and the destination could collude with some intermediate ASes to frame another AS on the path by not forwarding packets to it.

In the paper [132] potential attacks against OPT as well as defense mechanisms are explored. Additionally, Zhang et al. have performed a formal analysis of the OPT protocol suite [263].

12.3 OPT Design Overview

We consider a setting in which source H_S in AS S sends a packet to destination H_D in AS D along a sequence of ASes AS_i. We refer to H_S, H_D, and the ASes AS_i as *tracing entities*.

One of OPT's crucial requirements is to avoid storing per-flow state on the intermediate routers; unlike prior approaches that require each router to maintain a secret key for each flow, our design enables routers to derive the secret keys on the fly using an efficient pseudorandom function in combination with local AS-level secrets that are stored at the routers.

In a nutshell, source authentication and path validation, without requiring routers to maintain per-source or per-flow state, are achieved as follows:

1. In the packet header, source H_S includes the hash of the packet payload $H(P)$ to help routers quickly verify or update OPT information, while avoiding a hash computation over the entire packet.
2. On demand, a router in AS_i generates a key $K^{\sigma}_{AS_i}$ using an efficient symmetric-cryptographic operation that requires as input only AS_i's local secret SV_{AS_i} and a special value called $\text{SESSIONID}_{\sigma}$, which the source has added to the packet header. Consequently, generating deterministic keys this way is not only stateless, but can also be faster than storing or retrieving secrets from main memory.[2]
3. Each AS_i extends a special authentication field in the packet header, the *Path Verification Field* (PVF), by performing one MAC operation using the key derived in the previous step. The resulting MACs are nested in a verification chain that is later used for path validation.

Not storing per-source or per-flow state on intermediate routers makes OPT robust against DoS attacks that are based on state exhaustion. Including the hash of the packet payload $H(P)$ in the packet header enables a second important optimization: intermediate routers can parallelize the computation of MACs and packet hashes, or sporadically validate $H(P)$.

12.3.1 OPT Protocol Overview

We provide a brief overview of the OPT protocol in this section to provide intuition on how the protocol works. In Section 12.4 a more detailed view of

[2]Computing an efficient pseudorandom function (PRF) is faster than fetching a byte from main memory: a deterministic key derivation using Intel's AESni takes around 50 cycles, whereas a main memory access requires on the order of 200 cycles.

the OPT protocol is provided. The underlying DRKey key derivation protocol is described in Section 12.5. The notation used in this chapter is summarized in Table 12.1.

We assume that the source and destination entities that perform the tracing of intermediate ASes can establish a secret key between themselves. Several approaches exist for setting up an end-to-end secret key: TLS if one of the end hosts is an HTTPS server, or through IPsec. SCION offers a shared symmetric key between end hosts (described in Section 12.5) to support such a key setup. Alternatively, trust on first use (TOFU) can be used in conjunction with SSH to set up a key, or self-certifying identifiers as public keys [7, 170, 179, 247], or self-validation using an anonymous communication service [93].

1. **Key setup:** OPT runs in *sessions*. In each session σ, source H_S selects a path $PATH_\sigma$ over which it intends to send packets to destination H_D. The source then generates a session with the identifier SESSIONID_σ.

   We assume that H_S and H_D share a secret key. To distinguish the key shared via an out-of-band mechanism such as TLS and the key shared via the DRKey infrastructure, we introduce two keys: $K^\sigma_{H_S}$ and $K^\sigma_{H_S H_D}$. Key $K^\sigma_{H_S}$ is the key shared out-of-band, which would not require trusting the DRKey infrastructure. Key $K^\sigma_{H_S H_D}$ is obtained via DRKey and derived as shown in Equation 12.5. If the DRKey infrastructure is fully trusted, then $K^\sigma_{H_S}$ can be equal to $K^\sigma_{H_S H_D}$.

   To obtain the keys $K^\sigma_{AS_i}$ for each AS on the path, H_S queries its local certificate server. We assume a mechanism for encrypted and authenticated communication between H_S and its local certificate server.
2. **Generation of verification fields:** H_S uses the path information to precompute for each AS_i on $PATH_\sigma$ an origin verification field OV_i. H_S generates an additional common path verification field PVF that is initialized using $K^\sigma_{H_S}$.
3. **Verification and update by intermediate routers:** The source inserts SESSIONID_σ into the OPT extension header of packets within session σ so that each intermediate AS_i on $PATH_\sigma$ can use the DRKey protocol to dynamically compute its symmetric key $K^\sigma_{AS_i}$ shared with H_S and H_D using SESSIONID_σ as shown in Equation 12.6.

   Using the symmetric key $K^\sigma_{AS_i}$ and PVF, each intermediate AS_i can recompute and validate its verification field OV_i — if it matches, AS_i has successfully validated the authenticity of the source and the content of the packet, and validated the path traversed so far. AS_i then updates the common PVF field in the header by applying a MAC operation with the corresponding key. This process helps subsequent ASes and the destination validate that each AS on the path has indeed processed the packet.

H_S	Source entity
H_D	Destination entity
K_A^σ	Symmetric key shared among H_S, H_D, and AS A for a single session σ, derived as shown in Equation 12.6
$K_{H_S}^\sigma$	Symmetric key shared between H_S and H_D for a single session σ, generated by H_S
$K_{H_S H_D}^\sigma$	Symmetric key shared between H_S and H_D for a single session σ, generated by AS as shown in Equation 12.5
$K_{A \to B}$	Symmetric key shared between AS A and AS B
$K_{A \to B}^p$	Symmetric key shared between AS A and AS B for protocol p
$K_{A \to B:H_B}^p$	Symmetric key shared between AS A and H_B in AS B for protocol p
$K_{A:H_A \to B:H_B}^p$	Symmetric key shared between H_A in AS A and H_B in AS B for protocol p
$K_{A \to B:H_B, C:H_C}^p$	Symmetric key shared between AS A, H_B in AS B and H_C in AS C for protocol p
SV_A	AS A's local secret value
SESSIONID_σ	Session identifier of session σ
$PATH_\sigma$	Session σ's path information
PVF	Field enabling H_D to verify the path
PVF^{H_S}	Field enabling AS_i and H_D to verify the path
PVF^{H_D}	Field enabling H_D to confirm the actual path
OV_i	Field enabling AS_i to validate the packet sender
OPV_i	Field enabling AS_i to verify both the packet sender and path
$PRF_K(\cdot)$	Pseudorandom function using key K
$MAC_K(\cdot)$	Message authentication code using key K
$H(\cdot)$	Cryptographic hash operation
P	Network packet payload
DATAHASH	Hash of the packet's payload (i.e., $H(P)$)

Table 12.1: Notation used in this chapter. The arrow notation in the subscript of keys indicates the secret value used to derive the key — the secret key associated with the entity on the left side of the arrow is used for key derivation. For all the keys with an arrow subscript listed in the table, AS A's secret value SV_A is used for their derivation. The detailed key derivation operations are explained in Section 12.5.

4. **Verification by the destination:** The destination finally recomputes the verification fields using all the symmetric keys shared with the ASes on the path. Successful verification indicates source and packet content authentication as well as path validation.

OriginValidation
DATAHASH (128 bits)
SESSIONID (128 bits)
OV_1 (128 bits)
OV_2 (128 bits)
⋮
OV_D (128 bits)

PathTrace
DATAHASH (128 bits)
SESSIONID (128 bits)
PVF (128 bits)

Figure 12.1: The packet header formats for *OriginValidation* (left) and *PathTrace* (right).

12.4 OPT Protocol Description

The DRKey protocols and the techniques we introduce in this section span a protocol family of source authentication and path validation with varying assumptions and properties. Unfortunately, exploring the entire design space is out of scope for this chapter; we thus present only three protocol instantiations:

1. **OriginValidation:** for source authentication by each AS (§12.4.1) (H_S and H_D trust each other)
2. **PathTrace:** for path validation by the destination (§12.4.2) (H_S and H_D trust each other)
3. **Origin and Path Trace (OPT):** for source authentication and path validation by each AS and the destination (§12.4.3) (H_S and H_D may not trust each other)

12.4.1 OriginValidation

Origin validation enables each intermediate AS and the destination to perform source authentication using MACs computed over the hash of the packet. For efficient authentication, the source includes the following fields in the packet header (see left-hand side of Figure 12.1):

- DATAHASH : a hash $H(P)$ of the packet payload P.
- SESSIONID : a value chosen by H_S.
- OV_i and OV_D : the origin verification fields, message authentication codes that the source creates for each intermediate AS and the destination. We let $OV_i = MAC_{K^{\sigma}_{AS_i}}(H(P))$ be computed over DATAHASH using key $K^{\sigma}_{AS_i}$ that AS_i shares with H_S; and similarly $OV_D = MAC_{K^{\sigma}_{H_S}}(H(P))$.

Origin validation provides efficient MAC verification using the DATAHASH field without requiring each intermediate AS to compute the hash over the entire packet.

When intermediate AS_1 receives a packet from source H_S, it computes the symmetric key $K^{\sigma}_{AS_1}$ that it shares with H_S using $\text{SESSIONID}_{\sigma}$ from the packet header and its local secret SV_{AS_1}. AS_1 then computes $MAC_{K^{\sigma}_{AS_1}}(\text{DATAHASH})$ and checks whether it is the same as OV_1 (as contained in the packet header). If so, AS_1 is assured that the packet indeed originated from the claimed source H_S, and forwards the packet to AS_2. The other intermediate ASes AS_i and the destination H_D perform similar operations.

12.4.2 PathTrace

PathTrace helps the source and destination validate that a received packet indeed traversed the source-selected path. This main objective is achieved by the path validation field (PVF), a packet header field containing a nested MAC that intermediate ASes update as they forward the packet. As indicated by the right-hand side of Figure 12.1, only the DATAHASH, SESSIONID, and PVF fields are used for PathTrace. Therefore, the packet overhead does not depend on the path length.

We next describe how PathTrace enables the source and the destination to validate the path. We start with the destination.

PathTrace for Destination

To enable *only* the destination to validate the path, the source generates the initial PVF value, PVF_0, which is a MAC of DATAHASH using the shared symmetric key between the source and the destination. The source initializes the PVF value in the packet header with the initial PVF_0:

$$\boxed{\text{PVF}} = \text{PVF}_0 = MAC_{K^{\sigma}_{H_S}}(\text{DATAHASH}) \tag{12.1}$$

Each intermediate AS_i on the path overrides the PVF field in the packet header with its own augmented PVF_i value. To this end, the AS first fetches the previous PVF value PVF_{i-1} from the packet header and then applies a MAC using its local key $K^{\sigma}_{AS_i}$. More precisely, the AS updates the PVF field in the packet header as:

$$\boxed{\text{PVF}} = \text{PVF}_i = MAC_{K^{\sigma}_{AS_i}}(\text{DATAHASH} \parallel \text{PVF}_{i-1}) \tag{12.2}$$

The symmetric key $K^{\sigma}_{AS_i}$ is shared with *both* the source and the destination according to the key setup protocol in Section 12.5. Hence, upon receiving a packet, the destination first recreates the nested MACs (here shown for a path

Origin and Path Trace

DataHash (128 bits)
SessionID (128 bits)
PVF (128 bits)
Timestamp (32 bits)
OPV_1 (128 bits)
OPV_2 (128 bits)
⋮
OPV_D (128 bits)

Figure 12.2: OPT header. Source H_S initializes all fields; intermediate ASes update only the PVF field.

of two ASes with keys $K^{\sigma}_{AS_1}$ and $K^{\sigma}_{AS_2}$, respectively):

$$\text{PVF}' = MAC_{K^{\sigma}_{AS_2}}(\text{DATAHASH} \parallel MAC_{K^{\sigma}_{AS_1}}(\text{DATAHASH} \parallel MAC_{K^{\sigma}_{H_S}}(\text{DATAHASH}))) \tag{12.3}$$

If PVF′ is the same as PVF in the packet header, the destination can be sure that the packet was indeed delivered on the source-selected path. Otherwise, the destination drops the packet.

PathTrace for Source

To help the source authenticate that its packet is delivered to the intended destination using the source-selected path, the destination forwards the final PVF value and hash from the received packet header back to the source, authenticated with a key shared between the source and destination:

$$D \rightarrow S: \quad \text{DATAHASH}, MAC_{K^{\sigma}_{H_S H_D}}(\text{DATAHASH} \parallel \text{PVF}) \tag{12.4}$$

Upon receiving this information, the source performs the validation by reconstructing the nested MACs using DataHash as shown in Equation 12.3 and by comparing it with the received value. A successful validation indicates that the packet was indeed delivered on the source-selected path to the destination.

Retroactive-PathTrace

Retroactive-PathTrace supports path validation without the apparent key setup process in advance. Instead, it utilizes Retroactive-DRKey (see Section 12.5), which can run *after* the session has started. This allows the source and destination to immediately start sending traffic without initial latency caused by the key setup.

Retroactive-PathTrace requires the destination to store information for each received packet to enable later checking. More precisely, the destination stores the tuple (SESSIONID$_\sigma$, DATAHASH, PVF) for each packet. When the destination wants to validate the path, it requests the source to initiate Retroactive-DRKey so that intermediate ASes reveal the keys that were used for the received packets. Then the destination can check the PVF fields and detect coward attacks. The source can independently initiate the retroactive process as well.

12.4.3 Origin Validation and Path Trace

In this section, we introduce OPT, which combines *OriginValidation* and *PathTrace* such that all entities (including intermediate ASes) on the path can perform both source authentication and path validation when they trust the source. We assume that all the ASes in a session are loosely time synchronized (within a few milliseconds), e.g., using NTP [175] or Roughtime [101]; see also Section 7.7 on Page 159.

Figure 12.2 illustrates the OPT header format. In addition to DATAHASH, SESSIONID, and PVF, an OPT header includes the following fields to enable each intermediate AS to perform path validation.

- TIMESTAMP : the time when H_S creates the OPT packet. This mitigates timing-based attacks (such as replay attacks).
- OPV$_i$: Origin and Path Verification field. OPV$_i$ is a MAC that enables all entities on the path to perform path validation.

The SOURCE INITIALIZATION function in Algorithm 9 on the next page describes how the source initializes the OPT header fields. Each OPV field includes the following as inputs.

- **Previous OPV:** Including OPV$_{i-1}$ in the computation of OPV$_i$ supports the detection of malicious intermediate ASes that forward the packet to a benign AS, which is not specified by the source but follows the protocol.
- **Previous AS identifier:** the OPV field by itself cannot support entities in detecting the packet injection attack. Hence, we include the identifier of the previous AS from which each entity receives the packet.
- **Timestamp:** This field mitigates authenticator cloning attacks. Consider an example where packet P_{crt} is expected to be sent along the source-selected path $PATH_{crt}$, the source previously sent packet P_{old} on $PATH_{old}$, and P_{crt} and P_{old} have the same payload. Consider AS AS_{bad} that is in both $PATH_{crt}$ and $PATH_{old}$ such that $PATH_{crt} = \{AS_1, AS_2, \ldots, AS_{bad}, AS_{bad+1}, \ldots, AS_n\}$ and $PATH_{old} = \{AS'_1, AS'_2, \ldots, AS_{bad}, AS'_{bad+1}, \ldots, AS_m\}$. In this scenario, AS_{bad} can replace $\{AS_{bad+1}, \ldots, AS_n\}$ with $\{AS'_{bad+1}, \ldots, AS_m\}$ in $PATH_{crt}$ and all the corresponding fields in the P_{crt} header with those in P_{old}. Therefore, without TIMESTAMP, the destination cannot detect the misbehavior and ends up validating path

Algorithm 9 OPT header initialization and validation pseudocode.

1: **function** SOURCE INITIALIZATION
Require: $K^{\sigma}_{AS_i}$ and $K^{\sigma}_{H_S}$, which AS_i's and H_D share with H_S, respectively after running key setup. SESSIONID_σ chosen by H_S.
2: $\boxed{\text{DATAHASH}} \leftarrow H(P)$
3: $\boxed{\text{SESSIONID}_\sigma} \leftarrow \text{SESSIONID}_\sigma$
4: $\boxed{\text{PVF}} \leftarrow \text{PVF}_0 = MAC_{K^{\sigma}_{H_S}}(\text{DATAHASH})$
5: $l =$ source-selected path length
6: **for** each intermediate AS_i, where $1 \leqslant i < l$ **do**
7: $\text{PVF}_i = MAC_{K^{\sigma}_{AS_i}}(\text{PVF}_{i-1})$
8: $\boxed{\text{OPV}_i} \leftarrow MAC_{K^{\sigma}_{AS_i}}(\text{PVF}_{i-1} \parallel \text{DATAHASH} \parallel AS_{i-1} \parallel \text{TIMESTAMP})$
9: **end for**
10: $\boxed{\text{OPV}_D} \leftarrow MAC_{K^{\sigma}_{H_S}}(\text{PVF}_{l-1} \parallel \text{DATAHASH} \parallel AS_{l-1} \parallel \text{TIMESTAMP})$
11: $\boxed{\text{TIMESTAMP}} \leftarrow$ current time
12: **end function**

13: **function** VALIDATION AND UPDATE BY AS_i
14: (Note PVF in OPT header = PVF_{i-1})
15: Compute $\text{OPV}'_i = MAC_{K^{\sigma}_{AS_i}}(\text{PVF}_{i-1} \parallel \text{DATAHASH} \parallel AS_{i-1} \parallel \text{TIMESTAMP})$
16: **if** $\text{OPV}'_i == \text{OPV}_i$ and TIMESTAMP not expired **then**
17: $\boxed{\text{PVF}} \leftarrow \text{PVF}_i = MAC_{K^{\sigma}_{AS_i}}(\text{PVF}_{i-1})$
18: Forward the packet to AS_{i+1}
19: **else**
20: Drop the packet
21: **end if**
22: **end function**

23: **function** DESTINATION VALIDATION
24: (Note PVF in OPT header = PVF_{l-1})
25: $l =$ source-selected path length
26: Compute $\text{PVF}' = MAC_{K^{\sigma}_{AS_{l-1}}}(\ldots(MAC_{K^{\sigma}_{AS_1}}(MAC_{K^{\sigma}_{H_S}}(\text{DATAHASH}))))$
27: Compute $\text{OPV}'_D = MAC_{K^{\sigma}_{H_S}}(\text{PVF}_{l-1} \parallel \text{DATAHASH} \parallel AS_{l-1} \parallel \text{TIMESTAMP})$
28: **if** $(\text{PVF}' == \boxed{\text{PVF}})$ **and** $(\text{OPV}'_D == \boxed{\text{OPV}_D})$ **then**
29: Validation succeeds
30: Prepare packet using Equation 12.4 and forward to source
31: **else**
32: Drop the packet
33: **end if**
34: **end function**

$\{AS_1, AS_2, \ldots, AS_{bad}, AS'_{bad+1}, \ldots, AS_m\}$ for P_{crt}. By setting the TIMESTAMP field when the source sends out a packet, authenticator cloning attacks are mitigated with loose time synchronization between the source and ASes on the path.

The function VALIDATION AND UPDATE BY AS_i and DESTINATION VALIDATION in Algorithm 9 describe the OPT procedure that an intermediate AS_i and the destination performs, respectively.

Distrusting Source and Destination

The previous protocols assume that the source and the destination are honest and trust each other. We now relax this assumption and present an extension that handles distrusting entities. In OPT, the source can generate all PVFs by itself since it knows all $K^{\sigma}_{AS_i}$'s. Consequently, a malicious source can collude with an intermediate AS (e.g., AS_2) and forward the packet on a path $H_S \rightarrow AS_2 \rightarrow H_D$ without going through AS_1.

To prevent such an attack and address the problem of a distrusting source and destination, we use the key setup protocol in Section 12.5 so that intermediate ASes generate two separate shared keys for the source and the destination. Unlike OPT, the Extended-OPT header requires two PVF fields: PVF^{H_S}, which enables intermediate ASes and the destination to validate the source, and PVF^{H_D}, which enables the destination to confirm the *actual* path, even if the source is malicious and colludes with (at least) one intermediate AS. More details are presented in our technical papers [132, 263] on OPT.

12.5 Dynamically Recreatable Keys (DRKey) 

When SCION border routers need to authenticate data-plane packets, highly efficient authentication mechanisms are needed to avoid opportunities for DoS attacks. For efficient high-speed data-plane processing, only simple operations can be performed, ruling out asymmetric cryptographic operations such as per-packet digital signature generation or verification. A tight processing budget only permits efficient symmetric cryptographic operations, such as the computation or verification of MACs — but a challenge remains on how to obtain the keys for the MAC computations.

Generating keys for all destinations would require much effort for the router, and would require excessive state — both aspects could be exploited to mount DoS attacks. For instance, Schuchard et al. show how exhausting router state can be used to mount attacks to paralyze the Internet [219].

Our approach is to offload the key setup to the AS's certificate server, and to enable SCION border routers to efficiently derive keys on the fly. As a result, SCION border routers can operate at high speed for data-plane operations, while the complex control logic is outsourced to the AS's services.

Efficient on-the-fly key establishment is useful in several SCION contexts. For example, due to the efficient key establishment on intermediate routers, SCION facilitates the first network control protocol (SCMP) that supports the authentication of network control messages (see Section 7.6 on Page 155 and Section 4.2.5 on Page 82). The naïve approach of adding digital signatures to control messages could create a processing bottleneck at routers when many SCMP messages are created in response to a link failure. Thus, efficient symmetric cryptographic keys are necessary and constitute an important building block for the efficient and authentic propagation of network control messages.

A second example is SCION's OPT protocol as described in this chapter. OPT not only benefits from an *efficient* key setup, but also from a *retroactive* setup: the retroactive nature of the key setup allows the key setup to happen on a path P even *after* the first data packet has been sent along P. This enables the detection of sophisticated attacks such as coward attacks [161], in which malicious entities try to eschew detection.

The remainder of this chapter describes the Dynamically Recreatable Key (DRKey) setup protocols, which enable routers to derive symmetric cryptographic keys on the fly from a single local secret. More precisely, an AS uses *one* local secret key (known only to SCION border routers and servers in that AS) to derive a symmetric key for another AS or end host on the fly (without keeping per-AS/end-host state) using an efficient pseudorandom function (PRF). Hardware implementations of modern block ciphers enable faster key computation than memory lookup from DRAM, and therefore such dynamic key derivation based on a single secret can even result in a speedup over fetching keys from main memory.

For the presented DRKey variants, we assume that each entity E has a public/private key pair (PK_E, PK_E^-), and that the public keys are correctly distributed to all other parties. We use the SCION control-plane PKI (Section 4.2) to obtain authentic AS public-key certificates. Moreover, we assume that each AS E has a local secret value SV_E, which is renewed regularly. The current implementation envisions a daily renewal.

12.5.1 DRKey Suite

Both OPT and SCMP require DRKeys to be shared between different entities. To facilitate the exchange, SCION introduces the DRKey suite, which provides a unified way of requesting and exchanging desired DRKeys. The certificate servers in each AS form the backbone of the DRKey infrastructure.

The first-order DRKey $K_{A \to B}$ is the basis for the entire DRKey suite as higher-order keys are derived from it. The key $K_{A \to B}$ is derivable by AS A and only depends on the secret value SV_A that is known only to infrastructure nodes in AS A.

$$K_{A \to B} = PRF_{SV_A}(B)$$

Since AS B also needs to know $K_{A \to B}$, we will later describe a key exchange protocol where a certificate server of AS B fetches the key from a certificate server in AS A. Similarly, AS A will need to fetch the key $K_{B \to A}$ from AS B. These first-order symmetric keys shared between A and B are then used to derive higher-order DRKeys. Certificate servers in each AS are responsible for all aspects of DRKey management, for instance also for distribution of second-order keys to local hosts.

Since the secret values SV_i change regularly (in the current version they are valid for 24 hours), certificate servers periodically prefetch first-order keys with all other ASes. With an increasing number of SCION ASes, an AS would prefetch the most commonly used keys, and adopt a lazy key exchange approach for infrequently used keys.

To fetch $K_{A \to B}$, certificate server CS_B in AS B starts the key exchange with a certificate server CS_A in AS A by sending the request

$$\begin{aligned} signature = \; & \{A \parallel val\_time \parallel timestamp\}_{PK_B^-} \\ CS_B \to CS_A \; : \; & A, B, val\_time, timestamp, signature \end{aligned}$$

where *timestamp* denotes the current time, and *val_time* specifies a point in time at which the requested key is valid. The requested key may not be valid at the time of request, either because it is already expired or because it will become valid in the future.

If the request has a recent timestamp, CS_A replies with the encrypted DRKey, where *exp_time* denotes the actual expiration time of the key:

$$\begin{aligned} K_{A \to B} = \; & PRF_{SV_A}(B) \\ ciphertext = \; & \{A, K_{A \to B}\}_{PK_B} \\ signature = \; & \{ciphertext \parallel exp\_time \parallel timestamp\}_{PK_A^-} \\ CS_A \to CS_B \; : \; & ciphertext, exp\_time, timestamp, signature \end{aligned}$$

After CS_B has received $K_{A \to B}$, it is shared among all certificate servers in B to ensure a consistent view of the shared keys. Each CS can now answer requests by local end hosts in the AS.

To avoid explicit key revocation, the shared DRKeys are short-lived. Each key has a validity period, which is based on the lifetime T of the secret value. Certificate servers prefetch keys ahead of the expiration of the current key, to allow for seamless transition to the next key. To avoid a key request storm, the requesting certificate servers randomize their prefetch times.

Second-order DRKeys are derived from first-order DRKeys. End hosts can request second-order DRKeys from their local certificate servers. A secure channel between the end host and the certificate server is assumed, to ensure authenticity and secrecy of the request and response messages. The derivation and availability of the second-order DRKeys is specific to the protocol.

However, the certificate server provides a standard key request mechanism for any protocol. A request has the form $\{type, requestID, protocol, source, destination, additional$ (optional)$\}$. Following are the requests corresponding to the second-order DRKeys:

DRKey	request
$K^p_{A\to B}$	$(0, requestID, p, A, B, \_)$
$K^p_{A\to B:H_B}$	$(1, requestID, p, A, H_B, \_)$
$K^p_{A:H_A\to B:H_B}$	$(2, requestID, p, H_A, H_B, \_)$
$K^p_{A\to B:H_B,C:H_C}$	$(3, requestID, p, A, H_B, H_C)$

The exchange is initiated by the end host H_A in AS A. It creates a request according to the desired key and sends it to a local certificate server CS_A

$$H_A \to CS_A \;:\quad request, val\_time, timestamp$$

where val_time denotes a point in time at which the requested key is valid. Similarly as above, the requested key may not be valid at the time of request, either because it is already expired or because it will become valid in the future. The timestamp is used to prevent message replay attacks.

CS_A verifies that the request is valid and not expired. Additionally, it checks that H_A is authorized to request the key. This is only the case if either *sender* or *destination* is equal to $A : H_A$. On success, CS_A computes the requested key K for protocol p and responds with the requested key K and its associated expiration time:

$$CS_A \to H_A \;:\quad requestID, K, exp\_time, timestamp$$

12.5.2 DRKey Derivation Protocol for OPT

OPT requires that for source H_S in AS S and destination H_D in AS D: (a) each AS_i on the path shares symmetric key $K^\sigma_{AS_i}$ with both H_S and H_D, and (b) H_S and H_D directly share symmetric key $K^\sigma_{H_S H_D}$. To accomplish this, the DRKey suite is used. The key setup is done independently for each direction of communication and comes with the *retroactivity* property: H_S and H_D can request keys shared with the intermediate ASes to enable subsequent path validation even *after* communication has begun.

The DRKeys used in OPT can be computed by certificate servers in AS S. To minimize the number of requests, the source H_S first creates a path descriptor $PATH_{H_S\to H_D} = \langle H_S, AS_1, AS_2, \ldots, H_D\rangle$ for the source-selected path from H_S to H_D. The source H_S sends an OPT DRKey request to its local certificate server

CS_S, where *val_time* and *timestamp* are as before:

$$H_S \rightarrow CS_S: \quad \text{SESSIONID}_\sigma, PATH_{H_S \rightarrow H_D}, val\_time, timestamp$$

The request is handled by the local certificate server CS_S by first checking the timestamp and the value of H_S in $PATH_{H_S \rightarrow H_D}$. On success, CS_S derives and returns all keys, along with their expiration time (the expiration times of all ASes are aligned):

$$K^{opt}_{S:H_S \rightarrow D:H_D} = PRF_{K_{S \rightarrow D}}(\text{"}OPT\text{"} \parallel H_S \parallel H_D)$$

$$K^{opt}_{AS_i \rightarrow S:H_S,D:H_D} = PRF_{K_{AS_i \rightarrow S}}(\text{"}OPT\text{"} \parallel H_S \parallel H_D)$$

$$K^{\sigma}_{H_S H_D} = PRF_{K^{opt}_{S:H_S \rightarrow D:H_D}}(\text{SESSIONID}_\sigma) \quad (12.5)$$

$$K^{\sigma}_{AS_i} = PRF_{K^{opt}_{AS_i \rightarrow S:H_S,D:H_D}}(\text{SESSIONID}_\sigma) \quad (12.6)$$

$$CS_S \rightarrow H_S: \quad K^{\sigma}_{H_S H_D}, K^{\sigma}_{AS_1}, K^{\sigma}_{AS_2}, \ldots, exp\_time, timestamp$$

After H_S has received the list of DRKeys, the key exchange with destination H_D can be initiated. The list is encrypted and authenticated using $K^{\sigma}_{H_S H_D}$. Note that this key can be fetched by H_D from its local certificate server CS_D without interaction with AS S.

$$ciphertext = \{K^{\sigma}_{AS_1}, K^{\sigma}_{AS_2}, \ldots\}_{K^{\sigma}_{H_S H_D}}$$

$$mac = MAC_{K^{\sigma}_{H_S H_D}}(\text{SESSIONID}_\sigma \parallel ciphertext \parallel timestamp)$$

$$H_S \rightarrow H_D: \quad \text{SESSIONID}_\sigma, ciphertext, timestamp, mac$$

Both H_S and H_D are now in possession of all DRKeys in the direction from H_S to H_D. The key setup for the other direction works analogously.

Running this key setup before starting the communication between H_S and H_D will result in increased latency. However, the setup can be run retroactively, such that the keys are deterministically chosen and can be requested even after the communication has started. As described in Section 12.4.2, there is a retroactive instantiation of PathTrace, which will directly benefit from this property.

Attentive readers might notice that the presented key setup differs from the OPT paper [132]. There are multiple advantages over the original design: Introducing two tiers of keys allows derivation of all DRKeys by the local certificate server. This reduces latency and stress on certificate servers in core ASes.

A protocol supporting a distrusting source and destination can be implemented similarly. More details on this point can be found in the original OPT paper [132].

12.5.3 DRKey Derivation Protocol for SCMP

Authenticating network control messages is an important component of a secure Internet architecture. To prevent DoS attacks on SCION infrastructure devices, the generation of authenticated SCMP messages needs to be efficient. Using DRKey, SCION infrastructure devices can use efficient symmetric cryptography without the need for key lookups. End hosts benefit from the added security of authenticated SCMP messages, but have to perform an inexpensive key lookup on the local certificate server in case the validation key is not locally cached.

SCMP messages can be sent either by infrastructure nodes or by end hosts. To prevent attacks, infrastructure nodes use a DRKey, which can be derived using only the local secret value. This entails different types of SCMP authentication keys, depending on the creating and verifying entity.

SCMP messages created by an AS S addressed to another AS D are authenticated using the DRKey $K^{scmp}_{S\to D}$. It is only dependent on $K_{S\to D}$, which can be dynamically computed by S:

$$K^{scmp}_{S\to D} = PRF_{K_{S\to D}}(\text{“}SCMP\text{”})$$

SCMP messages created by an AS S addressed to host H_D in AS D are authenticated using the DRKey $K^{scmp}_{S\to D:H_D}$. This also is only dependent on $K_{S\to D}$ and can be dynamically computed by S. To verify, end host H_D has to request the key from its local certificate server as described in Section 12.5.1, if it is not present. Additionally, H_D sending an SCMP message to S will use the same key as well. This allows for fast verification on the AS side.

$$K^{scmp}_{S\to D:H_D} - PRF_{K_{S\to D}}(\text{“}SCMP\text{”} \parallel H_D)$$

SCMP messages created by end host H_S in AS S designated for end host H_D in AS D are authenticated using the following key:

$$K^{scmp}_{S:H_S\to D:H_D} = PRF_{K_{S\to D}}(\text{“}SCMP\text{”} \parallel H_S \parallel H_D)$$

Both H_S and H_D have to request this key from a local certificate server. The induced overhead is compensated by the security gain that is achieved by authenticating SCMP messages.

As we envision local networks using network address translation (see Section 10.8.2), the addresses used in SCMP packets may be modified by NAT devices, causing authentication failures. To address this issue, a NAT device has to re-authenticate an SCMP packet by replacing the original authentication information with new information created with a new key.[3] To enable this, certificate servers are configured to return keys for local IP ranges [210] with loosened access control rules. The exact details need to be worked out, if NAT turns out to find adoption in SCION.

[3] If the SCMP message contains the address and layer-4 headers, the NAT device has to rewrite them as well.

12.5.4 Optimizations

As border routers need to be able to create SCMP messages at line rate, the DRKey derivation has to be efficient. Thus, a hardware implementation of AES can be used for the pseudorandom function. However, the input to derive $K^{scmp}_{S \to D:H_D}$ does not fit into one AES block if H_D has an IPv6 address. To address this issue, we introduce a separate secret value SV^{scmp}_S into the DRKey suite, which is used to derive $K^{scmp}_{S \to D}$ directly. $K^{scmp}_{S \to D}$ is turned into a first-order key and is shared alongside the original first-order key $K_{S \to D}$ between certificate servers. Furthermore, the second-order key is redefined as follows:

$$K^{scmp}_{S \to D:H_D} = PRF_{K^{scmp}_{S \to D}}(H_D)$$

Consequently, the input to derive $K^{scmp}_{S \to D:H_D}$ does now fit into one AES block even if H_D has an IPv6 address. Thus only one AES block operation instead of two is needed for this derivation step.

Part IV

Analysis and Evaluation

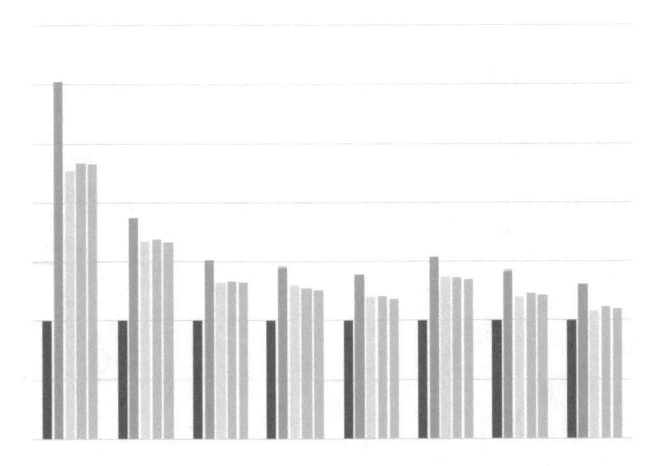

13 Security Analysis

DAVID BARRERA, TOBIAS KLENZE, ADRIAN PERRIG,
RAPHAEL M. REISCHUK, BENJAMIN ROTHENBERGER,
PAWEL SZALACHOWSKI

Evaluating the security of a network architecture, including its routing protocols and supporting infrastructure, is an ambitious undertaking. Indeed, the security guarantees afforded by the architecture, as well as the security of the architecture itself, depend on a number of factors. For example, should we assume correct implementation and configuration of the protocol at all deploying nodes? Do we consider an adversary that can eavesdrop on a large portion of network links? Can malware on end hosts send arbitrary traffic?

Even considering the multitude of factors, the vulnerability of one service or component in the architecture may impact the security of another. It is therefore important not only to analyze all major components independently, but also to analyze their interactions.

This chapter aims to shed light on the overall security of SCION. We explain how the architecture defends against *existing* network attacks (e.g., DDoS or source address spoofing), and how its design is resilient to *new* attacks that become possible in the SCION architecture. The comparison with today's Internet is slightly unbalanced as many of today's flaws and attacks (for instance with BGP and DNS) are not possible in SCION and will thus not be discussed in this chapter. The attacks that we will discuss, however, are mostly *not* possible in today's Internet. This mismatch could create the illusion that SCION enables more attacks than the current Internet. However, for each SCION attack we mention, we also provide either an argument why the attack is not possible, or we show how the attack can be countered. In the current Internet, countermeasures for many of the ongoing attacks are either less likely to be deployed or lead to further problems themselves. In the end, SCION offers dramatically improved security compared to today's Internet.

In this chapter, we analyze the security of the control and data planes, as well as the security of supporting infrastructure elements. The analysis does *not* cover the extensions SIBRA, RAINS, OPT, or DRKey, nor the PKIs or ISD coordination. The security of these systems is discussed in their respective papers. As a methodology, we assume an attacker's point of view and emulate

A. Perrig et al., *SCION: A Secure Internet Architecture*, Information Security and Cryptography, https://doi.org/10.1007/978-3-319-67080-5_13

a real-world adversary to analyze the design specification and source code of SCION. For each attack, we describe the adversary's goal and how SCION can defend against the attack. We also discuss several *doomsday* scenarios where private keys for critical services are leaked, or powerful state-level adversaries attempt to monitor, or even disable, global and local communication.

While simplicity was one of the design goals of SCION, protocol design is an error-prone process and an informal analysis alone cannot provide the same assurance as formal proofs. We thus aim to formally verify security properties at the protocol level. Some components of SCION have already been verified, such as ARPKI and OPT. We refer interested readers to the research papers [23, 24, 132], as reproducing the results here would go beyond the scope of this book. The verification of the core protocol (beaconing and data-plane forwarding) is still underway and does not feature in this book. Similarly, we are currently in the process of verifying the implementation, so we only briefly report on our approach in Section 13.3.

Besides our own analysis, we refer the reader to evaluations of SCION by third and independent parties. Ding et al. [70] have evaluated five popular future Internet architectures (FIAs) with a particular focus on the security properties provided. After comparing the five FIAs, the authors conclude that SCION offers more security properties than the other four (see Section VII and Table 2 of the study [70]).

Chapter Contents

13.1 Security Goals

There are two groups of actors in SCION, and they have their own security goals. The first are autonomous systems (ASes), which have routing policies based on business contracts and technological constraints. The second are end hosts, either machines offering certain services (e.g., web servers) or clients making use of those services. In this section, we briefly describe the security properties that SCION should provide for both groups.

13.1.1 Overall Security Goals

Besides the particular concerns of actors, some security goals apply to the entire architecture. These are resilience to failures, availability, and support for a heterogeneous but global trust environment.

Availability is one of the most important security goals in today's Internet, and is of concern to both ASes and end hosts in SCION. We show how SCION achieves a high level of availability, and provides tools to protect against denial-of-service attacks.

Today's heterogeneous trust environment is reflected in SCION through the concepts of isolation and transparency. Routing within an ISD is independent of other ISDs. Nevertheless, global connectivity is achieved for packets that need to traverse ISD bounds.

13.1.2 Autonomous Systems

The overarching goal of ASes is to reliably and effectively achieve network connectivity to other entities connected to the Internet, and in case of ISPs, provide connectivity to their customers. In the control plane, an AS wants to verify routing information it receives and ensure that routing information it propagates cannot be altered by malicious entities. In the data plane, an honest AS will forward packets only along locally valid segments. We define a segment as locally valid if the hop fields used for forwarding in the AS (the hop field leading to its provider, plus additional hop fields in the case of peering or change of segment) correspond to a PCB in the control plane.

13.1.3 End Hosts

The security goals of end hosts are diverse and manifold. Many of them should be addressed on different levels of the network stack. We only discuss the properties that are important at the network layer. End hosts have the following control-plane security goals:

- **Reachability:** The common case is for hosts to be reachable by any other host on the Internet and be able to reach other hosts.
- **Path diversity:** As disjoint paths improve availability, there should be a diverse set of paths available to choose from.

The following security goals concern the data plane:

- **Truthful forwarding:** The path selected by the source is the path traversed during packet forwarding.
- **Path transparency:** End hosts should know or be able to infer the path that the packet takes.
- **Packet integrity:** Receiving hosts should be able to verify that a packet, including its path, is the same as the one sent by the source.
- **Source authentication:** The receiving host should be able to authenticate the origin of a packet.
- **Weak and strong detectability:** An on-path attacker, as an intermediate node, cannot effectively disguise his own presence on the path, even when changing the path information in the packet's header. This property is strong when the destination is able to tell whether a disguise took place. Localization of the attacker is not required under this definition.

We note that packet integrity, source authentication, truthful forwarding, and strong detectability are only provided when using the OPT extension described in Chapter 12.

13.2 Threat Model

In the context of the control and data planes, we consider the following threat model. We assume that there exist hostile participants at arbitrary locations within the network. The adversary can not only passively eavesdrop on messages, but also actively tamper with the communication, i.e., drop, delay, or alter packets that it should forward, or inject packets into the network. For almost all attacks we will look at the network topology on an AS level, and assume that the adversary has compromised entire ASes (as opposed to just certain routers within an AS). All hostile nodes (ASes) are assumed to share a channel for information exchange outside of the current network. When discussing cryptographic primitives, we assume that the adversary is computationally bounded and has no efficient way of breaking cryptographic primitives.

Regarding SCION-specific capabilities, we assume the adversary is able to register as an AS with the ISD core and perform regular operations. However, we expect registration operations in the core to be throttled and visible to other nodes within the ISD. In particular, we assume a mechanism that prevents large numbers of malicious ASes from joining the ISD rapidly or automatically. For increased security, we expect that for each new AS registration, the necessary amount of due diligence and verification is performed by core ASes, as we explain in Section 4.2.3.

When an AS acts maliciously, we assume that the adversary can eavesdrop on all control and data messages traversing the AS. By compromising an AS, the adversary learns all cryptographic keys and settings. He can also control how the AS behaves including redirection of traffic, fabrication, replay, and modification of packets. Data signed by the AS will be valid until the corresponding certificates for that AS expire or are revoked.

Assuming the above-mentioned capabilities, the goal of the adversary is to prevent communication availability of other ASes, or eavesdrop on traffic that is not traversing a malicious AS.

13.3 Software Security

A critical aspect of security in a system is software security. Exploiting software defects allows a malicious entity to take over entire systems. Recently, the "Vault 7" revelations by WikiLeaks disclosed that over 300 types of Cisco switches are vulnerable to remote code execution [55], and MikroTik's routers can be exploited to gain root privileges [251]. Another example is Juniper Networks discovering unauthorized backdoors in its firewalls such that an external adversary could decrypt traffic encrypted by the firewall [48, 246, 261, 262]. These examples show that beside programming errors, rogue software developers must also be considered.

In this section, we discuss software verification and analysis as a tool to detect vulnerabilities in the SCION codebase. Further, we outline how code correctness is ensured in the SCION implementation.

Threat Model

For software security, we consider a more specific threat model and assume the following adversary capabilities. Since the attacker has full access to the public source code, he can therefore identify possible flaws in the implementation. He can run his own SCION nodes and infrastructure services (e.g., a beacon server), incorporate entire ASes and participate in the SCION network. The adversary is also aware of other publicly available exploits to compromise end hosts. By exploiting an end host, the adversary gets full control over that host and learns all private keys stored on the host.

Verification Techniques

We use static and dynamic analysis to find implementation flaws and errors that have not been detected by our unit and integration tests. Such code analysis tools are designed to analyze source code or compiled versions of code in order to find security flaws. The tools contribute to a secure codebase by identifying security-relevant portions of code, but are not enough to claim that code is secure. The SCION codebase has been analyzed using state-of-the-art analysis software. All detected issues have been fixed. Dynamic testing is an automated technique based on invalid, unexpected, or random input data, which we use to reveal unexpected behavior such as failing assertions, crashes, or memory leaks. We employ a mutation-based approach by altering existing data samples to create test data. We will continue to adjust our test suite as the codebase matures.

Apart from software analysis, we are working on formal verification techniques to ensure the correctness of SCION components. We have started with the verification of the SCION border router code and will expand our techniques to other components such as certificate, beacon, and path servers.

Strengths of SCION Regarding Software Security

Regarding software security and correctness, SCION offers several advantages compared to the current Internet. We discuss two specific advantages.

First, the design of SCION is publicly available and described in detail in this book and in various scientific papers [20, 22, 49, 266]. Moreover, the implementation of SCION is also freely available under the Apache License Version 2.0 [89]. Interested researchers, developers, network operators, governments, non-profit organizations, and other stakeholders can thus inspect SCION's specification and source code in detail to convince themselves of its correctness and backdoor-freeness. This is a tremendous benefit in comparison to today's Internet, as, in particular, the semantics of commercial routers is opaque in its details, and correct behavior cannot easily be verified. By allowing a wide public to see and test the code, we hope to publicly expose any flaws and potential backdoors.

The second benefit stems from SCION's simple data plane, leaving complex operations to the control plane. For a high-speed network architecture, an efficient data plane is critical — consequently, a simple data plane is beneficial to facilitate construction of efficient code. In fact, SCION's border router implementation in the Go programming language is about 10,000 lines of code. By comparison, modern network operating systems, such as CISCO's IOS or Juniper's JUNOS, are believed to have millions of lines of code [53, 73] as they implement numerous different protocols as well as control-plane operations. The SCION control plane is implemented as services. Since their operation is not as performance-critical as the data plane, the services can be implemented in higher-level languages that facilitate simple development. The current version of SCION utilizes Python as the programming language for the services, but more performance-critical services, such as the path servers, are planned to be implemented in the Go programming language.

13.4 Control-Plane Path Manipulation

Path manipulation attacks try to alter the path chosen by a sender. The goal of the adversary is to attract traffic towards himself, towards another target, or to manipulate the paths available to and chosen by hosts in some other way (possibly without the sender or receiver noticing). For instance, he may want to attract downstream traffic for financial gain (since as a provider, he will be compensated based on traffic volume) or to perform man-in-the-middle attacks on transit traffic. On the control plane, he can attempt to attract traffic by selectively disseminating PCBs or by forging new PCBs. As shown in the following examples, path manipulation attacks either have limited impact or require efficiently breaking the cryptographic primitives used in SCION.

In today's Internet, researchers and network operators are noticing that the Border Gateway Protocol (BGP) has numerous shortcomings [25, 117, 216, 219] and especially lacks integrity protection for routing update messages. Maliciously acting routers can advertise IP prefixes from address spaces that are unused or that belong to other ASes, and effectively redirect traffic to hosts under the control of the attacker. To address these problems, specific improvements (such as BGPsec) have been proposed [3, 44, 100, 116, 157, 158, 248], although none of these improved protocols have seen widespread deployment.

SCION control-plane messages, unlike those of BGP, are authenticated and integrity protected, paving the way for solving issues such as path hijacking. Moreover, SCION's routing plane converges instantaneously, in stark contrast to BGP whose iterative refinement of routes can result in slow convergence. As such, recovery and path discovery times are shorter and more predictable in SCION.

In the following, we examine several approaches to manipulating paths in the SCION control plane, and show for each case how SCION's design prevents the corresponding attack, or helps to mitigate it.

13.4.1 Path Hijacking Through Interposition

To become on-path, an adversary might try to manipulate the path creation or beaconing process. More precisely, as illustrated in Figure 13.1, provider AS A sends two beacons to customer ASes B and M. A malicious AS M who can eavesdrop on links between AS A and AS B could try to intercept and disseminate the "better" beacon meant for B by injecting its own hop fields into the PCB toward downstream ASes. This could offer B an attractive up-path traversing M to the core.

Similarly, assuming that the attacker has control over the path between A and B, he can block downstream PCBs from A such that the malicious AS M can position itself as the only upstream path to the core for B.

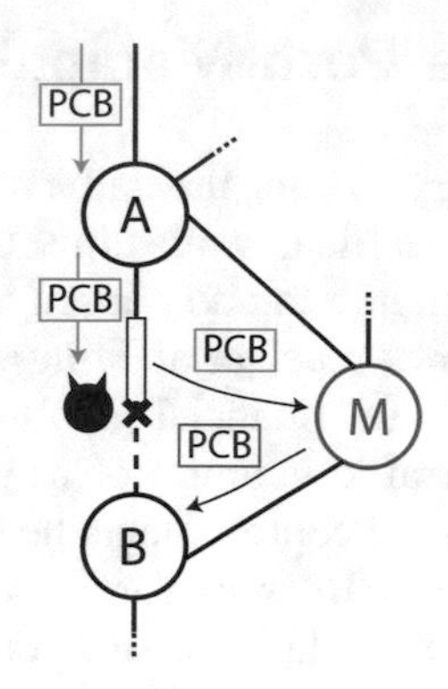

Figure 13.1: Interposition attack.

The attack is detectable by downstream ASes, because the PCBs disseminated by A towards B contain B as an egress AS identifier. Therefore, verification of inbound PCBs will fail, because the adversary's PCBs are not signed with the expected key.

Assuming that an adversary wants to interpose an AS by modifying an already existing path, he would need to modify the corresponding hop fields. As hop fields are integrity protected and include the previous hop field in the MAC calculation, malicious modifications of hop fields are prevented. However, if the adversary can block the traffic between A and B, then indeed he can force traffic redirection through M. This attack is fundamental and generally cannot be prevented.

13.4.2 Creation of Spurious ASes

An adversary could try to spoof other ASes by introducing nonexistent entities. If he succeeds and traffic is sent upstream with the spoofed entity as a source, the traffic will appear to originate downstream from beyond the malicious AS. This allows the adversary to plausibly deny the misbehavior and complicate detection of this attack.

However, this attack is difficult to execute, because spoofing a new AS requires a registration of that AS with the ISD core. However, if a malicious AS M obtains a valid certificate, we cannot prevent it from announcing malicious paths traversing M. Therefore, each AS needs to be checked thoroughly during the registration process. In case the malicious AS M does not obtain a valid certificate, the adversary cannot craft valid PCBs and HFs.

Similarly to creating a fake AS, if an adversary wanted to introduce a new ISD (possibly spoofed), it would need to generate its own TRC, and all cross-verifying ISDs would need to verify its legitimacy (see Chapter 5).

13.4.3 Peering Link Misuse

Downstream beacons may be recorded by on-path attackers (e.g., eavesdrop on a link between ASes). By re-injecting that beacon into another link, the adversary can extend paths as long as the beacon is correctly forwarded.

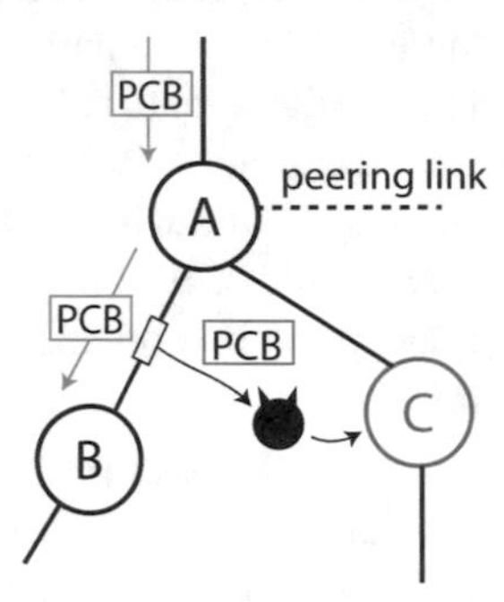

Figure 13.2: Beacon theft. AS *A* wants to selectively share access to the peering link with *B*, but not with AS *C*. An eavesdropping adversary reads the beacon intended for AS *B* and re-injects it at his own AS *C* to gain access to the peering link.

Consider the example in Figure 13.2. AS *A* wants to share its peering link only with one of its downstream neighbors, *B*, and therefore decides to selectively announce the peering link in PCBs sent to *B*. The monitoring adversary misuses this beacon to gain access to the peering link by prepending it to his own path. Apart from eavesdropping on the link, the adversary is able to obtain the necessary hop fields by querying a path server and extracting them from registered paths.

SCION successfully mitigates this attack by including specific "next hop" information in the PCB before disseminating it further downstream (see Equation 7.6). Furthermore, each hop field contains an egress interface. If a malicious entity tries to misuse a stolen PCB by adding it to its own segments, verification will fail upstream as the egress interface mismatches. Therefore, the peering link can only be used by the intended AS.

13.4.4 Manipulation of the Path Selection Process

Path selection is one of the main benefits of SCION compared to the current Internet, where hosts have no control over the forwarding paths that their packets traverse. With the benefits of freedom regarding path selection, however, comes the risk for hosts to choose non-optimal paths. In this section, we demonstrate how an attacker can trick hosts downstream into choosing non-optimal paths. We argue that, compared to the current Internet, the path selection in SCION offers higher security overall since (a) *path transparency* enables a path-selecting end host to identify the potentially malicious ASes on the path, and (b) *path control* enables the host to avoid such malicious ASes, even if the cost of such attacker-free paths appears higher.

In SCION, path selection is used in three cases. First, a beacon server selects which PCB to announce downstream. Second, the beacon server chooses which paths it wants to register at the local and core path servers. Third, the end host performs path selection from all available path segments. We now describe path selection attacks (or path preference attacks), which aim at influencing the path selection process in SCION. The goal of such attacks is to make paths that are controlled by the attacker more attractive than other available paths. A simple example is a low or even negative price in a pricing system, or announcing high bandwidth and low latency for a path.

The following attacks are only successful if the attacker is located within the same ISD and upstream relative to the victim AS. It is not possible to attract traffic away from the core as traffic travels upstream towards the core. Furthermore, the attack may be discovered downstream (e.g., by seeing large numbers of paths become available), but also during path registrations. After detection, paths traversing the adversary AS can be identified and avoided by regular ASes.

Fake Up-Link Announcement

In a fake up-link announcement attack, the attacker receives PCBs from its upstream provider and announces the PCBs downstream (see Figure 13.3). If such an adversary-announced path complies with the policy of the downstream ASes, the corresponding fake PCBs may be added as one of the k paths available to the end host. At this point, the end host may select such a fake link for communication.

The attacker controls a set of consecutive ASes and can also advertise fake links as possible upstream links to downstream entities. The advertised paths may selectively be crafted with good properties such as high bandwidth, long lifetime, low price, or any other desirable path property. This increases the chance for such a bogus path to be selected by the communicating parties. In Figure 13.3, an attacker floods $m \times n$ path segments, which become available to D, and possibly to the victim.

While such bogus paths can have some desirable properties, they will need to traverse at least three malicious ASes. As there might exist other shorter paths, this decreases their chance of being chosen by a downstream server for PCB dissemination or by a host for construction of a forwarding path. Furthermore, without creating fake ASes, there would be no AS-level diversity.

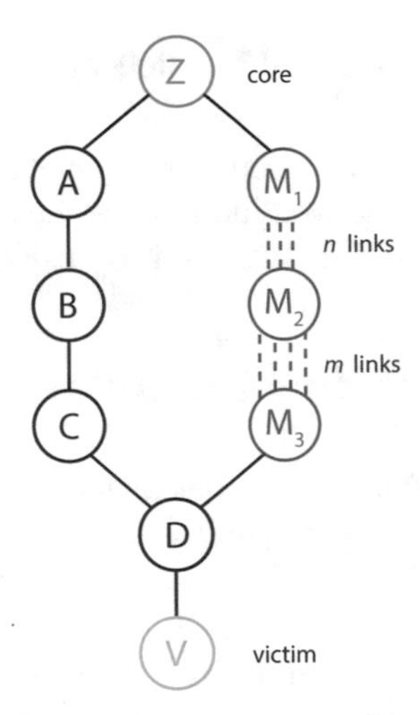

Figure 13.3: Fake up-link announcement. Malicious ASes M_i advertise $m \times n$ path segments to increase the chance of selection by victim V.

Wormhole Attack

A malicious node M_1 can send a PCB not only to his customers, but also out of band to another, colluding malicious node M_2. This creates new segments to M_2 and his customers, which may not correspond to actual paths in the network topology. Similarly, a fake path can be announced through a fake peering link and attract traffic even across ISDs. Without specific prevention mechanisms, such a wormhole attack is unavoidable in routing [115]. To detect wormhole attacks, latency measurements with per-link timestamps are one potential approach. Each ISP would announce the latency of links in the PCB. In combination with a timestamping extension, this would help reveal the wormhole.

Fake Peering Link Announcement

As an instance of a wormhole attack, an adversary advertises fake peering links, thus offering short routes to many different destination ASes within and outside its own ISD. Downstream ASes will likely have a policy of preferring paths with many peering links and thus are more likely to disseminate PCBs from the adversary. Similarly, hosts are more likely to choose short routes that make use of peering links. However a peering link can only be used if the neighboring AS also announces it. If the attacker is colluding with an external AS, a wormhole becomes possible. On the data plane, whenever a packet containing a fake peering link is received by the adversary, he can transparently exchange the fake peering link hop fields with valid hop fields to the colluding AS (see Section 13.5). To avoid detection of the path alteration by the receiver, the colluding AS can replace the added hop fields with the fake peering link hop fields the sender inserted.

To defend against this attack, methods to detect the wormhole attack are needed [115]. As discussed above, link latency measurements can help reveal the wormhole and render the fake peering link suspicious or unattractive.

13.5 Data-Plane Path Manipulation

Besides manipulating the routing decisions on the control plane, adversaries can also attempt to influence forwarding in the data plane. Because the forwarding path selection has already been made in the control plane, an off-path attacker is limited in his abilities. The adversary can merely attempt to disrupt the connectivity of the chosen path and force the host to select a new path. We discuss availability attacks in Section 13.7 and concentrate on the case of an on-path attacker in this section.

To differentiate these attacks from path manipulation attacks in the control plane, we assume a static control plane. This means that path servers have a constant set of paths available, and attackers are restricted to engaging in attacks by receiving, manipulating, and sending data traffic as opposed to sending control messages. Adversaries may try to attract or divert traffic from certain points in the network, craft new hop fields and segments, or combine existing ones in order to create new paths to influence the way outgoing data packets are forwarded, to manipulate the routing history of the packet, and to cover up their own actions.

While some of these attacks seem to be quite severe, they are under the strong assumption of an on-path attacker. This is in contrast to most attacks considered in protocols such as BGP, where attackers are less restricted. In addition, we will show that the attacks presented here are, if not entirely preventable, at least *detectable*, especially when using particular SCION extensions for path integrity protection.

We briefly review the SCION extensions that are relevant for this section. Using the end-to-end *SCION packet security extension* (see Section 15.1.4), the sender protects the header of a SCION packet with either a MAC or a signature, which can be checked by the receiver. All fields of the header are protected, except for the `CurrHF` and `CurrINF` pointers, which are changed by routers. Thus, the destination can detect tampering with the sender's *intended* path. *Origin and Path Trace* (OPT; see Chapter 12) on the other hand provides guarantees for the path that the packet *actually* took. Each intermediate router adds cryptographically secured information that malicious routers cannot produce, thus making attacks detectable. It should be noted that neither of these extensions are used by default for data packets.

13.5.1 Source Address Spoofing

A recent study has shown that more than 40% of all global ASes allow some level of IP address spoofing [33]. These ASes do not perform appropriate levels of ingress filtering, which means that they forward outbound traffic that claims to have originated in another network or AS. In today's Internet, source address spoofing is often used to hide the true origin of a packet. As shown in Figure 13.4, spoofing the origin of a packet allows an adversary, say `1.1.1.1`, to redirect (or even amplify) traffic to the victim `3.3.3.3` by first sending a request to an arbitrary host (or vulnerable service), say `2.2.2.2`, whose answer is then sent to the *alleged* originator of the request `3.3.3.3` (instead of to the *true* originator `1.1.1.1`).

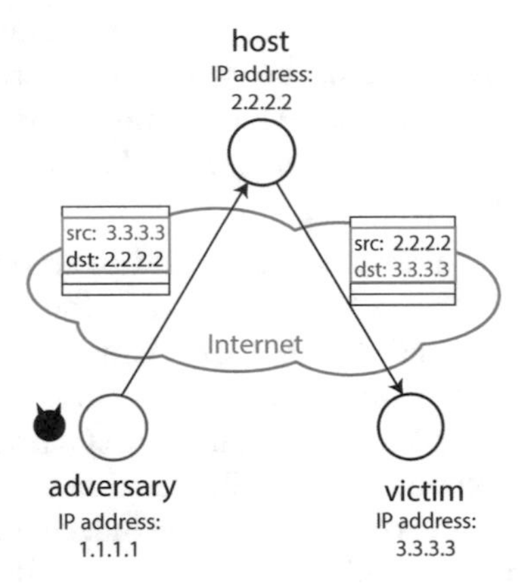

Figure 13.4: Source address spoofing in today's Internet. A malicious entity spoofs the source address of the packet, which will be reflected towards the victim by an arbitrary host, or even amplified by a vulnerable service.

The address fields in a SCION packet header can be arbitrarily picked by an adversary as they are not integrity protected. However, as SCION does not forward the packet based on the destination address until the packet reaches the destination AS, the effect of address spoofing is limited. Since SCION packet forwarding follows the hop fields, a packet needs a correct sequence of hop fields to be delivered to the destination.

In the simplest source address spoofing attack in SCION, the adversary simply embeds the source address of a different host v that is located in the same AS. The adversary embeds the hop fields required to reach the destination host h and sends off the packet. Since in SCION the receiver reverses the path for the response packet, v will receive the response.

In a more complicated case, the adversary wants the victim (who happens to be in a different AS) to obtain the response packet. Because of the way

SCION forwarding operates, the victim host will need to be located in an AS on the path traversed by the hop fields, either before or after the adversary's AS. If the adversary is located in AS A, the victim in AS V, and the host h in AS H, then the adversary can select a path $V - A - H$ (potentially with additional intermediate ASes), set the current hop field pointer to the hop field corresponding to AS A, and send the packet to the local egress border router which will forward the packet on the path toward AS H. When h inverts the path, the response will be delivered to v. This attack works if the adversary is located in an AS on the up-segment of AS V, in an AS on a core-segment, or in an AS on the down-segment of AS H.

Since SCION border routers check whether the destination ISD and AS numbers match their own ISD and AS numbers to determine if the packet should be locally delivered, an adversary can also create a path $A - V - H$ to mount this attack. The response packet will be delivered to v if the hop field corresponding to AS V does not have the forward-only flag set. Furthermore, to hide its traces and frame an innocent AS F, the adversary can use a path of the form $F - A - V - H$.

The fact that the adversary has to be located in an AS encoded by the hop fields in the packet facilitates adversary localization. Moreover, the SCION packet security extension and the OPT extension can prevent these attacks. The SCION packet security extension in conjunction with the DRKey system described in Section 12.5 enables that even the first packet sent to a destination provides source authentication. The OPT extension enables the destination to verify that all the claimed ASes on the path were indeed traversed, preventing the attacks where the adversary's initial packet only partially traverses the path. In case a malicious AS creates many fake hosts to overwhelm a destination server, source authentication enables the server to perform per-AS load balancing.

In a special case where the destination needs to perform a path lookup to return a packet (e.g., if a path has just expired or if a uni-directional path was used), the destination can insist on source authentication to prevent reflection attacks as described in the beginning of this section.

Armed with these countermeasures, we conclude that source address spoofing is not effective in SCION.

13.5.2 Modification of Packet Metadata

Packet metadata (such as the header and its path) is only partially integrity protected and thus vulnerable to unauthorized modifications. These modifications might have unwanted consequences in packet forwarding.

As part of the packet header, the SCION *common header* (see Figure 15.2 on Page 343) contains pointers to the current *info field* and to the current *hop field*. These pointers are updated as the packet traverses the network. If an unauthorized entity changes these pointers, there is a high probability that HF verification will fail. An adversary can also extend the packet with arbitrary content, set the packet length accordingly, and adjust the pointers to a location of his choice. This leads to a *path extension attack*, where an adversary adds arbitrary hop fields of his choice, modifies the pointer, and sends the packet further downstream. Similarly, modification of the destination type will make the border router in the victim's AS incapable of delivering the packet to the correct end host.

The packet metadata is protected by the SCION packet security extension and by OPT, so the use of either is sufficient to defend against metadata manipulation attacks.

Info-Field Manipulation

The metadata for each path segment is stored in the *info field* inside the path in the SCION header (see Figure 15.5 on Page 347). They include a timestamp, which is set by the initiator of the PCB. This timestamp cannot be modified by an attacker as it is included in the calculation of the MAC for each hop field. Otherwise, the timestamp could be backdated (and make a path appear invalid) or set to a later date (and extend the validity of the path).

Hop-Field Manipulation

Hop fields are protected with MACs and if the corresponding key is unknown, the attacker can at best attempt to perform a brute-force attack to determine the key. Candidate keys can be validated by checking the MAC contained in sample hop fields. As SCION uses 128-bit keys by default, such an off-line attack is computationally infeasible in practice. Furthermore, the keys for the MAC computation are short-lived, with a validity period of 24 hours.

MAC schemes are not generally specified by SCION and may thus be individually chosen by each AS. A hop field's MAC is only checked by the AS that created it, which enables algorithm agility for the MAC scheme. If a certain MAC algorithm is discovered to be weak or insecure, ASes can quickly switch to a secure algorithm without the need for coordination with other ASes.

The adversary might also try to directly brute-force a MAC (instead of the MAC's key). However, one packet would need to be sent to verify each guess. For our ℓ-bit MAC, the adversary is expected to generate 2^{ℓ} packets to forge one correct MAC. For the relatively short 3-byte MACs currently used, the attacker would need to try $\approx$ 17 million different MACs to successfully forge the MAC of one hop field. For each incorrect hop field, the corresponding AS returns an SCMP packet. Even though an attacker can observe whether a hop field has been accepted, each incorrect guess is visible to a monitoring entity and thus the attack can be easily detected.

13.5.3 Path Truncation and Extension

In the beaconing process, each AS extends the encoded SCION path with its own information and authenticates it before forwarding the beacon. The MAC used for authentication also takes the information of the previous hop field into account (see Equation 7.8). Therefore, it seems possible to *shorten* a path by removing hop fields from the end of the path. As long as the previous hop field is included (if it exists), verification in the current AS will succeed. Similarly, an adversary could try to *extend* an existing valid path by using its own hop fields.

Both these attacks are not possible in SCION since source and destination hop fields contain an empty ingress or egress interface identifier (see Equation 7.12), and delivery of packets using a hop field marked with the forward flag is prohibited. Hence a router can detect that a packet terminates at its AS.

13.5.4 Path Splicing

In a path-splicing attack, an adversary takes valid path segments of different paths and splices them together to obtain a new valid path.

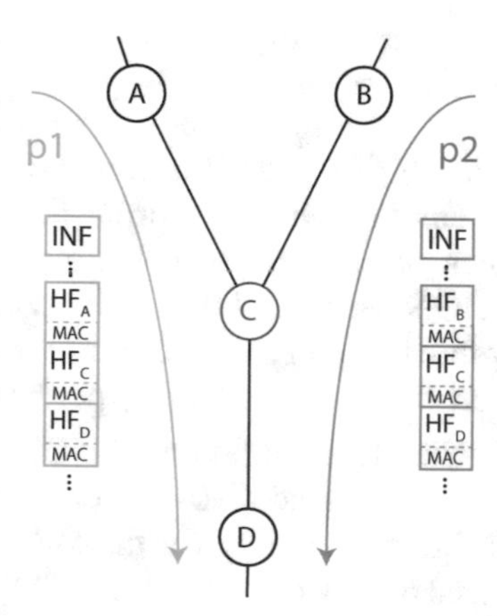

Figure 13.5: Path-splicing attack. A malicious AS C combines multiple paths to trick a downstream entity.

As illustrated in Figure 13.5, assume there exist two valid paths p_1 and p_2, each containing an info field (INF) and a sequence of hop fields (HFs). All hop fields are chained by appending the previous hop field to the current hop field and then calculating the MAC. The integrity of the previous hop field cannot be checked, since the MAC verification key is not available to entities other than the AS that generated the hop field.

Assume further that AS C is acting maliciously (active and on-path) and replaces HF_A with HF_B (and the source AS if necessary) for an incoming message and forwards it to AS D. From D's point of view, the MAC validation is successful. D cannot determine that the path has been modified by C and thinks that the traffic is coming from B. On the reversed path, C reverts its path changes and forwards the packet to A. MAC verification is still successful because after reversing the path (indicated by the 'up' flag) the integrity is verified with the next hop field instead of the previous one.

SCION, nevertheless, provides weak path integrity to guard against errors and misconfigurations, as well as simple path alterations. Without chaining of hop fields, any end host could simply splice valid paths by combining hop fields. The chaining prevents such attacks by end hosts, but does not prevent active path alterations by malicious routers *on* the path. However, an off-path entity cannot attract or influence the flow.

Using OPT, SCION can provide a strong notion of path integrity. The destination can verify that the path has not been modified while traversing the intended hops.

13.5.5 Path Segment Replacement

In a manner similar to path splicing, an on-path attacker may replace one or more entire path segments of a packet with different, valid segments, which he can obtain for instance from path servers. The whole segment must be replaced as otherwise the next hop's verification of its hop field will fail and the packet will be dropped. This leads to an interesting property of on-path forwarding attacks, which we call *weak detectability*: an attacker cannot manipulate a packet's path information to disguise her own presence on the path.[1] This property is considered to be weak, since the receiving host does not know *which* of the hops on the manipulated path is the attacker, and cannot even tell *whether* a segment replacement attack has taken place. We can guarantee a stronger detectability property by using the SCION packet security extension or OPT. Since the SCION packet security extension provides end-to-end path integrity protection, and OPT provides origin and path validation, any segment replacement will be noticed at the receiving end host. Strong detectability in this sense does not imply the ability to tell *which* of the nodes were acting maliciously, only that *some* of them were (OPT provides no fault localization mechanism).

This attack's restriction that the path may only be replaced by a combination of valid path segments can be partially lifted. The adversary can modify the already traversed portion of the current segment and past segments in a completely arbitrary manner (for instance, deleting and adding valid and invalid hop fields). There is some risk of detection for the attacker, since such a fake

[1] While hop fields do not contain globally valid AS identifiers, publicly available segments at path servers can be used to map hop fields to AS names, and thus to extract the sequence of ASes that the packet claims to have traversed.

segment will not be registered with the path servers. An unknown segment being used implies that either that segment is non-registered (hidden) or a segment replacement attack has taken place. If the receiving host knows that the sender does not make use of any hidden segments, then such an attack is strongly detectable.

In segment replacement attacks, the attacker is able to transparently revert any changes to the segments on replies by the destination. For instance, if an attacker M is an intermediate AS on the path of a packet from A to B, then M can replace the packet's past path (leading up to, but not including M). The new path may not be a valid end-to-end path. However, when B reverses the path and sends a new packet, that packet would reach M, who can then transparently change the invalid path back to the valid path to A. The SCION packet security extension and OPT make this attack strongly detectable.

Wormhole Attack

Similarly to wormhole attacks on the control plane (see Section 13.4.4), two colluding attackers in the network topology can create a variety of different wormhole attacks, if at least one of the attackers is on-path. We note that there are fundamental limits to path validation with respect to wormhole attacks, which also apply to SCION even when using the SCION packet security extension and OPT. We refer the reader to Chapter 12 for more details.

13.6 Censorship and Surveillance

In this section, we discuss attacks that target the integrity and confidentiality of communication traffic. If a malicious AS resides on the path between source and destination, it is able to inspect the traffic between the communicating endpoints. On-path adversarial traffic inspection is inevitable in any network infrastructure, including today's Internet. Therefore, endpoints need to use encryption to hide sensitive data. Otherwise, a cautious on-path adversary can easily eavesdrop passively on the link. An active adversary can take a step further and alter data of passing packets, or just drop them.

Increased Confidentiality Against Censorship and Surveillance

On-path traffic inspection is often referred to as *surveillance*. The scope of such an attack can be scaled to target a large number of victims. A powerful attacker can either passively monitor traffic (possibly at multiple observation points) or actively try to redirect traffic for inspection.

In contrast to today's Internet, SCION has a number of built-in features that natively protect the confidentiality of data, even without employing encryption of the data.

First, SCION's ISDs prevent entities in remote ISDs from manipulating the local control plane, preventing adversaries from tricking other networks into sending traffic through them. Inside ISDs, the core may appear to be a good vantage point for performing surveillance. Oppressive states may collude with or compel their local ASes to perform surveillance. However, peering links offer an alternative path out of the ISD without traversing the core. Due to source-selected paths and path transparency, senders can select paths through ASes they trust.

Second, SCION's native multipath communication hampers surveillance in that an adversary would have to be present on *all* paths that a traffic flow is using concurrently, which would result in a costly effort since path selection in SCION is very agile and unpredictable, meaning that end hosts can select paths based on an arbitrary selection process and change this selection at any time. Moreover, even if the adversary could eavesdrop on all the paths, it would need to correlate the packets that traversed different links but originate from the same user, which causes much higher overhead for the surveillance.

In addition to these basic mechanisms, SCION offers advanced techniques such as APNA [153], an architecture that provides strong source accountability and privacy-preserving communication. APNA appoints ISPs to authenticate hosts and their packets in the network and anonymize the identities of communicating partners. Moreover, state-of-the-art encryption mechanisms such as TLS 1.3 are provided in SCION for encapsulation of data traffic.

Anonymous Communication Against User and Host Identification

The act of communication on the Internet inevitably leaks information. In particular, network headers reveal information (e.g., source address, flow information), which might threaten anonymous communication and privacy. Based on this information, a state-level adversary could be enabled to enforce censorship. To counteract identification of users and hosts, SCION proposes the extensions HORNET [49] and OTA [156].

HORNET is a low-latency onion routing system that operates at the network layer and makes use of symmetric cryptography for data forwarding. It offers payload protection by default using a shared secret key between endpoints and routers and can defend against attacks that exploit multiple network observation points. Instead of keeping state at each relay, connection state (including, e.g., onion layer decryption keys) is carried within packet headers, allowing intermediate nodes to quickly forward traffic without per-packet state lookup.

OTA uses per-packet *one-time addresses*, which are issued by ASes to their customer hosts. Each one-time address is only used once as either a source or a destination address. This eliminates flow information from packet headers — implicitly (e.g., the standard 5-tuple in TCP/UDP packets) and explicitly (e.g., flow identifier) — while still allowing demultiplexing of seemingly unrelated packets to flows.

13.7 Attacks Against Availability

In this section, we discuss the effect of attacks targeting the availability of the SCION infrastructure using popular attacks also known in the current Internet, as well as SCION-specific attacks.

13.7.1 (Distributed) Denial-of-Service Attacks

Today's Internet lacks native mechanisms to defend against denial-of-service (DoS) attacks. The configuration of many of today's networks even facilitates the execution of DoS attacks. For instance, many ISPs on the Internet today do not enable protection against source address spoofing [7, 18, 33, 87]. The Internet does not offer victims inbound path selection, precluding path agility (i.e., quickly switching to a better-provisioned path) in the event of an attack. In addition, due to the revenue model, ISPs lack incentives to address DoS attacks, which contributes to the pervasiveness and increasing occurrences of DoS attacks on end hosts, servers, and network infrastructure. In fact, DoS attacks have become such a common occurrence [139] that they have spawned an entire industry of content distribution networks, cloud-based DoS mitigation systems, and a wide range of middleboxes aiming to help reduce the impact of attacks. Unfortunately, even when these systems are used, networks may be unavailable if the adversary can generate sufficient traffic to cause congestion in the core of the network, or can exploit a flaw or limitation in the defense system.

SCION's core defense against DoS is to enable inter-domain traffic management and resource allocation. The five core mechanisms are: (a) path announcements with a short lifetime, (b) non-registered (or hidden) paths, (c) multipath communication, (d) source authentication using OPT, and (e) SIBRA. These mechanisms can guarantee communication for two benign communicating entities and can mitigate the power of network-level congestion even for the case of publicly accessible services. We will discuss these mechanisms in more detail.

DoS on a Host and Application-Layer DoS

As an inter-domain routing architecture, SCION does not aim at providing specific defenses for application-layer DoS attacks [163], such as exploiting a vulnerability on a service. That is, if a publicly accessible SCION down-segment to the destination exists, any remote end host will be able to send traffic to that destination — just as in today's Internet. However, the destination can announce paths with limited lifetime that will not be renewed once an attack is detected. Alternatively, the destination could insist on source authentication to filter out unwanted hosts or source ASes, and preferentially handle requests by known clients, or requests originating from trusted ASes.

DoS on a Domain and Volumetric DoS

Attackers may attempt to attack an entire AS by sending large amounts of traffic, specifically attacking the AS's upstream links. On the Internet today, victims are forced to over-provision upstream links to account for peak bandwidth utilization, or alternatively must be able to rapidly add additional capacity when an attack is noticed. In SCION, victim ASes have control over inbound paths through which they are reached, thus enabling the following options:

- The victim has the ability to end the attack, along with inbound connectivity, by removing the public down-segments from the path servers. Once the paths to the victim are no longer available, the attacker can only continue to send traffic until the path expires, since an end-to-end path can no longer be built or used thereafter. To shorten the amount of time a path can be misused, the domain can announce paths with short lifetimes (e.g., 10 minutes).
- The previous option effectively terminates the attack, but also (deliberately) terminates connectivity to the victim AS. In cases where the victim wants to continue to make paths available, but only to a smaller set of authorized senders, the victim ceases publishing the paths to a path server, and temporarily distributes *non-registered* (or hidden) paths out of band to authorized senders (see Section 7.2.5 on Page 137).
- High availability can be achieved through *multipath* communication: as with today's Internet, ASes may peer with multiple ISPs to increase availability in case of attack. However unlike today's Internet, SCION allows ASes to make use of multiple paths *concurrently*, requiring the adversary to simultaneously and continuously flood *all* upstream links of a victim AS. This requirement greatly increases the attacker's effort. Further, SCION makes it easy to be *multi-homed*.
- DoS attacks in today's Internet often use packet reflection to redirect large volumes of traffic to a specific entity and hide the origin of their attack. Using the SCION packet security extension or OPT extension, SCION can authenticate the source address of each packet, prioritize authenticated traffic, and thus effectively mitigate DoS attacks using source address spoofing.

- Finally, SIBRA offers differentiated inter-domain resource allocation for fine-grained bandwidth control of individual flows, as described in Chapter 11.

DoS on a Link

Attacks such as Coremelt [231] and Crossfire [129] generate seemingly benign traffic that traverses specific inter-AS links. These links are usually high-capacity and frequently used, so that their saturation (and hence lack of availability) can impact large portions of Internet traffic. SCION's source-selected paths may allow attackers to target specific links more effectively than in today's Internet since the sender has more control over the end to end path. However, the current Internet route optimization process also introduces network bottlenecks [128].

SCION can defend against DoS by having core path servers keep track of path requests and by balancing responses across links as necessary. If SIBRA is actively deployed, then balancing is not strictly necessary. SIBRA will allocate a fair share of bandwidth to all authorized senders.

DoS on Essential Infrastructure Services

SCION's operation depends on the availability of beacon servers, path servers, name servers, and certificate servers for regular network operations; if any one of these services is not available within a domain, paths may fail to be created, disseminated, or authenticated. As such, the availability of these infrastructure services is of paramount importance to SCION.

The simplest way to provide high availability of these services is by replication. SCION's current codebase uses a distributed consistency service based on Apache's ZooKeeper [11]. We use this service to share state amongst the various infrastructure elements, so that, for example, several path servers all know the up-to-date set of paths. Once the service is operational, a *master* server is elected automatically, while all *standby* servers replicate the master state. If the master does not respond within a period of time, a new master becomes active. As additional capacity is needed, new standby servers can be added to the pool without downtime. Executing a DoS attack against infrastructure services thus requires sufficient resources to continually exhaust resources of all standby servers that are part of the pool. Moreover, for all requests, load-balancing amongst all the active servers is deployed by default. The details of SCION server failure resilience are described in Section 7.4 on Page 146.

In addition to replication, ASes within an ISD can rate-limit inbound requests originating from outside the ISD. This can be done generally for all external ISDs based on internal server capacity, or on a per-source-ISD basis. For requests that initiate from inside the ISD, SCION provides visibility and thus sources of excessive traffic requests can be identified.

Packet Replication Attacks

In packet replication attacks, the adversary resends packets previously received from other nodes. Compared to sending random packets, this has the advantage that replayed packets might bypass simple firewalls or intrusion detection systems. The replicated packets can be fresh or “stale” (packets that are delayed past a certain time). With large numbers of packets replayed, both the bandwidth of the network and the computing power of the involved nodes are uselessly consumed. This can be used to perform a DoS attack on victims. Furthermore, if a network has an accounting mechanism deployed (e.g., for fair resource sharing), a malicious entity could use packet replication to perform a framing attack. To prevent such replay attacks, we have designed a high-speed in-network replay detection system [155], which we plan to integrate into SCION in a future version.

In today’s Internet, the attacker typically would need to attract traffic from the victim. The observed traffic could then be replicated and re-injected into the network. In SCION, such attacks are harder since (a) an attacker must be on-path to successfully spam a victim with a replicated packet, and (b) SCION’s path transparency helps to identify the malicious entity.

13.7.2 Forwarding Loops

Forwarding loops occur when packets traverse the same subset of ASes multiple times (see the dashed red line in Figure 13.6). Each AS in this subset is traversed twice. The optimal path (green dotted line) would not include the loop but instead use a shortcut. The impact of this attack might be limited when considering a single attacker, but if a collaboration of malicious entities exploits this possibility, the availability of the intermediate ASes (e.g., on a network link between source and destination ASes) can be affected.

Apart from enabling attacks, forwarding loops are a design flaw and should be avoided in a clean-slate design. SCION prevents forwarding loops through two mechanisms. First, loops within a path segment are prevented in the beaconing process, where an AS ensures not to send a beacon to another AS that is already part of the path. Second, a valid SCION path is constructed by at most one path segment of each type (up-segment, core-segment, and down-segment). Since the up-segment and down-segment contain at most one core AS, and the core-segment is loop-free, no loop containing core ASes can be created by the combination of the three path segments. The combination of an up-segment and a down-segment, however, can result in a single loop if the sender is malicious and does not create a shortcut. This case is illustrated in Figure 13.6. Not only

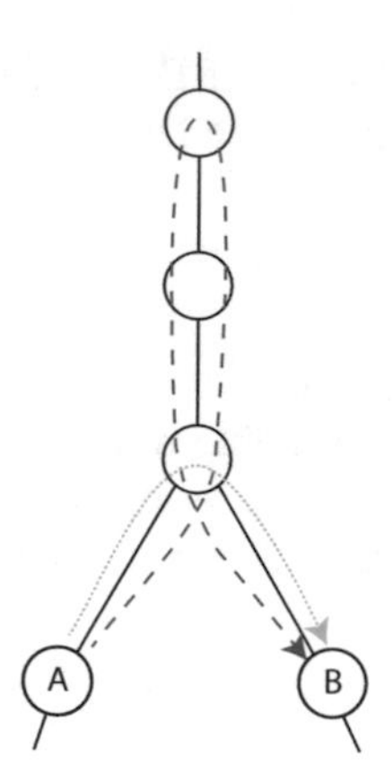

Figure 13.6: A forwarding loop occurs if a packet is sent along the dashed red path to get from source A to destination B. Using a shortcut as represented by the dotted green path avoids the loop.

can such maliciously constructed paths be detected, but also legitimate nodes can avoid such loops by constructing shortcut paths.

13.7.3 A Note on Resource Exhaustion in General

Today's Internet suffers from various cases of resource exhaustion, some of which stem from design decisions that no longer meet the requirements of growth (e.g., the exhaustion of the IPv4 space), from misconfiguration errors (e.g., the advertisement of bogus IP prefixes that flood BGP border tables), or from malicious activities that explicitly aim at exhausting resources (e.g., DoS attacks against vulnerable services).

To circumvent resource exhaustion, SCION generally avoids keeping state in performance-critical infrastructure wherever possible. For instance, to prevent state-exhaustion attacks and state inconsistencies, routers in SCION do not keep any forwarding state. However, resource exhaustion at any point in the infrastructure cannot fully be excluded. Hardware and computational power are limited and will be exhausted at some point, if the queuing rate of incoming requests is higher than the processing rate.

SCION extensions that use asymmetric cryptographic operations during session setup are susceptible to resource exhaustion, considering that asymmetric cryptographic operations are several orders of magnitude slower than symmetric ones. Too many sessions initiated in a short period of time by an adversary controlling multiple endpoints can effectively exhaust memory and computation resources of a victim. Apart from extensions, SCION uses signatures in the control plane. However, they are used only between infrastructure elements and the number of messages is limited by the number of ASes.

Asymmetric cryptography not only suffers from a (relatively) high overhead, but also requires the public keys of ASes and end hosts to be available. SCION uses the Dynamically Recreatable Key (DRKey) protocol, as described in Section 12.5, which enables routers to derive symmetric cryptographic keys on the fly from a single local secret. A sample use of DRKey in SCION is SCMP (see Section 4.2.5), the SCION alternative to ICMP, where it is used to efficiently create authenticated SCMP messages.

13.8 Absence of Kill Switches

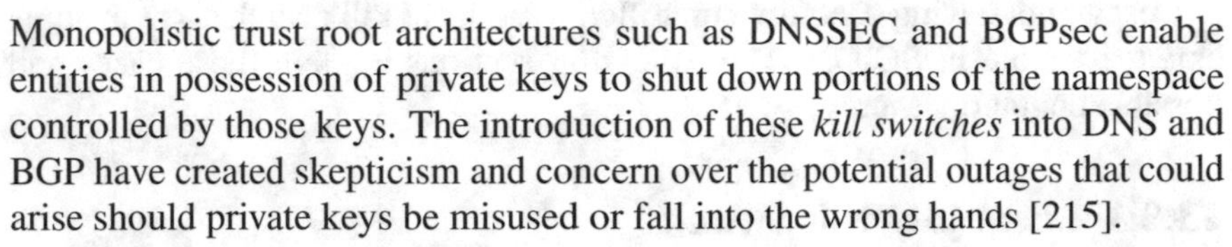

Monopolistic trust root architectures such as DNSSEC and BGPsec enable entities in possession of private keys to shut down portions of the namespace controlled by those keys. The introduction of these *kill switches* into DNS and BGP have created skepticism and concern over the potential outages that could arise should private keys be misused or fall into the wrong hands [215].

Briefly, monopolistic trust architectures work by delegating trust from a top-level key to lower-level keys. Due to this hierarchical trust structure, each key has full control over keys beneath it in the hierarchy. Kill switches work by revoking or maliciously substituting a public-key certificate at a specific point of the hierarchy (e.g., the certificate for the root zone in DNSSEC). The incorrect certificate will cause downstream signature validations to fail. A compromise of the top-level zone is the worst-case scenario; the entire namespace can be shut down. However, targeting specific zones lower in the hierarchy is also possible.

In DNSSEC and BGPsec, root keys are secured through multiple layers of physical security accompanied by a key rollover schedule. However, the physical location and key management processes must still be performed in a known jurisdiction; for DNSSEC, key management ceremonies are held quarterly at Verisign facilities in the United States. Even though the key rollover ceremony is monitored and logged, it may still be possible that a state-level adversary can gain access to root-level keys.

We note that private keys can be leaked through several means. Attackers might exploit a software vulnerability in a system with access to the key, or use social engineering for access to that system. Employees may go rogue or be threatened/extorted to reveal the keys. Short asymmetric cryptographic keys (e.g., RSA keys shorter than 1,024 bits) are likely breakable today by well-sponsored nation states actors, or can be broken in the future as computing speeds continue to increase. The broad attack surface makes it challenging to implement comprehensive protection mechanisms.

Even without key compromise, centralized architectures risk being unavailable if one or more of their critical services becomes unavailable. For example, a denial of service on services publishing revocation lists will cause clients to receive an incorrect view of currently valid certificates.

SCION's trust architecture is fundamentally different from that in the cases described above. In SCION, each ISD manages its own trust roots instead of a single global entity providing those roots. This structure gives each ISD autonomy in terms of key management (i.e., all key management operations can take place without contacting a parent authority) and in terms of trust. All entities inside the ISD already subscribe to the ISD's policies. What SCION enables is trust transparency for entities to know what additional roots need to be trusted for a given communication.

Despite not having centrally controlled trust, local kill switches are to some extent possible in SCION. The following sections explain these cases and possible countermeasures.

13.8.1 Local ISD Kill Switch

As in the case of DNSSEC and BGPsec, executing a kill switch inside a local ISD can be done at different levels of the AS-level hierarchy. One difference in SCION is that core ASes cannot be switched off by a parent authority since they manage their own cryptographic trust roots (see below). Another difference is that the attack vector of intra-ISD kill switches has only two entry levels; the core certifies all ASes in the ISD, but ASes do not certify ASes below them. A special case is the situation of nested ISDs (see Section 3.6 on Page 56), where a non-core AS also acts as a core for another ISD, resulting in an even more limited scope due to the isolation.

If the core's root keys are compromised, or the core is acting maliciously, then it is trivial to shut down communications traversing the core. Moreover, the core might stop propagating PCBs, precluding the discovery of new paths. In this case, downstream ASes will notice that PCBs are no longer being propagated, but all previously discovered (and still valid) paths are still usable for data-plane forwarding until they expire.

Perhaps a more stealthy kill switch would be to shut down path servers in victim ASes. While this cannot be done remotely, an adversarial entity controlling an ISD (e.g., a government) might compel core and non-core ASes to stop replying to path requests. Alternatively, the compelled ASes might return only a subset of all available paths. If this attack were used in conjunction with blackholing, senders in the ISD would have difficulty getting traffic out of the ISD. We would like to emphasize, however, that such attacks are even easier to perpetrate in today's Internet. In SCION, existing paths can continue to be used in the data plane as long as the traversed ASes allow the forwarding.

13.8.2 Remote ISD/AS Kill Switch

Since SCION ISDs independently manage their own cryptographic keys and namespace, it is not possible for a remote attacker (outside the target victim's ISD) to cause a kill switch in a different ISD. That is, without access to the private trust root keys in the remote ISD, the attacker is limited to data-plane attacks. Even if private keys became available to a remote attacker, they would need access to an AS inside the remote ISD to inject faulty information.

13.8.3 Recovery from Kill Switches

In the event of a non-core AS kill switch, the impacted AS needs to obtain a new certificate from the core. This process will vary depending on internal issuance protocols.

If a core AS's offline root key is compromised, the TRC must be re-issued, which could be time-consuming since it needs to be certified by a quorum of core ASes and cross-signed by neighboring ISDs. AS certificates may not need to be re-issued as long as the core AS's online root key was not breached, in which case the existing online root key can be re-signed by the new offline root key.

If the core AS has not been compromised, but is instead acting maliciously (e.g., by not propagating beacons downstream or tampering with responses for paths or certificates), one way to recover is for downstream ASes to self-organize and form a new ISD. By now operating autonomously, the new ISD can begin path discovery and traffic forwarding.

SCION, unlike BGP, has no notion of routing convergence. Instead, the flooding of beacons disseminates topology information. This means that in the worst case, if all paths must be re-created, fresh paths are established after a single flood has reached all ASes.

13.9 Resilience to Path Hijacking

In this section, we analyze the capabilities of an attacker hijacking IP tunnels that are operating as an overlay on top of the legacy Internet (see Section 10.1.2 on Page 194). More precisely, we discuss the properties that an incrementally deployed SCION network can achieve, where different SCION ASes need to connect via inter-site tunnels with traffic traversing the current Internet. Our results show that SCION even deployed as an overlay-only network still provides better resilience to path hijacking than the legacy Internet. Further analysis of partially deployed SCION is presented by Lee et al. [152].

13.9.1 Deployment Simulation

To evaluate the deployability and availability benefits of SCION under partial deployment, we perform several BGP simulations by extending the BSIM simulator [130], where the BGP paths are computed using route selection based on the standard BGP routing policies (Gao-Rexford Model [91]). As our topology dataset, we use a recent snapshot of the CAIDA Inferred AS Relationship dataset.

Tunneled Path Resilience

In this section, we investigate the potential benefit (i.e., resilience against prefix-hijacking attack) that a path constructed using a series of short tunnels can provide over a single BGP path. For our study, we use the following notation:

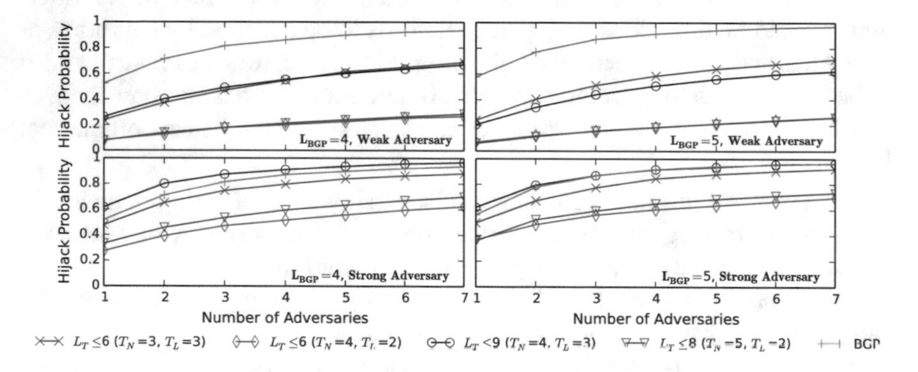

Figure 13.7: Probability of hijacking BGP and SCION paths under four tunnel settings while varying the number of adversaries. The upper and lower halves show the results for weak and strong adversary models respectively; and the left and right halves show the results for $L_{BGP} = 4$ and $L_{BGP} = 5$, respectively.

- (AS_x, AS_y): BGP path (list of ASes) between AS_x and AS_y,
- $|(AS_x, AS_y)|$: length (expressed in AS-level links) of path (AS_x, AS_y),
- T_N: number of deploying ASes $AS_1, AS_2, \ldots, AS_{T_N}$ that form the overlay end-to-end tunnel,
- T_L: length of the longest tunnel segment, i.e., $\max|(AS_i, AS_{i+1})|$ for $i \in [1, T_N - 1]$,
- L_{BGP}: length of the BGP path between AS_1 and AS_{T_N}, i.e., $|(AS_1, AS_{T_N})|$,
- L_T: length of the tunneled path between AS_1 and AS_{T_N}, i.e., $\sum_{i=1}^{T_N-1} |(AS_i, AS_{i+1})|$.

We assume that the first node of an end-to-end tunnel path (AS_1) is the source while the last (AS_{T_N}) is the destination. Then, L_{BGP} expresses the length of the BGP path between source and destination. We also assume that traffic from source to destination over the end-to-end tunnel traverses overlay nodes $AS_2, AS_3, \ldots, AS_{T_N-1}$ in that order. The tunnel's path on the AS-level is a concatenation of BGP paths:

$$(AS_1, AS_2), (AS_2, AS_3), \ldots, (AS_{T_N-1}, AS_{T_N})$$

For our simulation, we consider two adversary strategies designed to hijack traffic from source to destination: (a) a weak adversary, which announces only the destination's prefix; and (b) a strong adversary, which announces all prefixes of $AS_2, AS_3, \ldots, AS_{T_N}$. In both cases an adversary launches attacks from a randomly compromised AS, however he cannot compromise ASes on the path between source and destination.

Our experiment simulates eight scenarios by varying T_N, T_L, and L_{BGP} to analyze the resilience that tunnels can provide against prefix-hijacking attacks. For each scenario, while incrementing the number of adversarial ASes from one to seven, we construct and simulate 1,000 random and unique tunnel deployments, where AS_1 and AS_{T_N} are randomly chosen from *multi-homed* leaf ASes and the other tunnel nodes are chosen from all other ASes. We focus on multi-homed ASes, as they are more likely to start deploying an availability-enhancing technology (according to a study [152], about 57% of all leaf ASes are multi-homed).

Figure 13.7 summarizes the results of our simulation. In each graph, the x-axis represents the number of adversaries (varied from one to seven) and the y-axis represents probability values that an attack on the tunneled path will be successful. The two figures on the left show the hijack probability of the tunneled path for source and destination AS pairs that are four BGP hops apart ($L_{BGP} = 4$); and, on the right five BGP hops apart ($L_{BGP} = 5$). Moreover, the upper two figures show the results against weak adversaries while the lower two figures show the results against strong adversaries. Lastly, each plot also shows hijack probability for the BGP paths (green line with plus markers).

As expected, our results show that the hijack probability increases as the number of adversaries increases, and that the probability is higher for the strong adversary model than the weak adversary model. Furthermore, against the weak adversary model, the probabilities of hijacking the tunneled paths are similar for the two cases that have the same tunnel segment length (i.e., T_L) but different total length (i.e., L_T) — on the upper two figures, the purple line with diamond markers and the red line with inverted triangle markers almost overlap with each other. This is because the weak adversary model can only attack the last tunnel segment, i.e., (T_{N-1}, T_N), as the weak adversaries only announce the prefix of the destination (i.e., T_N).

The results show that the tunneled paths have lower hijack probability than the BGP paths even if the length of the tunneled path (i.e., L_T) is longer than that of the BGP paths (i.e., L_{BGP}). However, the result also shows that if the length of the tunneled paths becomes too long (e.g., twice the length of the BGP paths), the tunneled paths become more susceptible to hijacking attacks. However, in practice, it is highly unlikely that the tunneled paths would be twice the length of the BGP paths.

Lastly, the results show that the composition of the tunneled path affects the resilience against prefix-hijacking attacks: a tunneled path that is composed of shorter individual segments is more resilient than a path that is composed of longer individual segments. For the two cases where $L_T \leqslant 6$, the hijack probability is significantly lower when the length of the individual tunnel is kept shorter. In Figure 13.7, this can be seen by comparing the blue line with 'x' markers and the purple line with diamond markers. Moreover, the result shows that the tunneled paths that have longer total length but shorter individual tunnel segments (i.e., $L_T \leqslant 8$, red lines with inverted triangle markers) are more resilient than the tunneled paths that have shorter total length but longer individual segments (i.e., $L_T \leqslant 6$, blue lines with 'x' markers).

13.10 Summary

In this chapter, we have compared various security aspects of today's Internet with their SCION counterparts. We have shown that SCION provides built-in security defenses against many well-known network attacks that plague today's Internet operators and users. We have also examined specific attack vectors on SCION and showed that they either result in minimal impact or can easily be detected and mitigated, especially when using the SCION packet security extension or OPT.

We finally refer our readers to external evaluations of future Internet architectures (FIAs), for instance to a recent study by Ding et al. [70], which also demonstrate that the security properties achieved by SCION are stronger than those of other FIAs.

14 Power Consumption

David Barrera, Chen Chen, Adrian Perrig

The Internet, including user equipment, data transmission media, data centers, and access networks, requires a considerable amount of power, consuming nearly 1% of annual electricity production worldwide in 2010 [111]. Around 50 GW of power is consumed by network equipment, and this number is expected to double by 2020 [245]. Increased power consumption not only implies greater monetary cost, but also has an expanding environmental impact in the form of carbon footprint and pollution [96]. Reversing the trend is imperative and would pay off massively.

As a first step towards measuring and comparing the power consumption of IP networks and FIAs including SCION, we focus on the power consumption of the data plane. Since data-plane traffic represents 83% of the total power consumed by the Internet (compared to 17% consumed by the control plane [38]), our analysis covers the largest component of power consumption in IP networks and FIAs.

We observe that the designs of several candidate network architectures mainly vary in two dimensions: the design of packet-forwarding methods and the design of content cache methods. With respect to packet forwarding, SCION and NEBULA [9] both use packet-carried forwarding state (PCFS), which embeds forwarding information into each packet. In comparison, IP and NDN [184] routers maintain route tables and perform a route table lookup (RTL) to forward each packet. With respect to content-caching designs, NDN proposes *pervasive caching*, which equips each NDN router with a content cache. IP, SCION, and NEBULA enforce no requirement for content caching and therefore can use both end-to-end communication and *edge caching*, which is often seen in content delivery networks (CDNs). We summarize the two design dimensions in Table 14.1.

Accordingly, in addition to evaluating SCION's power consumption, this chapter seeks to answer a more fundamental question: what impact do design choices for packet forwarding and content caching exert on a network architecture's power consumption? Answers to the question not only help us to consider SCION's power consumption, but also guide us in designing a power-efficient network architecture.

A. Perrig et al., *SCION: A Secure Internet Architecture*, Information Security and Cryptography, https://doi.org/10.1007/978-3-319-67080-5_14

Architecture	Forwarding Technique		Cache Placement	
	PCFS	RTL	Edge	Pervasive
IP		TCAM	✓	
NDN		SRAM-BF		✓
NEBULA	PoC & PoP		✓	
SCION	Hop Field		✓	

Table 14.1: Methods used by network architectures for making forwarding decisions and caching content. PoC and PoP stand for *proof of consent* and *proof of provenance*, respectively.

The text in this chapter is based on the paper "Modeling Data-Plane Power Consumption of Future Internet Architectures" by Chen Chen, David Barrera, and Adrian Perrig, which was published in the Proceedings of the IEEE Conference on Collaboration and Internet Computing (CIC) 2016 [50]. From our analysis in the paper we were able to draw several observations: (a) the use of PCFS in SCION can be more power efficient than RTL in today's Internet (despite the larger packet size and cryptographic computations of PCFS); (b) based on our workload assumptions, end-to-end communication consumes less power than using in-network caches; and (c) there is no substantial difference between energy footprints of networks with edge caching as compared to ones with pervasive caching.

Chapter Contents

14.1 Modeling Power Consumption of an FIA Router

To model power consumption, we propose a generic router model that captures the forwarding behavior of both IP routers and FIA routers, as Figure 14.1 shows. In accordance with the two design principles (forwarding method and caching method) that we are investigating, we separate the content cache module and the forwarding-decision module from other router components. Table 14.1 summarizes the design choices that FIAs use for these two different modules.

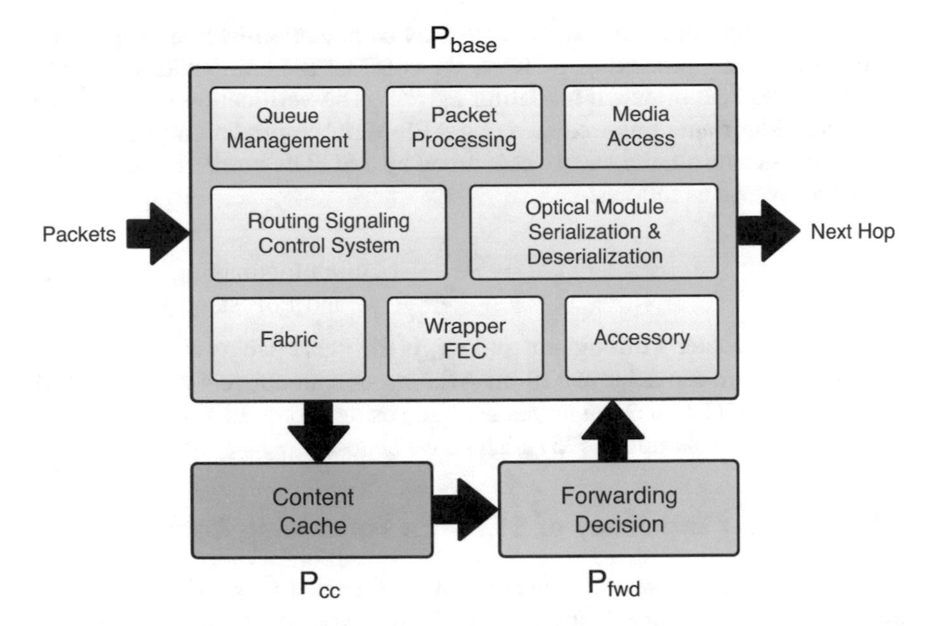

Figure 14.1: Abstraction of the forwarding behavior of an FIA router. We present a similar router-component dissection to Tamm et al. [236].

We group the rest of the router components, such as queue management and switching fabric, which are common components for both IP and FIA routers, in Figure 14.1, and treat their power consumption as a baseline for our analysis.

We denote the total power consumed by an IP or FIA router (measured in watts) to forward packets as P^{arch}, the power consumption of the local content-caching system as P^{arch}_{cc}, the power consumption of making forwarding decisions as P^{arch}_{fwd}, and the baseline power consumption of all the other components as P_{base}. The superscript "*arch*" is substituted by IP, NDN, SCION, or NEBULA. Accordingly, $P^{arch} = P_{base} + P^{arch}_{fwd} + P^{arch}_{cc}$.

Example. We take P^{SCION}_{fwd} as an example and refer the interested reader to our paper [50] for a complete description of packet forwarding and content cache models of IP, NDN, and NEBULA.

SCION uses packet-carried state for finding the interface via which to forward a packet. The forwarding decisions reside in the packet header and thus no inter-domain routing table needs to be stored on routers. The lack of routing tables (and thus lack of relatively expensive table lookup operations) helps reduce the power consumption of packet forwarding, P^{SCION}_{fwd}. However, SCION routers use cryptographic primitives to verify the integrity of the routing decisions embedded in the packet headers, which again adds to P^{SCION}_{fwd}.

Since the hop-field verification is the only computation-intensive operation in the forwarding process of SCION, we consider the computation of cryptographic verification when modeling P_{fwd}^{SCION}. The verification process on a SCION border router only requires one AES-MAC computation to verify the hop field. As a result, the energy consumed by a SCION border router P_{fwd}^{SCION} can be expressed as follows:

$$P_{fwd}^{SCION} = \frac{I \cdot E_{AES}}{s_{pkt}} \tag{14.1}$$

where I is the router's throughput and s_{pkt} is the mean packet size. A typical Helion AES core can achieve 128-bit AES throughput above 40 Gbps (320M AES ops/s) [108] based on an implementation using Virtex-7 FPGA with 6.6 W on-chip power consumption [257]. Thus, we choose $E_{AES} = 20nJ/op$.

14.1.1 Power Efficiency of SCION's Forwarding Logic

To investigate the power efficiency of SCION packet-forwarding logic, we compare the packet-forwarding logic of SCION with those of IP, NDN, and NEBULA using the complete model and parameter settings in the full paper [50].

Figure 14.2 illustrates the results when the link speed varies from 1 Gbps to 40 Gbps. In general, regarding power consumption, the packet-forwarding logic using PCFS holds advantages over that with RTL (except that NEBULA routers consume more power than NDN routers with small FIBs). Specifically, SCION's packet-forwarding logic consumes 2–3 orders of magnitudes less than those of IP, NEBULA, and NDN with different FIB sizes. We can attribute SCION's power efficiency to both usage of PCFS and its simple forwarding logic.

14.2 Simulation

Based on the model of FIA routers in Section 14.1, we now use a holistic method to demonstrate the power consumption of SCION and other FIAs in content distribution scenarios. We conduct our experiments by simulating the forwarding behavior of the IP network and FIAs when used for content distribution. We provide a sensitivity analysis that evaluates the impact of changing parameters in our paper [50].

14.2.1 Simulation Setup

The topology used in our simulations is based on the router-level topology of education backbone networks (Abilene and Geant) and Rocketfuel (Telstra,

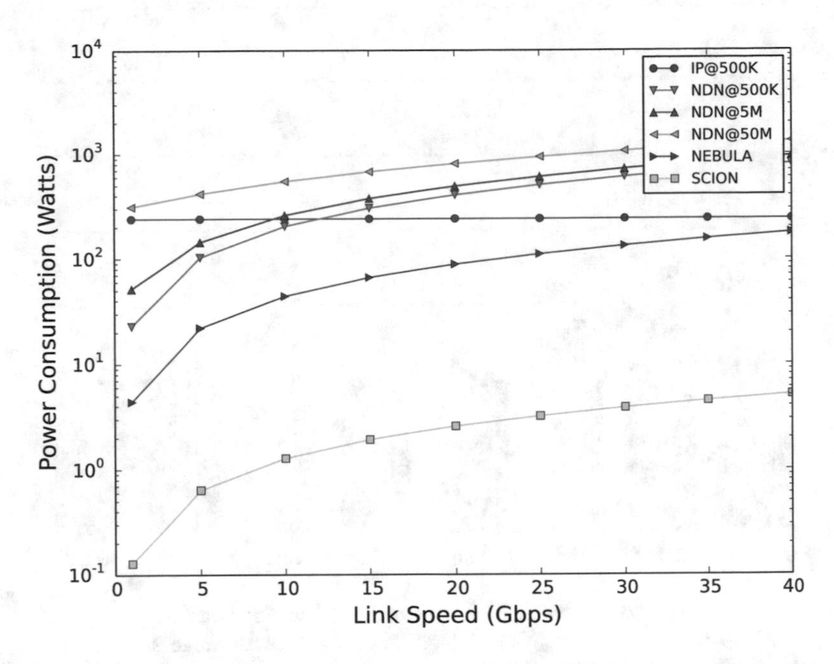

Figure 14.2: P_{fwd} under different link speeds for border routers. For the FIB size of NDN, Perino and Varvello have suggested 20M entries [197]. Accordingly, we vary NDN's FIB size from 500K entries to 50M entries to demonstrate the influence of routing-table size on routers' power consumption.

Sprint, NTT, Verio, Level3, AT&T) [227]. We follow the methods proposed by Fayazbakhsh et al. [86] to approximate access networks by trees appended to each point of presence (PoP). The internal nodes of the trees are border routers.

For content access patterns, previous work has suggested that a Zipf distribution closely approximates real-world content access from end hosts [86]. We use synthesized content access traces with $\alpha = 0.99$ (which approximates US users' behavior, as pointed out by Fayazbakhsh et al. [86]). For the query distribution, we assume in our simulation that the leaf PoP generates queries for content, where the number of queries generated is linear in the population of the city where the PoP is located.

To control the content cache capacity, we define a cache budget ratio. Let R be the number of routers capable of caching, C be the average cache capacity of each router, O be the total number of pieces of content that would benefit from caching, and s be the average size of each piece of content. We define the cache budget ratio $c = \frac{R \times C}{O \times s}$. We choose $c = 5\%$ as a baseline, which is

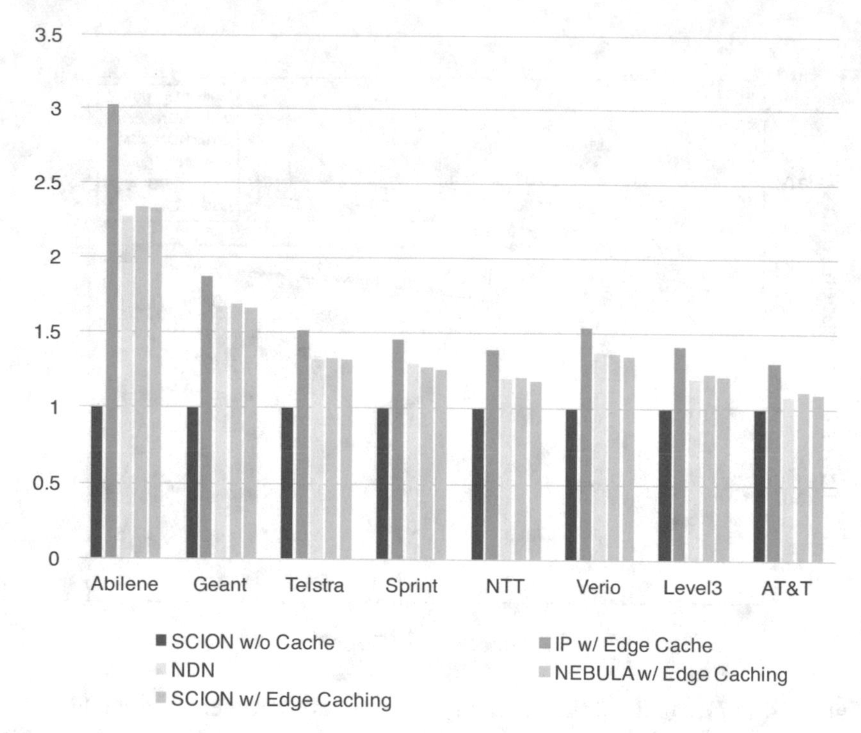

Figure 14.3: Power consumption of end-to-end communication, edge caching, and pervasive caching with capacity ratio $c = 0.05$. All results are normalized by SCION without content caching.

observed as a relationship between the CDN cache provisioning and the total requested objects seen by a CDN server per day [86].

For edge caching used in IP, NEBULA, and SCION, we simply assume that content requests are only served by each standalone cache server. We call this strategy *simple edge caching*. For pervasive caching, we assume *on-path cache discovery*, in which only content cached by the on-path routers is returned. For cache eviction, we select the least-recently used (LRU) method as our baseline strategy. Note that designing an optimal or high-performance cache replacement strategy is out of scope for this investigation.

14.2.2 Simulation Results

Figure 14.3 shows the power consumed by different network architectures with and without content caching. In particular, *SCION without content caching consumes 15-50% less power than other network architectures*. On one hand, SCION's advantage arises from its efficient packet forwarding. On the other hand, SCION's design benefits from the fact that the small real-world locality of the content access pattern determines that caching content is less power efficient.

Compared to SCION with edge caching, NDN with pervasive caching saves only a marginal amount (~2%) of power. Compared to IP with edge caching, NDN consumes up to 16% less power. The result implies that pervasive caching helps reduce power consumption, but the power reduction is limited. The reason is twofold: (a) multiple-layer caching provides limited improvement over single-layer caching, as indicated by previous work [86]; (b) pervasive caching requires more power-consuming caching devices, which further reduces the small advantage in power consumption.

We summarize our main observations as follows:

1. Network architectures that use packet-carried forwarding state instead of routing-table lookups exhibit lower power consumption.
2. FIAs without caching consume less overall power compared to those that use caching.
3. Under our workload assumptions, the use of pervasive caching results in marginal reductions in power consumption.

Although the variability of these results is nontrivial and depends on future technology innovations, which cannot be predicted, our results show that SCION's design not only results in a more secure and transparent operation, but at the same time can be implemented without negative impact on power consumption.

Part V

Specifications

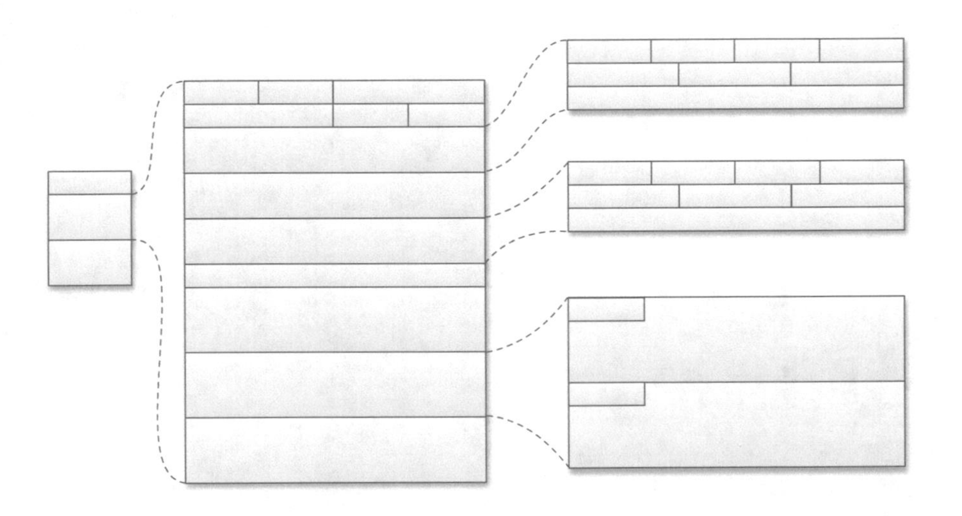

15 Packet and Message Formats

ADRIAN PERRIG, STEPHEN SHIRLEY, PAWEL SZALACHOWSKI

In this chapter, we describe the header formats of SCION control and data packets. We start with the description of the generic SCION header, which consists of four parts: a common header, a forwarding path, an extensions chain, and a layer-4 protocol header.

Chapter Contents

15.1 SCION Packet

A high-level layout of a SCION packet is presented in Figure 15.1. Before discussing its components in detail in the following sections, we give a brief overview.

- **Common header (Section 15.1.1):** Every SCION packet contains a mandatory *common header* in the first 8 bytes of the packet. The most important information encoded within the common header is the length of the packet, the types of the source and destination address, the current position in a path, and the type of the next header (an extension or layer-4 protocol).
- **Addresses (Section 15.1.2):** Source and destination addresses are placed right after the common header. A SCION address consists of an ISD

A. Perrig et al., *SCION: A Secure Internet Architecture*, Information Security and Cryptography, https://doi.org/10.1007/978-3-319-67080-5_15

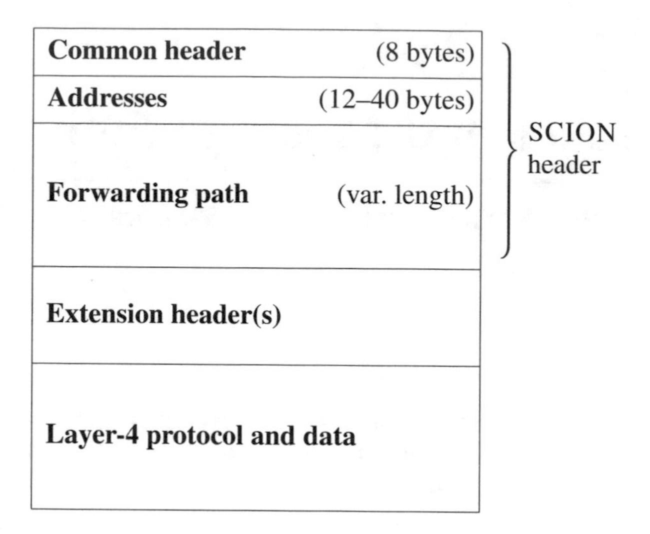

Figure 15.1: High-level layout of a SCION packet.

identifier, an AS identifier, and an end-host address. As end-host addressing in SCION is local to an AS, it allows ASes to use different address spaces (e.g., IPv4, IPv6, or MAC addresses) to address their end hosts. Consequently, SCION permits *hybrid addressing*, that is, source and destination can have addresses of different types. The concatenation of the source and destination addresses has to be aligned to a multiple of 8 bytes.

- **Forwarding path (Section 15.1.3):** The *forwarding path* consists of a sequence of *info fields* (INF) and *hop fields* (HF), which contain information required by border routers for packet forwarding. The forwarding path has to be aligned to a multiple of 8 bytes. The current INF and HF are indicated by pointers in the common header. Note that the forwarding path is empty for SCION packets that do not leave the origin AS.

  The concatenation of common header, addresses, and forwarding path constitutes the *SCION header*.
- **Extension headers (Section 15.1.4):** A SCION packet can additionally contain multiple *extension headers*. These are constructed as a chain (i.e., one extension points to the next), and the final extension points to the layer-4 header. There are two types of extension headers:
  - *hop-by-hop* extensions, processed by source and destination end hosts, as well as by every border router on the path; and
  - *end-to-end* extensions, processed only by source and destination end hosts.

  Extensions have to be aligned to a multiple of 8 bytes.

- **Layer-4 protocol and data (Section 15.1.5):** The remaining data of the packet is the *payload*, which is usually encapsulated within a *layer-4 protocol*.

In the remainder of this section, we describe the elements of a SCION packet in detail.

15.1.1 SCION Common Header

The common header is an essential element that every SCION packet must contain. The common header contains various fields such as the length of the entire packet, the length of the SCION header, types of source and destination addresses, next header (an extension or upper-layer protocol), and the current position of the packet on the path. The detailed format is presented in Figure 15.2; its fields are described below.

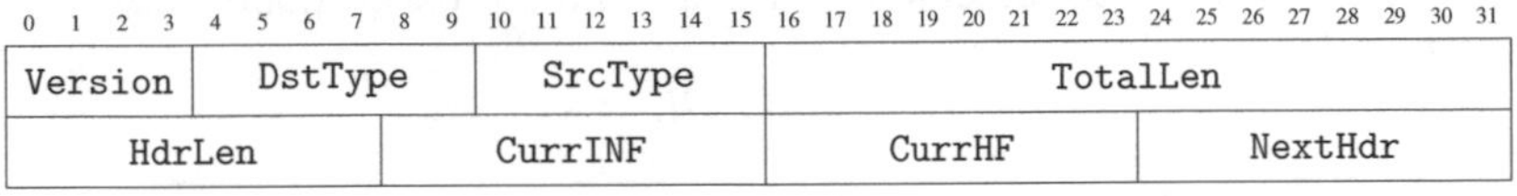

0 1 2 3 4 5 6 7 8 9 10 11 12 13 14 15 16 17 18 19 20 21 22 23 24 25 26 27 28 29 30 31

Version	DstType	SrcType	TotalLen
HdrLen	CurrINF	CurrHF	NextHdr

Figure 15.2: The SCION common header (8 bytes).

- `Version`: The first common header field is the version number of the SCION protocol. The field is 4 bits long; the current version of SCION has version number 0.
- `DstType` / `SrcType`: As SCION is agnostic to the addressing scheme used within an AS, the `DstType` and `SrcType` fields define the types of addresses that communicating end hosts use within their ASes. The address type implicitly defines the address length, and within a single packet, the source and destination addresses can have different types (hybrid addressing). The current implementation supports the following three end-host address types:
  - `IPv4`: IPv4 host address (4 bytes),
  - `IPv6`: IPv6 host address (16 bytes),
  - `Service`: The service address (2 bytes) is used to indicate the desired SCION service (see Sections 7.4.6 and 15.1.2).

  The complete SCION address, besides an end-host address, includes ISD and AS identifiers, which are encoded within 4 bytes (the ISD identifier is encoded within the most significant 12 bits, and the AS identifier is encoded within the remaining 20 bits). Thus, the minimum size of a single SCION address is 6 bytes, while the maximum is 20 bytes (see Section 15.1.2). Each address type field is encoded within 6 bits, which supports up to 64 different address types.
- `TotalLen`: total packet length in bytes. This field is 16 bits long, hence the maximum length of a SCION packet is 65,535 bytes.

- **HdrLen**: length of the SCION header in bytes (i.e., the sum of the lengths of the common header, the source and destination addresses, and the path). All SCION header fields are aligned to a multiple of 8 bytes. The SCION header length is computed as

$$\texttt{HdrLen} \times 8 \tag{15.1}$$

The 8 bits of the HdrLen field limit the SCION header to a maximum length of 2,040 bytes.

- **CurrINF**: pointer to the current info field of a forwarding path. This field is used by a border router to identify (and verify) the current series of hop fields (see Section 8.2). The pointer is updated by a router whenever the last hop field of a path segment has been processed, and thus the next router will start processing the first hop field of the next path segment.

The absolute byte offset to the current info field is computed as

$$\texttt{CurrINF} \times 8 \tag{15.2}$$

For every SCION packet with non-empty forwarding path, we require

$$8 + \texttt{SrcLen} + \texttt{DstLen} + \texttt{AddrPad} < \texttt{CurrINF} \times 8 < \texttt{HdrLen} \times 8 \tag{15.3}$$

where 8 denotes the length of the common header, SrcLen and DstLen are the lengths of the source and destination addresses respectively, and AddrPad is an optional padding added to align the concatenation of source and destination addresses to an 8-byte boundary (see Section 15.1.2). If the forwarding path is empty, the CurrINF field is set to 0.

- **CurrHF**: pointer to the current hop field of a forwarding path. This field is used by a border router to identify the current hop field, which in combination with the corresponding info field allows packets to be forwarded. For more information, see Section 8.2 on Page 164. An end-to-end example of packet forwarding, including updates of the CurrHF and CurrINF fields, is provided in Section 10.8 on Page 223.

The absolute byte offset to the current hop field is computed as

$$\texttt{CurrHF} \times 8 \tag{15.4}$$

As above, for every SCION packet with non-empty forwarding path, we require

$$\texttt{CurrINF} \times 8 < \texttt{CurrHF} \times 8 < \texttt{HdrLen} \times 8 \tag{15.5}$$

As info and hop field pointers (CurrINF, CurrHF) are expressed within 8 bits, the maximum offset is 2,040 bytes. If the forwarding path is empty, the CurrHF field is set to 0.

- `NextHdr`: field that encodes the type of the first header after the SCION header. This can be either a SCION extension or a layer-4 protocol such as TCP or UDP. Values of this field respect and extend IANA's assigned internet protocol numbers [120], (e.g., TCP and UDP have numbers 6 and 17, correspondingly).

15.1.2 SCION Addresses

SCION addresses are placed in the SCION packet directly after the common header (see Figure 15.1 on Page 342). Every SCION address consists of ISD and AS identifiers, and an end-host address. The ISD identifier is globally unique, the AS identifier is locally unique within the ISD, and the end-host address is routable within the AS. We note that the local address assignment should respect the rules of the address type being used, e.g., in the case of IPv4 the address space should either be RFC 1918 [210] private address space, or public address space that is owned by the network. See Section 10.3 for more details on the implications of the choice of end-host addresses.

The current SCION implementation uses 12 bits for the ISD identifier and 20 bits for the AS identifier. This allows up to 4,096 ISDs globally, and 1,048,576 ASes per ISD. However, the value 0 for both ISD and AS identifiers is reserved. The size of the end-host address is variable and depends on the address type. The length of a complete SCION address is thus also variable and is determined as follows.

Type	Size
`IPv4`	8 B
`IPv6`	20 B
`Service`	6 B

Table 15.1: Size of a SCION address, which depends on the type of the end-host address.

The types IPv4 and IPv6 are used for standard unicast end-to-end communication. The combination of IPv4 or IPv6 as a source address and `Service` type for a destination address is used for sending control-plane requests to a SCION service. The type `Service` is introduced to inform the destination AS that a given packet should be sent as an anycast packet to an instance of the correct service. However, SCION does not dictate how such an anycast mechanism should be implemented. (Our reference implementation uses the discovery service — see details in Section 7.4.7 on Page 152.) As a concrete server is queried through an anycast mechanism, a response to the request is sent as a standard data packet (i.e., the server uses its own IPv4 or IPv6 address in the source SCION address). The defined types of `Service` addresses are presented in Table 15.2.

SCION source and destination addresses are placed in a SCION packet in a way that enables border routers easy access to ISD-AS identifiers and the destination address (which is read more frequently than the source address). Namely, first destination and source ISD-AS identifiers are placed, which are then concatenated with destination and source host addresses. If necessary, this concatenation is padded to align the address block to an 8-byte boundary. For uniform addressing with IPv4 or IPv6 end-host addresses, the padding is not necessary as the concatenation of two SCION addresses is 16 and 40 bytes long, respectively. However, hybrid addressing, e.g., IPv6 as source and IPv4 as destination host address, requires 4 bytes of padding. An example of how addresses with different types are placed in a SCION packet is presented in Figure 15.3.

0	11	31	43	63
DstISD	DstAS	SrcISD	SrcAS	
DstHostAddr (IPv6)				
SrcHostAddr (IPv4)		Padding		

Figure 15.3: Example of hybrid addressing: SCION addresses with different types for end-host addresses.

15.1.3 Forwarding Path

To construct a forwarding path, an end host needs to extract and concatenate info and hop fields, created during the path-segment construction process (see Section 7.1). The forwarding path (i.e., in the data-plane format) is a list of such fields placed in the SCION packet after the addresses. A high-level illustration of a path is presented in Figure 15.4.

A single path can be composed of maximally three series of hop fields: series of HFs extracted from an up-segment (that leads toward a core AS), HFs from a core-segment (used for routing between core ASes), and HFs obtained from a down-segment (to forward a packet from a core AS to the destination). None of the series is mandatory, and the forwarding path is empty if a packet is sent within a local AS. However, if a packet is sent to a remote AS, the order of the series has to be preserved (e.g., HFs from a down-segment cannot precede HFs from a core-segment, and HFs from a core-segment cannot precede HFs from an up-segment).

Each list of HFs starts with an info field and at least two consecutive hop fields. The current position of a packet on its path is determined through pointers placed in the common header (see Section 15.1.1 for details).

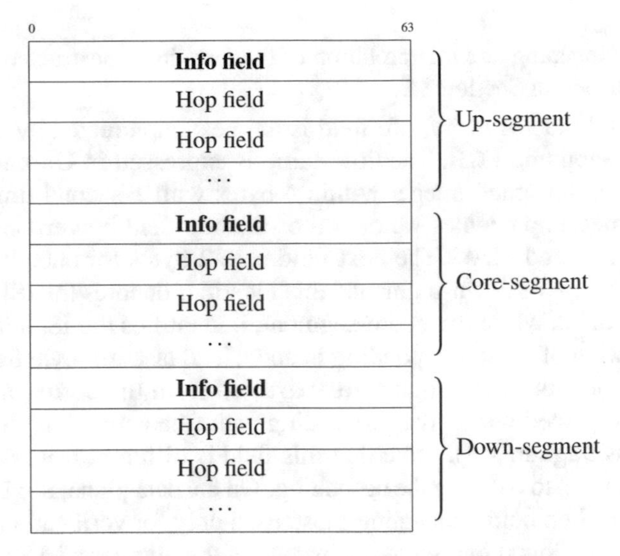

Figure 15.4: High-level layout of a path.

Info Field (INF)

An info field (INF) starts every series of HFs from a given path segment in the data-plane path format. It identifies the type of the path, and contains information required for hop-field validation. It also contains the length of the corresponding path segment, and it carries the identifier of the ISD which initiated the propagation of the path. The format of the info field is depicted in Figure 15.5.

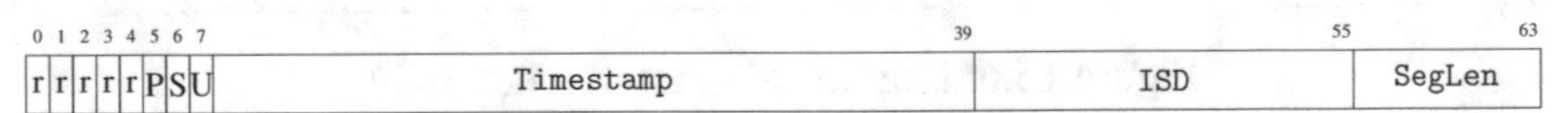

Figure 15.5: Format of an info field.

The first field of every info field is called `Flags` and is eight bits long. The bits of this field identify the type of a path as follows:

- `r`: unused and reserved for future use.
- `P`: peering-shortcut flag. If set to true, then the forwarding path is built as a peering-shortcut path (see Section 8.2.1). It is only valid if the shortcut flag is also set.
- `S`: shortcut flag. If set to true, then the forwarding path is built as a shortcut path (see Section 8.2.1).
- `U`: direction (up-segment) flag. If set to true, then the HFs from the corresponding path segment have an up-segment orientation, otherwise the HFs have a down-segment orientation.

The exact meaning and interpretation of the flags in a constructed forwarding path is described in Section 8.2.

The second field of every info field is a timestamp created by the initiator of the corresponding PCB. The timestamp is expressed in Unix time, and is encoded as an unsigned integer within 4 bytes with 1-second time granularity. This timestamp enables validation of the hop field by verification of the expiration time and MAC. The next field uses 2 bytes for encoding the ISD identifier. For up- or down-segments, this identifier denotes the ISD that hosts the path segment, while for a core-segment, it identifies the ISD that initiated the propagation of the corresponding beacon. The last, a 1-byte field `SegLen` encodes the length of HFs that were extracted from the corresponding path segment and placed within the forwarding path (the actual length in bytes is computed as `SegLen` × 8). Note that this field is different from `SegLen` in a PCB, which is set to 0 during the beaconing. On the data plane, `SegLen` denotes the number of hop fields (including those used only for verification, and those used for peering links) in a segment, instead of the number of ASes which are traversed. Its value is not changed during forwarding.

Hop Field (HF)

Hop fields contain information necessary for a border router to forward packets. They include encoded interfaces on which a packet is forwarded, the expiration time set by the AS, and a MAC (to authenticate the HF). A sample hop field is presented in Figure 15.6.

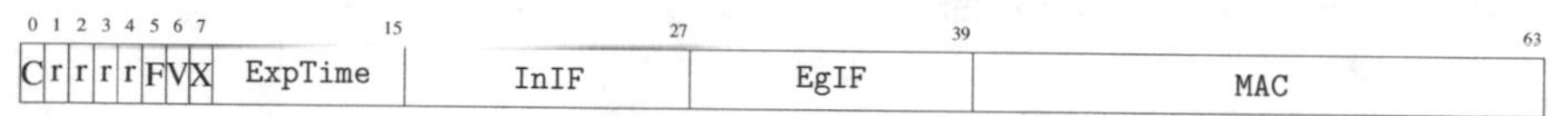

Figure 15.6: Example format of a hop field.

The format of a hop field is not unified and its layout may depend on the AS; however, the size of a hop field must always be a multiple of 8 bytes. The first field of every hop field is a 1-byte `Flags` field that describes the hop field.

- `C`: continue/stop flag, used to encode hop fields longer than 8 bytes. If a hop field consists of multiple 8-byte lines, then the flag is set on the first and the last lines, to encode the beginning and the end of the hop field accordingly (the intermediate lines have this flag set to 0). The other flags are set only in the first line.
- `r`: unused and reserved for future use.
- `F`: forward-only flag. It is set by the AS that created the HF to indicate that delivery of packets to the AS's end hosts is not permitted. This flag is set by an AS during beaconing and it is immutable (i.e., is included in the MAC calculation and cannot be modified by end hosts).

- `V`: verify-only flag. If set, it informs a router that the marked hop field is used only for MAC verification (not used in actual packet forwarding).
- `X`: cross-over flag. If set, it informs a border router that the path switches to a new path segment in this AS.

The exact meaning and interpretation of the flags in a constructed forwarding path is described in Section 8.2.

The next field inside the HF is the `ExpTime` field, which denotes when a hop field expires. The field is 1-byte long, thus there are 256 different values available to express an expiration time. The expiration time expressed by the value of this field is relative, and an absolute expiration time in seconds is computed in combination with the timestamp field (from the corresponding info field) as follows:

$$\mathtt{TS} + \left((1 + \mathtt{ExpTime}) \times \left\lfloor \frac{24 \times 60 \times 60}{256} \right\rfloor \right) \tag{15.6}$$

where `TS` is the value of the corresponding timestamp field. Hence, the minimal lifetime of a hop field is about 5 minutes, while the maximum validity of an HF is one day.

The following two fields, `InIF` and `EgIF`, represent ingress and egress interface identifiers between which a packet should be forwarded. The length of these fields is variable, and any encoding scheme can be implemented by the AS. In our implementation, these fields have a length of 12 bits each. This bounds the number of AS interfaces to 4,096.

The last item in the hop field is an authentication tag (`MAC`). The length of this field is variable (in our implementation, a length of three bytes was chosen). The AS itself can freely decide how to generate MACs and what size they should have.

15.1.4 Extensions

The forwarding path is the last element of the SCION header, and extension headers are placed after the path. Extensions are optional, and if they are present, they form a chain (one extension defines the type of the next one, and so on). The type of the first extension header is defined in the SCION common header, and the last extension in the chain points to the layer-4 protocol (e.g., UDP or TCP). For packets that do not contain any extension, the `NextHdr` field of the common header is set directly to the layer-4 protocol number.

SCION supports the following two classes of extension headers: (a) hop-by-hop extensions, and (b) end-to-end extensions. Every border router has to process hop-by-hop extensions, as do source and destination end hosts, and all hop-by-hop extensions must be placed before end-to-end extensions. End-to-end extensions are processed only by source and destination end hosts. The number of extensions in a packet is limited only by the maximum packet length. However, for efficiency reasons, border routers are required to process only the first three hop-by-hop extensions.[1]

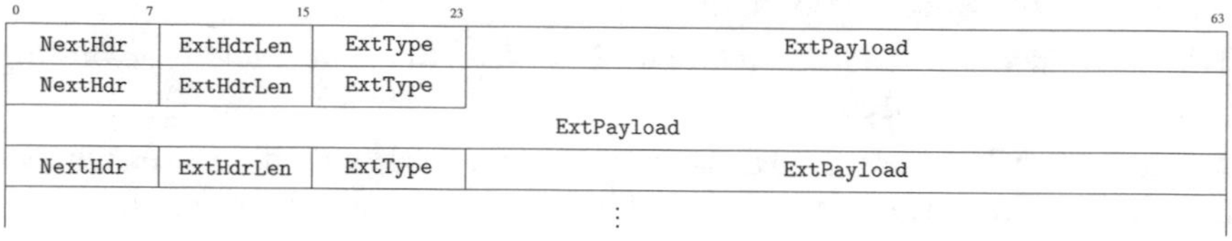

0 7	15	23	63
NextHdr	ExtHdrLen	ExtType	ExtPayload
NextHdr	ExtHdrLen	ExtType	
ExtPayload			
NextHdr	ExtHdrLen	ExtType	ExtPayload
⋮			

Figure 15.7: An extension chain formed by the three extension headers.

An example of an extension chain is presented in Figure 15.7. NextHdr is the first field of an extension header. It specifies the type of the extension or protocol header that follows the extension header. Values for these fields are chosen according to IANA's protocol numbers [120]. For hop-by-hop extensions, SCION uses the value 0x00 (that is the same value as IPv6's hop-by-hop extension) in the NextHdr field, while end-to-end extensions are indicated by the value 0xfd.

The second byte of every SCION extension header defines the length of that header. Extension headers have variable length, which must be a multiple of 8 bytes. The length of a given extension is computed as follows:

$$\texttt{ExtHdrLen} \times 8 \tag{15.7}$$

and $\texttt{ExtHdrLen} \neq 0$, which gives 8 bytes as a minimum length, and 2,040 bytes as a maximum. Space for an extension header must be allocated within a packet by the source, and intermediate routers may not increase the size of extension headers (or create additional ones).

The last common field of every SCION extension header defines the type of extension within a given class (i.e., type within the hop-by-hop or end-to-end extension class). The field is 1-byte long, thus it is possible to define 256 different SCION extensions per class.

SCION does not forbid multiple SCION extensions of the same class and type to be put into a single packet. Hence, as the length of a single extension is capped at 2,048 bytes, if necessary, an extension's payload can be split into multiple extension headers.

SCION by default supports the following extensions.

[1] An exception are SCMP packets, for which border routers may process up to four hop-by-hop extensions.

Path Transport Extension

The path transport extension is an end-to-end extension that allows communicating end hosts to pass or update a path. It is designed to transport a forwarding path (i.e., a path in the data-plane format, see Section 15.1.3). The transported path can be accompanied with a SCION address, to indicate that the end host can be reached via another address and path to support host mobility. This extension has a multitude of uses, for example for DoS defense, where a server can pass a secret SCION path to the client, after the user has been authenticated and identified as benign.

The path transport extension that conveys a path in the data-plane format (set of info (INF) and hop (HF) fields) is depicted in Figure 15.8. The extension

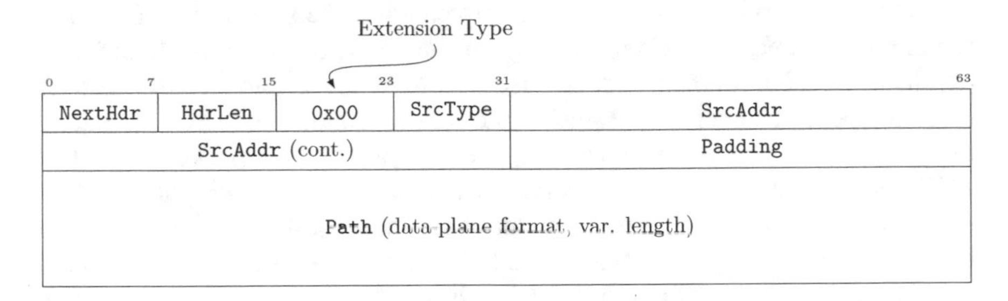

Figure 15.8: An example of a path transport extension. It is an end-to-end extension, thus the preceding `NextHdr` field has value `0xfd`.

contains the field `SrcType`, which indicates what type of SCION address the path is prepended with. A source can inform a destination that the source is reachable through another address via the `SrcAddr` field (ISD and AS identifiers concatenated with a host address). If necessary, padding is added so that the full path is aligned to an 8-byte boundary (Section 15.1.3 describes details of the data-plane path format). The `Path` field is not mandatory and an empty path can be sent, and through such a construction a source can indicate a new SCION address.

One-Hop Path Extension

The one-hop path extension is a hop-by-hop extension introduced to handle communication between two entities from neighboring ASes that do not have a forwarding path. Currently, it is used only by beacon servers during the beaconing process, and we present this extension in that context.

Beacons themselves are used for the creation forwarding paths, thus it is not possible to use a forwarding path for the first packet sent. Moreover, whenever a beacon server receives a beacon, it has to know from which ingress interface (i.e., from which border router) the beacon was forwarded. Without such information, the beacon server would not be able to create a valid hop field for the beacon. Hence, an ingress router which forwards the beacon to the beacon server needs to put its own interface identifier into the packet.

To enable communication in this case, the first packet sent (i.e., the TCP SYN packet with the beacon service address specified as the destination address) carries the one-hop path extension. This packet also has a forwarding path where the info field and the first hop field are created by the originating beacon server, while the second hop field is empty. The first router (i.e., an egress router of the sender AS), simply verifies and forwards the packet to its neighbor router (ignoring the extension), as it does for every data packet. The second router (i.e., an ingress router of the receiver AS), detects that this packet is special (i.e., it has the one-hop path extension and it is addressed to the beacon service), and contributes to the forwarding path by creating the second hop field and replacing the empty hop field in the original path. Note that the router can create hop fields as it knows the secret key used for MAC generation/verification. A receiving beacon server's TCP stack removes the one-hop path extension, and then treats the packet in a standard way, sending a SYN-ACK packet back. The TCP stack of the sender beacon server reads the completed forwarding path from the SYN-ACK packet. After that, an end-to-end one-hop path is known to the two beacon servers, and can be used to communicate between them. Any subsequent packet in the communication is sent using the created path. The header of the one-hop path extension is presented in Figure 15.9.

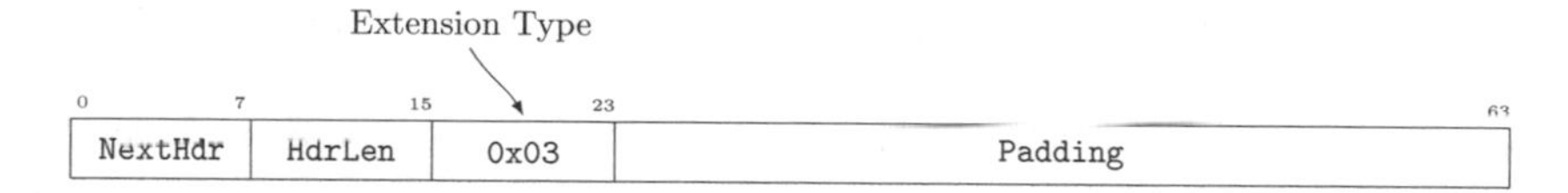

Figure 15.9: The header of the one-hop path extension. It is a hop-by-hop extension, thus the preceding NextHdr field has value 0x00.

SCION Packet Security Extension

The SCION packet security extension is an end-to-end extension that allows communicating end hosts to protect sent data on the packet level. It has a generic design that allows encryption and/or authentication of packets via either symmetric or asymmetric cryptography. The header of the SCION packet security extension is presented in Figure 15.10.

The SecMode field indicates what kind of protection is applied to the packet that carries the extension header. The content and size of the Metadata and Authenticator fields depends on the value of the SecMode field. The metadata field can include information such as timestamp or sequence number (the

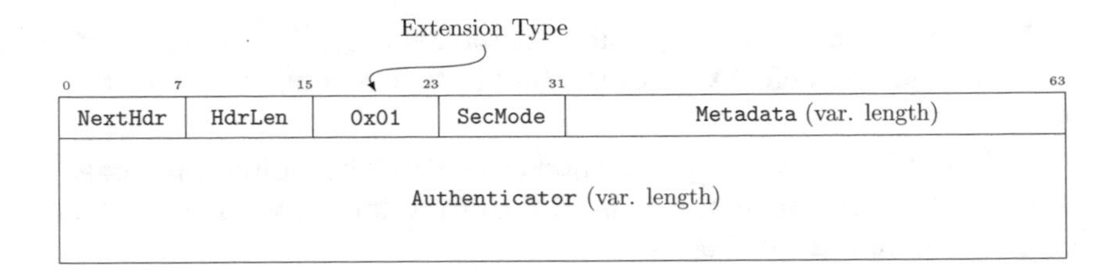

Figure 15.10: The header of the SCION packet security extension. It is an end-to-end extension, thus the preceding `NextHdr` field has value `0xfd`.

field is empty when metadata are unnecessary). The size of the authenticator field has to be aligned to a multiple of 8 bytes. Currently, we support the following security modes (i.e., values of the `SecMode` field):

- **AES-CMAC** (`0x00`): MAC scheme [226] producing 16-byte authenticators.
- **HMAC-SHA256** (`0x01`): MAC scheme [136] producing 32-byte authenticators.
- **Ed25519** (`0x02`): digital signature scheme [30] with 64-byte signatures.
- **GCM-AES128** (`0x03`): authenticated encryption scheme [76] producing 16-byte authenticators.

The metadata for all these modes is a Unix timestamp encoded as a 4-byte unsigned integer.

The SCION packet security extension can be used for encrypted and authenticated, or authenticated-only packets. However, as packets need to be processed and modified by intermediate routers the protection has to be realized in a special way:

- Before protecting a packet, the authenticator field is filled-in with zero-valued bytes.
- When a packet is only authenticated (through a MAC or a digital signature), then the entire packet is protected, except: the `TotalLen`, `HdrLen`, `CurrHF`, and `CurrINF` fields, and the sizes and payloads of hop-by-hop extensions.
- When a packet is encrypted and authenticated, then all end-to-end extensions (excluding the SCION packet security extension itself) and the layer-4 header with payload are encrypted, while the following are authenticated: SCION Header (except `TotalLen`, `HdrLen`, `CurrHF`, and `CurrINF` fields), types and order of all extensions, and all end-to-end extensions. (Hop-by-hop extensions are not protected, as they can be modified by intermediate border routers.)

Additionally, when the one-hop path extension is used, the second hop field has to be set to zero before verification (as that field was modified by a border router).

The SCION packet security extension can be used by various protocols. In our current implementation, it is used for protecting SCMP messages (see details in Section 4.2.5 on Page 82).

MTU Extension

The maximum transmission unit (MTU) extension is an end-to-end extension that helps communicating end hosts to coordinate the MTU used during transmission. As the MTU depends on the path used, the end hosts can pass the used MTU for the first packet sent via a given path.

The standard procedure is as follows. The source obtains path segments, creates a forwarding path, and learns its MTU (the relevant information is encoded within the path segments). For the first packet sent, the effective MTU is indicated within the MTU extension. For the responding packet, the destination indicates its MTU only if it is smaller than the received one (e.g., when the destination's LAN dictates a smaller MTU). Otherwise, the responding packet is sent without the MTU Extension.

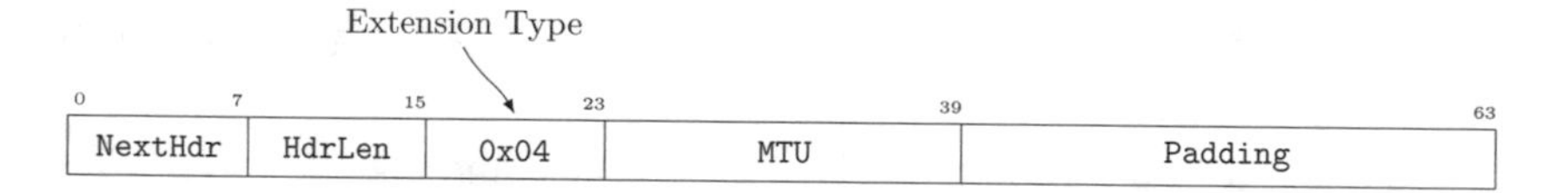

Figure 15.11: The header of the MTU extension. It is an end-to-end extension, thus the preceding `NextHdr` field has value `0xfd`.

The MTU extension can be integrated with transport protocols (such as TCP — see Section 9.3). The header of the MTU extension is presented in Figure 15.11.

AS-Level Anycast Extension

The AS-level anycast extension is a hop-by-hop extension that implements the AS-level anycast service mechanism presented in Section 7.5. The goal of the extension is to allow an anycast request to a service's server to be sent to any intermediate AS on a path to the core. Although the extension is hop-by-hop, it is processed only by ingress non-core border routers on the up-path (i.e., the forwarding path has the up direction).

The header of the extension is presented in Figure 15.12. The `Flags` field is reserved for future use, the `Service` field specifies the service that should

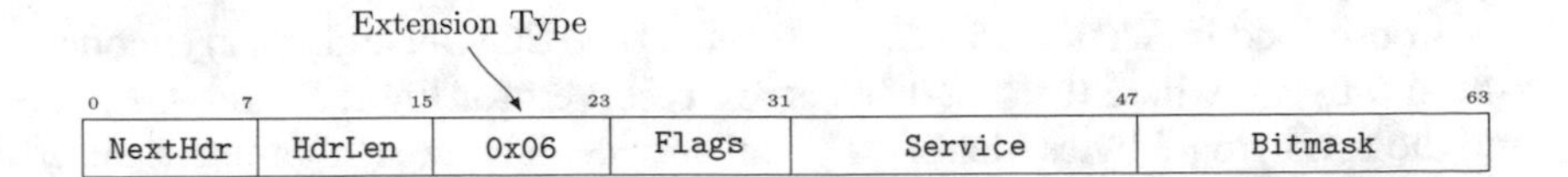

Figure 15.12: The header of the AS-level anycast extension. It is a hop-by-hop extension, thus the preceding `NextHdr` field has value `0x00`.

handle the request (SCION service address), and the `Bitmask` field specifies the ASes that should send the packet to the service's server. The ASes are specified by pointers to their corresponding hop fields. If the *i*-th most significant bit is set to 1 then the hop field starting at the byte

$$i \times 8 + \texttt{offset}$$

is set as an anycast hop field (`offset` is the length of the SCION common header and the addresses of the packet). An ingress router processing the anycast hop field tries to deliver the packet to the service's server. If this server does not operate, the packet is forwarded to the next AS.

SCMP Extension

SCION introduces the SCMP extension header, which is described along with the SCMP packet format in Section 15.6.1.

15.1.5 Layer-4 Protocols

The last element of a SCION packet is the layer-4 protocol header. Each protocol has a unique protocol number (assigned by IANA), which is encoded either within the last extension header (if any exists), or within the SCION common header (when a packet does not contain any extensions).

15.2 Control Plane

SCION control-plane messages are encapsulated within SCION packets, and control-plane protocols are implemented using TCP as the default transport protocol. There are two types of control-plane dispatching in SCION. First, in the network layer, control-plane packets can be anycast, i.e., assigned to a given service, and finally to an appropriate server (see details in Section 7.4.7). Second, in the application layer a message is assigned to a given processing logic that should be implemented by the server.

As described in Section 15.1.2, the service type SCION address is encoded within 6 bytes, where the first 4 bytes are reserved for the ISD and AS pair, and the following 2 bytes identify a given service. The most significant bit of a service address determines whether the address is multicast (the bit is set) or anycast. The addresses are assigned to the SCION services as presented in Table 15.2.

Service	Anycast Address	Multicast Address
Beacon Service	0x0000	0x8000
Path Service	0x0001	0x8001
Certificate Service	0x0002	0x8002
SIBRA Service	0x0003	0x8003
SIG Service	0x0004	0x8004
Discovery Service	0x0005	0x8005
RAINS Service	0x0006	0x8006
Time Service	0x0007	0x8007

Table 15.2: SCION services with corresponding service addresses.

The SCION network layer is able to dispatch control-plane messages to corresponding services, while control-plane messages are dispatched in the application layer to corresponding processing logics. Every SCION control-plane message starts with a 4-byte field that indicates the length of the message. The length field is followed by an encoded payload. For the encoding we use Cap'n Proto encoding [218], which enables language-neutral, platform-independent, extensible, and fast data serialization. As the exact encoding is determined by Cap'n Proto, message formats presented in this section are only logical.

The basic message types are presented in Table 15.3, while the corresponding control message formats are described in the rest of this chapter.

15.3 PCB and Path Segment

The format of a PCB is presented in Figure 15.13. It consists of the info field (created by a core AS that initiated the beacon creation) and a series of AS entries that follows the order of propagation (i.e., each AS during beacon propagation appends one entry). Details on the path construction process are presented in Section 7.1.

Message Type	Description
`Beacon`	Path-Segment Construction Beacon
`Iface`	Interface state
`PathSegRequest`	Path-segment request
`PathSegReply`	Path-segment reply
`PathSegReg`	Path-segment registration
`PathSegSync`	Path-segment synchronization
`PathRev`	Path revocation
`CertRequest`	Certificate request
`Certificate`	Certificate registration or reply
`TRCRequest`	TRC request
`TRC`	TRC registration or reply
`DRKeyRequest`	DRKey key exchange request
`DRKeyResponse`	DRKey key exchange response

Table 15.3: Message types in the application layer.

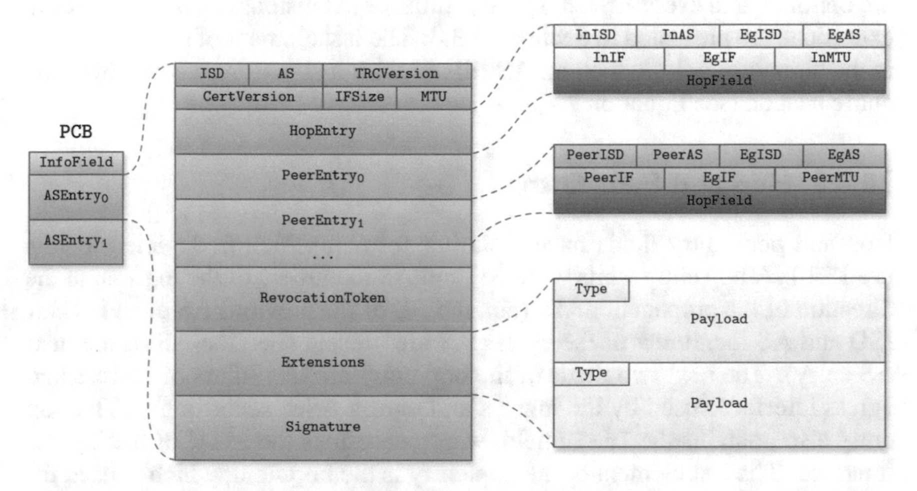

Figure 15.13: Format of the path-segment construction beacon (PCB), color scheme as in Section 7.1.

15.3.1 AS Entry

The AS entry is the main information appended to a beacon by every AS that has contributed to the beacon. Figure 15.13 depicts the format of the AS entry. It consists of metadata, a series of hop and peer entries, a revocation token, beacon extensions, and a signature.

Each AS entry starts with its ISD and AS identifiers. Next, the `TRCVersion` field informs all downstream ASes of the current version of the TRC possessed by the AS that created the entry. Similarly, `CertVersion` denotes the AS's certificate version. The `IFSize` field describes how many bits are used to encode the AS's interfaces within its hop fields, while `MTU` states the MTU of the AS's internal network.

After this metadata, the list of entries that specify routing decisions start. There are two types of entries, and they have the same structure. The first entry is interpreted as a *Hop Entry*. This entry carries routing information required to forward packets between an ingress interface (connecting to a parent AS), and an egress interface (connecting to a child AS). All other entries within an AS entry are called *Peer Entries* and they are optional. Every peer entry carries routing information needed to forward packets between a peer interface (connecting to a peer AS) and an egress interface (connecting to a child AS). The structure of the entries is presented in Figure 15.13, and their details are described in the following section.

AS entries include also the `RevocationToken` field (details on the revocation mechanism are presented in Section 7.3) and beacon extensions. Extensions are optional, and every AS can specify multiple extensions. Details on beacon extensions are presented in Section 15.3.4. The last element of every AS entry is the signature field, which an AS fills in with a signature created over the entire beacon (see Equation 7.4).

15.3.2 Hop and Peer Entry

Hop and peer entry fields have a similar format, which is depicted in Figure 15.13. A hop entry starts with ISD and AS identifiers of the ingress, in the direction of the propagation, AS (i.e., the AS of the previous AS entry). Then, ISD and AS identifiers of the egress AS are present (i.e., the AS of the next AS entry). The next two fields of the hop entry are identifiers of ingress and egress interfaces used by the ingress and egress ASes accordingly. The hop entry also contains the `InMTU` field, which describes the MTU of the ingress interface. The last element of the hop entry is the hop field, which is used for building the full data-plane path, and forwarding traffic between ingress and egress interfaces of the AS (interfaces are specified within the hop field). Note that interface identifiers of the hop field can be determined through the `IFSize` field of the AS entry. Using this field, end hosts can use paths at interface-level granularity (i.e., end hosts can determine which interfaces a packet traverses).

Peer entries have a similar format; however ISD and AS identifiers identify a peer AS of the AS that created the entry, the interface identifiers fields are assigned to the peer and egress interface, and the `PeerMTU` field denotes the MTU of the peering link. The hop field allows forwarding of traffic between peer and egress interfaces. Peer entries are optional for intra-ISD beacons and are not valid in core beacons.

15.3.3 Revocation Token

The RevocationToken field contains the information that allows authentication of revocation messages issued by this AS (details of the revocation mechanism are presented in Section 7.3). More specifically, it contains a descriptor of the hash algorithm used, and a root value of the revocation hash tree. The root is required to verify whether a revocation message is authentic (an example revocation message is presented in Figure 15.18 on Page 362).

15.3.4 Beacon Extensions

Beacons have their own extension mechanism as depicted in Figure 15.13. Beacon extensions are optional and are placed at the end of a beacon just before the signature field. Every AS entry can have multiple extensions, and each beacon extension starts with its Type encoded within 1 byte. The rest of an extension is its payload. Every extension is signed by the AS that created it, to protect its integrity. However, extensions can have unprotected fields that are not covered by the signature.

Announcement Extension

New ISDs must be announced in advance by their neighbors through a beacon extension (see Chapter 5). An announcement extension is presented in Figure 15.14. It contains the first TRC (version 0) of the newcomer with the TRC's quarantine flag set to true or false depending on whether the extension constitutes an *early announcement* or a *final announcement*. This field is optional, but the extension also contains an identifier of a hash algorithm and the hash of the TRC (these fields are mandatory). Each ISD is limited to making at most five early announcements at any point in time. The TRC field is not included in the signature calculation, although it is mandatory for PCBs (core ASes learn about new ISDs through PCBs). When a PCB, with an announcement extension, is transformed into a path segment (see Section 15.3.6), the TRC field is removed (this is done for efficiency reasons, to avoid sending large TRCs during the path lookup process).

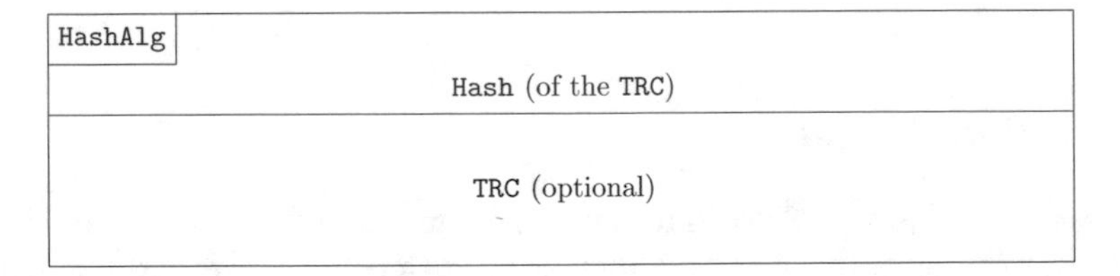

Figure 15.14: Format of announcement extensions.

Routing Policy Extension

A routing policy extension enables an AS to specify which ASes can (or cannot) use a beacon as a path segment. This mechanism allows expression of some source-based routing policies (see Section 10.9.2 on Page 232). The extension allows the permission to be set for either an entire beacon or an interface included in the beacon:

- If an AS is not permitted to use a beacon, the AS is not allowed to register the beacon as a path segment with the core ASes (a core path server has to reject such a registration), hence the path segment is not visible to other ASes.
- If an AS is not permitted to use an interface, the AS still can register the beacon as a path segment, however forwarding packets through the forbidden interface can be disabled for this AS (the border router can just sample and drop packets).

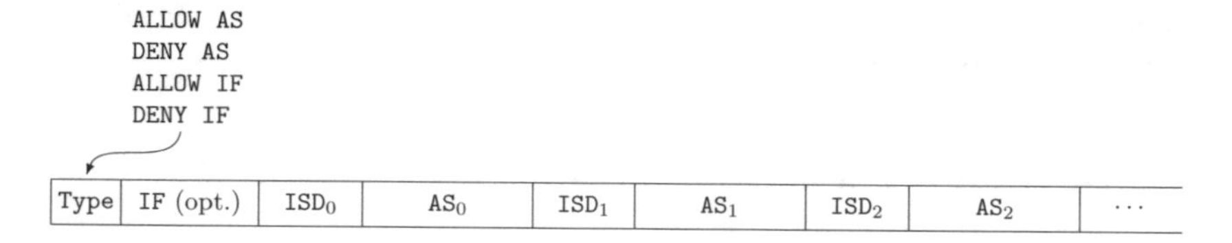

Figure 15.15: Format of routing policy extensions.

The format of a routing policy extension in presented in Figure 15.15. It lists ASes whitelisted or blacklisted to use the interface or the entire beacon as a path segment. Since this extension is signed, policy integrity is achieved.

15.3.5 Signature

The last field of a PCB is the signature. The signature is calculated over the entire PCB, as presented in Equation (7.4) on Page 121. The signing algorithm and the public key required to verify the given signature are indicated by the AS's certificate, which in turn is determined through the `CertVersion` field (see Figure 15.13).

15.3.6 Path Segments

Path segments are complete SCION paths containing all metadata required to build forwarding paths. They are registered with path servers and fetched by end hosts. Path segments use the same format as path-segment construction beacons, and the only changes are that: (a) unprotected fields of extensions are discarded, and (b) the last AS entry terminates the path (setting egress ISD, AS, and interface identifiers to zeros). To turn a PCB into a path segment, an AS conducts the following (note that this AS must be the last AS specified within this PCB):

1. Remove all unprotected fields of the PCB's extensions.
2. Add the new AS entry, which is also a termination entry (i.e., the PCB with this entry cannot be extended anymore).
3. Sign the PCB via the AS's private key, and set the signature field of the AS's (i.e., the last) AS entry.

15.4 Path Management Messages

There are several types of path management messages. Table 15.3 on Page 357 defines the different path management types. These types are encoded by three different message formats, which we explain in this section.

The format of a path request is depicted in Figure 15.16. It contains source ISD and AS identifiers, and destination ISD and AS identifiers. Through these values a requester requests a set of path segments from the source to the destination AS. The requester can specify the empty destination AS (i.e., AS identifier equals zero), which denotes a request for a path to any core AS within the destination ISD. The request messages contain the `CacheOnly` flag field. If the flag is set then only the responder's local cache is used to find requested path(s).

SrcISD	SrcAS	DstISD	DstAS	CacheOnly

Figure 15.16: Format of path request messages.

Path segments are encapsulated within a message that follows the structure from Figure 15.17. This message format is used in the following three cases:

- path replies (from a path server to a requester),
- path registrations (from a beacon server to a path server),
- path synchronization (between path servers from the same AS).

Payloads of these messages consist of a list of path segments, where every path segment is prepended with a field that represents the type of the path segment (type can be one of `UP`, `DOWN`, or `CORE`).

The final message format is used to propagate path revocation information among end hosts and the path infrastructure. The format of the message is presented in Figure 15.18. The revocation message consists of the ISD and AS identifier of the revocation message issuer, revoked interface identifier (`IF`), time period in which the interface is considered to be revoked (`Epoch`), and other fields (`Nonce`, `Proof`, `PrevRoot`, and `NextRoot`) that together prove that

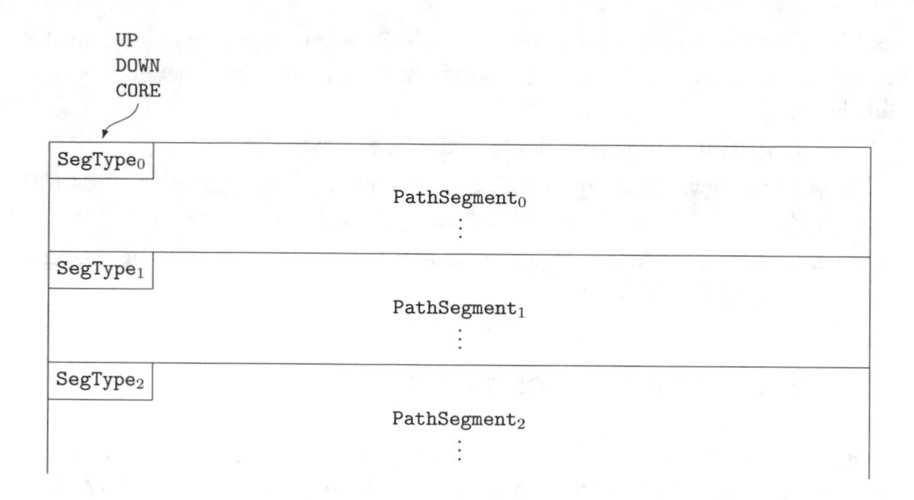

Figure 15.17: Format of path reply, registration, and synchronization messages.

the revocation is correct. Authenticity of the revocation message is verified against a revocation token contained in the path segment (see Section 15.3.3). The message also specifies a hash algorithm used to construct and verify the proof. The details of the SCION path revocation mechanism are described in Section 7.3.

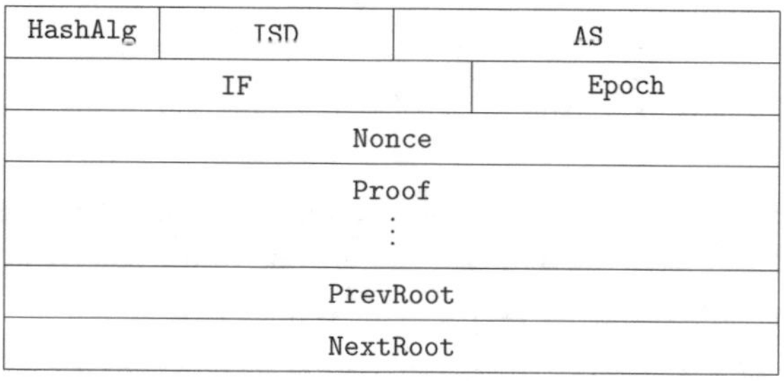

HashAlg	ISD		AS
IF			Epoch
Nonce			
Proof ⋮			
PrevRoot			
NextRoot			

Figure 15.18: Format of path revocation messages.

15.5 PKI Interactions

To interact with the control-plane PKI, SCION provides the following message formats.

The format of a certificate request message is depicted in Figure 15.19. It consists of ISD and AS identifiers and the certificate version field, which a requester specifies to ask for a given certificate. By setting the version field to *null*, the requester asks for the most recent certificate.

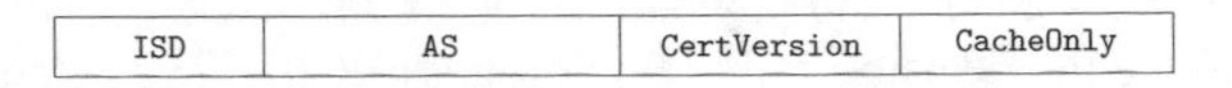

Figure 15.19: Format of certificate request messages.

Similarly, the format of a TRC request message (presented in Figure 15.20) includes an ISD identifier and the requested TRC version. Again, a version field set to *null* is used to request the most recent TRC.

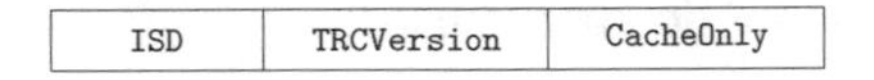

Figure 15.20: Format of TRC request messages.

Request messages contain the `CacheOnly` flag field. If the flag is set then only the responder's local cache is used to find a requested certificate/TRC (without contacting any remote server).

Replies to these requests contain only the requested certificate or TRC.

15.6 SCMP Packet

15.6.1 SCMP Extension Header

Each SCMP packet contains a mandatory hop-by-hop SCMP extension header. For efficiency reasons, the header is always the first extension header, however it does not count towards the limit of hop-by-hop extensions, so that a packet that is received with the maximum number of hop-by-hop extensions can still be replied to. By having SCMP as the first extension, any router on the path can efficiently check whether a packet is an SCMP packet (i.e., routers find it from the common header's `NextHdr` field).

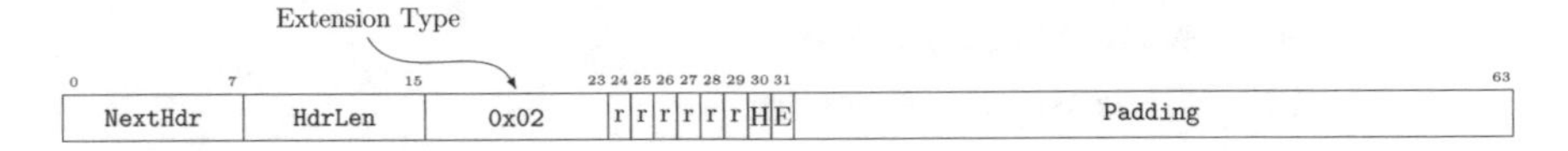

Figure 15.21: The header of the SCMP extension. It is a hop-by-hop extension, thus the preceding `NextHdr` field has value `0x00`.

The extension header is depicted in Figure 15.21. It contains a single byte encoding the following flags:

- `H`: hop-by-hop flag, set if this is a hop-by-hop SCMP message.
- `E`: error flag, set if this is an SCMP error packet.

- `r`: unused bit, reserved for future use.

The hop-by-hop flag tells the router on the path whether or not it needs to process the layer-4 header and payload of the SCMP packet. In both cases the router can make the decision only after processing the first extension header (which it can jump to directly through the `HdrLen` field in the common header).

The error flag indicates whether the SCMP message should be treated as an error. Depending on the type of a basic error that happens while parsing the packet, a router might not be able to reliably read the extension header, in which case it will treat all SCMP packets as if they were SCMP error packets. SCMP error packets must never generate SCMP error packets. This simple safeguard ensures that packet storms do not occur.

15.6.2 SCMP Layer-4 Header

An SCMP packet layout is presented in Figure 15.22. Each packet consists of an *SCMP Header* and *SCMP Payload*.

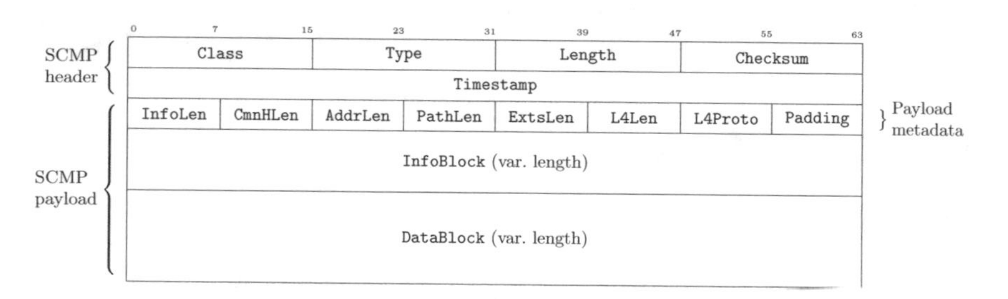

Figure 15.22: The layout of an SCMP packet.

SCMP Header

The SCMP header is presented in Figure 15.22. It contains the following fields:

- **`Class`**: SCMP message category.
- **`Type`**: SCMP message subcategory.
- **`Length`**: length (in bytes) of the SCMP header and SCMP payload.
- **`Timestamp`**: time when the SCMP header was generated. Time is expressed in microseconds since the Unix epoch.
- **`Checksum`**: Internet-style checksum [41] over the SCION address header, layer-4 protocol type, SCMP layer-4 header, and SCMP payload.

The classes and types of SCMP messages are described in Section 15.6.3.

SCMP Payload

The payload contains all the context required to interpret a message of the class and the type specified in the SCMP header. It has three parts (see Figure 15.22): *Payload Metadata* (mandatory field, 8 bytes), *Info Block* (optional field, variable length), *Data Block* (optional field, variable length).

The payload metadata field contains a series of length fields that describe the lengths of the relevant blocks (info block and data block, which can contain: quoted SCION common header, quoted address header, quoted path header, quoted extension headers, quoted layer-4 header). A length of zero means that the field is not present. The `L4Proto` field is the protocol number of the quoted layer-4 header, if it exists, and zero if it does not.

The class and type of the SCMP message defines the type of the info block, and which (if any) quoted blocks are present in the data block. The info block contains any additional information required to process the SCMP payload. The various classes and types, along with which info blocks and which quoted blocks are required are described in the next section.

15.6.3 Message Classes and Types

Below, we present classes and types of SCMP messages with their descriptions.

- **GENERAL** (`0x00`): General SCMP errors and echo request/replies.
  - **UNSPECIFIED** (`0x00`): Error that does not fall into any other category. Only to be used until a more specific error code can be allocated. `InfoBlock` includes: string describing the error. `DataBlock` includes: all headers.
  - **ECHO_REQUEST** (`0x01`): Echo request. `InfoBlock` includes: identifier (randomly generated per source app), sequence identifier (incremented for each packet). `DataBlock` can be user-supplied, set to a sequential byte string by default.
  - **ECHO_REPLY** (`0x02`): Echo reply. `InfoBlock` and `DataBlock` are copied from the request message.
  - **TRACEROUTE_REQUEST** (`0x03`): Traceroute request. `InfoBlock` includes: identifier (randomly generated per source app). Traceroute request is sent towards destination and every border router on the path detects it, generates a reply (see below), and forwards to the next hop.
  - **TRACEROUTE_REPLY** (`0x04`): Traceroute reply. `InfoBlock` is copied from the request message. `DataBlock` includes: interface identifier of the router that generated the reply.

- **FORWARDING** (`0x01`): SCMP forwarding and delivery errors. For each type, the `DataBlock` consists of quoted: common header, address header, and layer-4 header (if exists).
  - **UNREACH_NET** (`0x00`): Destination network unreachable: a layer-2 error, when there is no route to the destination host's network.
  - **UNREACH_HOST** (`0x01`): Destination host unreachable: a layer-2 error (e.g., the current machine is on the same segment as the destination host, and gets no response from it).
  - **L2_ERROR** (`0x02`): Layer-2 error not covered by other error types (e.g., TTL exceeded). `InfoBlock` includes: layer-2 error code(s).
  - **UNREACH_PROTO** (`0x03`): Destination host does not support the requested layer-4 protocol.
  - **UNREACH_PORT** (`0x04`): Destination host unable to parse layer-4 port number.
  - **UNKNOWN_HOST** (`0x05`): Destination host unknown (e.g., the destination address is an SVC address, for which the router cannot retrieve any service instances).
  - **BAD_HOST** (`0x06`): Destination host is invalid (e.g., the destination address is an unsupported SVC address).
  - **OVERSIZE_PKT** (`0x07`): Packet size is larger than MTU. `InfoBlock` includes: packet size and MTU.
  - **ADMIN_DENIED** (`0x08`): Communication with destination host administratively denied.
- **SCION COMMON HEADER** (`0x02`): SCION common header errors. For each type, the `DataBlock` consists of quoted: common header, address header, and layer-4 header (if exists).
  - **BAD_VERSION** (`0x00`): Invalid SCION version. (e.g., the SCION version is deprecated). Only versions that are known but not allowed are signaled, as with an unknown version the packet then cannot be processed.
  - **BAD_DST_TYPE** (`0x01`): Invalid destination address type. (e.g., the address type is deprecated). Only types that are known but not allowed are signaled, as with an unknown type the packet then cannot be processed.
  - **BAD_SRC_TYPE** (`0x02`): Invalid source address type. (e.g., the address type is deprecated). Only types that are known but not allowed are signaled, as with an unknown type the packet then cannot be processed.

  - **BAD_PKT_LEN** (`0x03`): `TotalLen` field in common header does not match the number of bytes received. `InfoBlock` includes: received bytes in packet.
  - **BAD_IOF_OFFSET** (`0x04`): Invalid `CurrINF` offset in common header (e.g., offset is non-zero for an empty path).
  - **BAD_HOF_OFFSET** (`0x05`): Invalid `CurrHF` offset in common header (e.g., offset is non-zero for an empty path).
- **PATH** (`0x03`): Forwarding-path-processing errors. For each type, the `DataBlock` consists of quoted: common header, address header, layer-4 header (if exists), and path (except the first type).
  - **PATH_REQUIRED** (`0x00`): Packet cannot be routed as it has no path.
  - **BAD_MAC** (`0x01`): MAC verification failed. `InfoBlock` includes: `CurrINF`, `CurrHF`.
  - **EXPIRED_HOF** (`0x02`): Hop field expired. `InfoBlock` includes: `CurrINF`, `CurrHF`.
  - **BAD_IF** (`0x03`): Invalid interface ID in HF. `InfoBlock` includes: `CurrINF`, `CurrHF`, direction flag.
  - **REVOKED_IF** (`0x04`): Revoked interface in path. `InfoBlock` includes: `CurrINF`, `CurrHF`, direction flag, revocation info.
  - **NON_FORWARD_HOF** (`0x05`): Current HF not valid for forwarding. (e.g., HF has the verify-only flag set). `InfoBlock` includes: `CurrINF`, `CurrHF`.
  - **DELIVERY_FWD_ONLY** (`0x06`): Delivery disallowed by the HF's forward-only flag. `InfoBlock` includes: `CurrINF`, `CurrHF`.
  - **DELIVERY_NON_LOCAL** (`0x07`): Delivery disallowed as destination is not local. `InfoBlock` includes: `CurrINF`, `CurrHF`.
- **EXTENSION** (`0x04`): SCION-extension-processing errors. For each type the `InfoBlock` includes the extensions index, while the `DataBlock` consists of quoted: common header, address header, layer-4 header (if exists), and extension headers.
  - **TOO_MANY_HOPBYHOP** (`0x00`): Too many hop-by-hop extensions.
  - **BAD_EXT_ORDER** (`0x01`): Invalid extension order. (e.g., SCMP extension is not the first one).
  - **BAD_HOPBYHOP** (`0x02`): Unsupported hop-by-hop extension.
  - **BAD_END2END** (`0x03`): Unsupported end-to-end extension.
- **SIBRA** (`0x05`): SIBRA errors. For each type, the `DataBlock` consists of quoted: common header, address header, layer-4 header (if exists), and

the SIBRA extension header. (In a future release of SCION we plan to add more SIBRA SCMP error types.)

- **SIBRA_BAD_VERSION** (`0x00`): Unsupported SIBRA version.
- **SIBRA_SETUP_NO_REQ** (`0x01`): Request flag not set in the setup packet.

16 Configuration File Formats

TOBIAS KLAUSMANN, STEPHEN SHIRLEY, PAWEL SZALACHOWSKI

This chapter describes the details of SCION configuration files. All SCION configuration files are represented in JSON [42] format.

Chapter Contents

16.1 Trust Root Configuration (TRC)

The TRC format is introduced in Section 4.2.1. This section provides a more detailed description of its structure. Every TRC consists of the following fields:

- `ISD`: ISD identifier as an integer from 1 to 4095.
- `Description`: human-readable description of an ISD.
- `Version`: version number of the TRC. An ISD assigns a consecutive version number (nonnegative integer) to every new TRC.
- `CreationTime`: Unix timestamp (integer) that expresses when the TRC was created.
- `ExpirationTime`: Unix timestamp (integer) that expresses when the TRC expires.
- `CoreASes`: dictionary of core ASes and their online and offline public keys (see Section 4.2.2). The keys are represented by Base64-encoded strings, and for each key a digital signature scheme is indicated.
- `RootCAs`: dictionary of root CAs and their public-key certificates. Certificates are implemented in the X.509v3 format [60] (as today), and

A. Perrig et al., *SCION: A Secure Internet Architecture*, Information Security and Cryptography, https://doi.org/10.1007/978-3-319-67080-5_16

represented in the DER format [119] encoded as a Base64 string. Each CA has also an associated online key used for signing TRCs of remote ISDs (the key is encoded in the same format as AS online keys), and a list of servers that handle TRC signing requests. Additionally, CAs specify keys and addresses of their ARPKI services.

- **`CertLogs`**: dictionary of end-entity certificate logs, and their addresses and public-key certificates (see details in Section 4.4). Certificates are implemented in the X.509v3 format, and represented in the DER format encoded as a Base64 string.
- **`ThresholdEEPKI`**: threshold number (nonnegative integer) of trusted parties (CAs and one log) that have to assert a domain's policy (see details in Section 4.4).
- **`RAINS`**: RAINS root public key (encoded as a Base64 string) and an online key used for signing TRCs of remote ISDs (the key has the same format as AS online keys). This section also includes a list of servers that handle TRC signing requests.
- **`QuorumTRC`**: quorum (i.e., threshold number, nonnegative integer) of core ASes that must sign a new TRC.
- **`QuorumCAs`**: quorum of root CAs required to change `RootCAs`, `CertLogs`, `ThresholdEEPKI`, and `QuorumCAs`.
- **`GracePeriod`**: period during which the TRC is valid after the creation time of a new TRC.
- **`Quarantine`**: Boolean flag defining whether the TRC is valid (`quarantine` = `false`) or an early announcement (`quarantine` = `true`). See Chapter 5 for more details.
- **`Signatures`**: dictionary of signatures of trust roots (core ASes, root CAs, and naming roots). This section also contains cross-signatures. The signatures are computed over the TRC, and are encoded as Base64 strings.

An example of the TRC is presented in Figure 16.1.

16.2 AS Certificates

Certificates are used to prove ownership of a public key. SCION employs certificates for routing, name, and entity authentication, and details of certificate management are given in Chapter 4. For end-entity authentication, standard TLS certificates (i.e., X.509v3 format [60]) are used, for name authentication DNSSEC's resource records [213] are deployed, while for routing authentication SCION proposes the following simple certificate format.

- **`Subject`**: string representing the entity that owns the certificate and the corresponding key pair. An AS is represented as a string *ISD-AS*.

```
{"ISD": 1,
 "Description": "The first (test) ISD",
 "Version": 2,
 "CreationTime": 1480927723,
 "ExpirationTime": 1483927723,
 "CoreASes": {
   "1-10": {"OnlineKeyAlg": "ed25519", "OfflineKeyAlg": "ed25519",
            "OnlineKey": "vDlNRHdRI4TZrdhfKn2MgV5A1ewoDv6h4osFdylEyCQ=",
            "OfflineKey": "S29jaGFtIEhlbmlhIGkgUGF1bGlua2UuIFBhd2VsLiA="},
   "1-11": {"OnlineKeyAlg": "ed25519", "OfflineKeyAlg": "ed25519",
            "OnlineKey": "5n33hhBRT86/lS6LOOhORUWweYranrnLkD8uqLzArB4=",
            "OfflineKey": "k0ScqpNRFMsal54sjlgbFxENWJq6ofdPOiazjiK9ta0="},
   "1-12": {"OnlineKeyAlg": "ed25519", "OfflineKeyAlg": "ed25519",
            "OnlineKey": "tuJOOW5bNrlzhoyohdifXo70Zc8zFl4nFyOT4JlgP1I=",
            "OfflineKey": "VYDONHZjckKqXHgprT9zmrDwGhL5dElakxNsGuxnd5I="},
   "1-13": {"OnlineKeyAlg": "ed25519", "OfflineKeyAlg": "ed25519",
            "OnlineKey": "cXRYKtY/L18KHs4dt8G6e4itodFhhj7f3LvBS5xo3as=",
            "OfflineKey": "wUw9f9wFov/kWykV/T94lJu6dfJ2aeQDOtzmnIbo32E="}},
 "RootCAs": {
   "VeriSign Class 3":   {"Certificate": "MIID3OwDQYJKoZIhvcNAQELBQA...",
          "OnlineKeyAlg": "ed25519", "OnlineKey": "F4tLPPhdEygoXidQK...",
          "TRCSrv": ["1-10 43.0.0.10"], "ARPKISrv": ["1-10 43.0.1.23"],
          "ARPKIKey": "LmvksbZlvZxYWv4qLusOGY..."},
   "GeoTrust Global CA": {"Certificate": "MIID1jCCAr6gAwIBAgIIUuuzQL...",
          "OnlineKeyAlg": "ed25519", "OnlineKey": "pW2wH8DzCRVw2KGH4...",
          "TRCSrv": ["1-21 1.23.117.88"], "ARPKISrv": ["1-21 1.23.119.8"],
          "ARPKIKey": "Pilk7MYtx5mow4VZEY2Nww..."},
   "DigiCert Root CA":   {"Certificate": "MIIE0zCCA7ugAwIBAgIQGNrRni...",
          "OnlineKeyAlg": "ed25519", "OnlineKey": "uppd7OMBMQGGHrNAk...",
          "TRCSrv": ["1-98 83.13.1.19"], "ARPKISrv": ["1-98 33.13.19.18"],
          "ARPKIKey": "/NI04nIa/3pBu/dlEv9rIe..."}},
 "CertLogs": {
   "ISD 1, Log1": {"1-11 1.1.2.3": "MIIH0zCCBbugAwI..."},
   "ISD 1, Log2": {"1-13 3.0.8.7": "MIIDbTCCAlWgAwI..."}},
 "ThresholdEEPKI": 3,
 "RAINS": {"RootRAINSKey": "fQRbxC1lfznQgUy286dUV4otp6F01vvpX1FQHKOt...",
           "OnlineKeyAlg": "ed25519", "OnlineKey": "VAsCtoEndLXAPtXVX...",
           "TRCSrv": ["1-5 13.3.33.210", "1-12 8.8.8.8"]},
 "QuorumTRC": 2,
 "QuorumCAs": 2,
 "GracePeriod": 18000,
 "Quarantine": false,
 "Signatures": {
   "1-11": "zQrFoqqaNfG62X5OyyraF8kQok4Ehh3POHooGemX+UwvhxhZnydw...",
   "1-12": "7DEAyG1ldO3jQqems22y9RZmD87VgBnbcvR7YxRIq58eLDkekV20...",
   "1-13": "D+Eg1O++oGfqKVXB/bxufdz5GbXY5CTQFGQbOSJCP07c8ebb3SzK...",
   "2-1": "ufTuR26sWp53MHu5suyQuChxWhWQM7gmgkLKJJIl2KJPAdK98Ki8a...",
   "ISD 2, RAINS": "2BwAtQ4mG9rdnpo1VGVIj96f/Ueq1TNgdXPI9YSlEREm...",
   "ISD 2, CA: TestCA": "ZO9NkrvTJ/Vec8X5T9ja1IV+o2xvhTQ6FZatnsO..."}}
```

Figure 16.1: Example of a TRC.

- **`Issuer`**: string identifying the entity that verified the binding between the public key and the subject entity. The issuer is the entity that signed the certificate. An AS is represented as a string *ISD-AS*.
- **`TRCVersion`**: version of the TRC that the issuer used at the time of signing the certificate.
- **`Version`**: version of the certificate. It has to be a unique (per subject) nonnegative integer.
- **`Comment`**: arbitrary and optional string used by the subject to describe the certificate.
- **`CanIssue`**: Boolean that describes whether the subject is allowed to issue certificates for other ASes.
- **`IssuingTime`**: point in time when the certificate was created (time is expressed as a Unix timestamp).
- **`ExpirationTime`**: expiration time of the certificate (Unix timestamp).
- **`EncAlgorithm`**: cryptographic algorithm that must be used to encrypt/decrypt a message with the subject's public/private key. An encryption algorithm is determined by a predefined string, and the list of supported algorithms is presented in Section 17.1.2.
- **`SubjectEncKey`**: subject's encryption key, encoded in Base64.
- **`SignAlgorithm`**: cryptographic algorithm that must be used to sign/verify a message with the subject's private/public key. A signing algorithm is determined by a predefined string, and the list of supported algorithms is presented in Section 17.1.2.
- **`SubjectSignKey`**: subject's signature key, encoded in Base64.
- **`Signature`**: issuer's signature over the certificate. The signature is encoded in Base64.

```
{"Subject": "1-16",
 "Issuer": "1-13",
 "TRCVersion": 2,
 "Version": 0,
 "Comment": "AS certificate",
 "CanIssue": false,
 "IssuingTime": 1480927723,
 "ExpirationTime": 1512463723,
 "EncAlgorithm": "curve25519xsalsa20poly1305",
 "SubjectEncKey": "Gfnet1MzpHGb3aUzbZQga+c44H+YNA6QM7b5p00dQkY=",
 "SignAlgorithm": "ed25519",
 "SubjectSignKey": "TqL566mz2H+uslHYoAYBhQeNlyxUq25gsmx38JHK8XA=",
 "Signature": "IdI4DeNqwa5TPkYwIeBDk3xN3605EJ/837mYyND1JcfwIOumhBK..."}
```

Figure 16.2: Example of an AS certificate.

An example certificate is presented in Figure 16.2. A private key that corresponds to the certificate's public key is encoded using Base64, and is stored as a single-line file.

A certificate chain is constructed as a simple concatenation of certificates, delimited by the numbers that order certificates in the chain. An example of a certificate chain that uses the presented certificate format is presented in Figure 16.3. The first certificate is a leaf certificate, and every subsequent certificate is signed by an owner (i.e., subject) of the next certificate. The last certificate is a self-signed (i.e., root) certificate. However, as presented in the example, it is allowed to exclude a root certificate from a chain, as an entity that validates the certificate chain is supposed to store correct root certificates in a corresponding TRC.


```
{"1":
 {"Subject": "1-16",
  "Issuer": "1-13",
  "TRCVersion": 2,
  "Version": 0,
  "Comment": "AS certificate",
  "CanIssue": false,
  "IssuingTime": 1480927723,
  "ExpirationTime": 1512463723,
  "EncAlgorithm": "curve25519xsalsa20poly1305",
  "SubjectEncKey": "Gfnet1MzpHGb3aUzbZQga+c44H+YNA6QM7b5p00dQkY=",
  "SignAlgorithm": "ed25519",
  "SubjectSigKey": "TqL566mz2H+uslHYoAYBhQeNlyxUq25gsmx38JHK8XA=",
  "Signature": "IdI4DeNqwa5TPkYwIeBDk3xN3605EJ/837mYyND1JcfwIOumhBK..."},
 "2":
 {"Subject": "1-13",
  "Issuer": "1-13",
  "TRCVersion": 2,
  "Version": 6,
  "Comment": "Core AS certificate",
  "CanIssue": true,
  "IssuingTime":  1442862832,
  "ExpirationTime": 1582463723,
  "EncAlgorithm": "curve25519xsalsa20poly1305",
  "SubjectEncKey": " Z8Kd0FTxwrPJtODRKFHtFJ5xAJejvpylHSYMbzaEbPQ=",
  "SignAlgorithm": "ed25519",
  "SubjectSigKey": " SKx1bhe3mh4Wl3eZ1ZsK1MwZwsSfcwvyn4FSI9yTvDs=",
  "Signature": "kKzkmxSszVGAHnjPfk8wo/hPSHgBIh8J5nHPXt+aCrnQi1SHeF2..."}}
```


Figure 16.3: Example of a certificate chain.

16.3 Discovery Service Configuration

Service discovery (described in Section 7.4.6) provides configuration files that are a central element of an AS's configuration. Although not all end hosts and servers of an AS receive all information from the AS's discovery service configuration file, some parts of it are necessary to enable operation of the control plane and data plane within an AS. A configuration file consists of the following fields:

- `ISD_AS`: ISD and AS identifiers in the *"ISD-AS"* format, where *ISD* is an integer from 1 to 4,095, and *AS* is an integer from 1 to 1,048,575.
- `MTU`: integer that describes (in bytes) the maximum transmission unit within the AS.
- `Overlay`: overlay protocol used within the AS to implement SCION. It can be one of the following:
  - `IPv4`: denotes an IPv4 overlay (i.e., SCION endpoints must have IPv4 addresses assigned).
  - `IPv6`: denotes an IPv6 overlay (i.e., SCION endpoints must have IPv6 addresses assigned).
  - `IPv4+6`: denotes that IPv4 and IPv6 overlays are deployed (i.e., SCION endpoints must have IPv4 and IPv6 addresses assigned).
  - `UDP/IPv4`: denotes a UDP/IPv4 overlay (i.e., SCION endpoints must have IPv4 addresses assigned and must listen on a pre-defined overlay UDP port).
  - `UDP/IPv6`: denotes a UDP/IPv6 overlay (i.e., SCION endpoints must have IPv6 addresses assigned and must listen on a pre-defined overlay UDP port).
  - `UDP/IPv4+6`: denotes that UDP/IPv4 and UDP/IPv6 overlays are deployed (i.e., SCION endpoints must have IPv4 and IPv6 addresses assigned and must listen on a pre-defined overlay UDP port).
  - `MPLS`: denotes an MPLS overlay (i.e., SCION endpoints must have MPLS labels assigned).
- `BorderRouters`: dictionary of router identifiers (arbitrary unique strings), and their local end-host addresses with an interface configuration. Each router's section can have an optional list (`InternalAddrs`) of its internally used addresses (i.e., addresses used when addressing the router itself, not for forwarding). An entry of the list contains an address (IPv4/IPv6 address or MPLS label) accompanied with the port.

  The next dictionary (`Interfaces`) specifies information about interfaces that the router supports (an interface connects with a border router of a neighboring AS). Each interface has assigned an interface identifier (positive integer) which must be unique within the AS. An interface configuration consists of the following:

  - `InternalAddrIdx`: index of the `InternalAddrs` list, which specifies the internal address of the router. This address is used for receiving local traffic (i.e., from the local AS).
  - `Overlay`: overlay protocol used to communicate with a neighbor AS's border router. It can be `IPv4`, `IPv6`, `UDP/IPv4`, `UDP/IPv6`, or `MPLS`.
  - `Bind`: address (and optionally port — if UDP encapsulation is used) that the router binds to.
  - `Public:` address (and optionally port) used for receiving remote traffic (i.e., from a neighbor AS's border router).
  - `Remote`: address (and optionally port) used by a neighbor AS's border router to receive traffic from the local AS.
  - `Bandwidth`: integer that describes bandwidth (in bits per second) of the interface.
  - `ISD_AS`: ISD and AS identifiers of the neighbor AS in the *"ISD-AS"* format.
  - `LinkType`: type of relation between local and neighbor ASes. The value of this field can be `PARENT`, `CHILD`, `PEER`, or `CORE` (if both ASes are core ASes).
  - `MTU`: integer that describes (in bytes) the maximum transmission unit of the interface.

- `ConsistencyService`: dictionary of consistency server identifiers (arbitrary unique strings), and their local (end-host) addresses. In our implementation, the consistency service is implemented with ZooKeeper [11], but this field is implementation-specific.
- `BeaconService`: dictionary of beacon server identifiers (arbitrary unique strings), and their address(es).
- `CertService`: dictionary of certificate server identifiers (arbitrary unique strings), and their addresses.
- `PathService`: dictionary of path server identifiers (arbitrary unique strings), and their addresses.
- `SibraService`: dictionary of SIBRA server identifiers (arbitrary unique strings), and their addresses.
- `RAINSService`: dictionary of RAINS server identifiers (arbitrary unique strings), and their addresses.

Addresses of servers can be specified within the two following sections. Each entity must have a `Public` address section, which is used by

- the entity as the source address for messages it sends, and

• all other entities as the destination address for packets sent to that entity.

A `Bind` section is optional. If specified, it contains the address(es) that the entity binds to. This is used in the case of NAT (a `Bind` section means that the entity will not bind to any addresses that are only listed in the `Public` section). Note that the address sections of servers can contain multiple addresses (in contrast to a router's interface address sections).

If multiple SCION entities are behind a legacy NAT device, then `Overlay-Port` should be specified in the `Public` section for each of them, and that port forwarded by the NAT device to the private address the entity runs on. (Note that this configuration only works if a UDP-based overlay is deployed within the AS.) If the NAT device is SCION-aware, then this is not needed, as the SCION-NAT dispatcher can forward incoming packets appropriately, and the overlay type does not matter.

Each address entry must contain `Addr` and `L4Port` fields. `OverlayPort` is optional (and only relevant if the UDP overlay is used), and typically only needed when running services behind a legacy NAT device as described above. If `OverlayPort` is not specified, then the default dispatcher port is used.

An example of a discovery service configuration file is presented in Figure 16.4. We emphasize that an AS can restrict visibility of some parts of the configuration file to some entities. For instance, end hosts do not need to know beacon servers or detailed border router interface information.

16.4 Router, Server, and End-Host Configuration

In this section, we describe how SCION infrastructure elements are configured and initialized.

Discovery Server

A discovery server needs an AS static configuration file that includes information on all SCION servers and border routers existing in the AS. As described in Section 7.4.6, the discovery service provides both static and dynamically generated information on the SCION services within its AS.

Since some of the information presented in the dynamic view cannot be learned from the consistency service, the values from the static configuration are mixed into the dynamic view. For example, the border routers do not register with the consistency service, and thus, information about them needs to be learned from the static configuration.

Border Router

In order to operate, a SCION border router requires the addresses of the local AS discovery servers, the AS's static configuration file (optionally), and a symmetric key used for hop-field verification. Distribution of symmetric keys within an AS is up to the AS. In particular, the AS can distribute a master key, which is then used to derive other keys, such as a hop-field creation/verification key or keys used in extensions (e.g., SIBRA or OPT).

Beacon Server

On startup, a beacon server must be provided with the addresses of the local AS discovery servers, the AS's static configuration file (optionally), a TRC (see Section 16.1), the AS's certificate (see Section 16.2), and a private key corresponding to the certificate (Section 16.2). The beacon server can additionally be given certificates of upstream ASes, older TRCs, or TRCs of other ISDs. To operate properly, beacon servers are also configured with the following parameters:

- **Hop-field key:** symmetric key used to authenticate hop fields. (The same key has to be used by border routers to verify hop-field MACs.)
- **Propagation period:** denotes how often a beacon server initiates path-segment construction beacon propagation.
- **Registration period:** describes the frequency of path-segment registration. Registration periods are configurable per path-segment type (i.e., up, down, or core).
- **Beaconing policy:** applied when a beacon server selects beacons for propagation. Details of the selection process are described in Section 7.1.4.
- **Registration policy:** applied when a beacon server selects beacons for up-, down-, or core-segment registration (each path-segment type has its own policy).

As presented, beacon servers are configured with beacon selection policies. The selection process and SCION path policies are described in Sections 7.1.4 and 10.9. In particular, the selection policy can implement the following factors:

- **Beacon store parameters**: internal configuration of a beacon store (see Section 7.1.4), such as number of stored beacons, number of beacons to select, number of stored selections, or how often the store is updated.
- **Selection filters**: factors to decide whether a given beacon is discarded. It can include a blacklist (or whitelist) of ASes or allowed ranges of the beacon's properties. In particular, these properties can include:
  - **Path length**: number of hops from an originator AS to the local AS.

  - **Last reception**: time that has elapsed since the PCB arrived at the AS's beacon store.
  - **Expiration time**: time when the given path segment expires (i.e., at least one of the hop fields expires).
  - **Last transmission**: last time at which the beacon server propagated the beacon or registered it as a path segment.
  - **Peering ASes**: number of peering ASes from the beacon.

  Disjointness can be another selection filter (an AS can specify how many common ASes/interfaces a candidate beacon can have with the previously selected beacons).
- **Property weights**: impact of different properties in the beacon selection process.
- **Policy parameters**: used to implement the routing-policy extensions (see Section 15.3.4), applicable to beaconing policies only. For example, an AS can specify a list of ASes that can or cannot be used in path segments. Similarly, the AS can specify a list of interfaces that can (or cannot) be used to forward traffic.

Path Server

On start, a path server has to be provided with the current versions of the AS's TRC, the certificate of the AS, addresses of discovery servers, and the AS's static configuration file (optionally).

Certificate Server

A certificate server must be initialized with all TRCs of its ISD and all certificates of its AS (with the corresponding private keys). Additionally, it can be provided with TRCs and certificates of remote ISDs/ASes. Besides TRCs and certificates, a certificate server requires the addresses of the local AS's discovery servers, the AS's static configuration file (optionally), and symmetric keys used in the DRKey and OPT protocols.

RAINS and SIBRA Servers

RAINS and SIBRA servers require the current TRC and the address(es) of discovery server(s), as well as their own configuration file.

End Host

An end host must possess the current TRC and the address(es) of discovery server(s).

```
{"ISD_AS": "1-11",
 "MTU": 1472,
 "Overlay": "UDP/IPv4+6",
 "BorderRouters": {
   "br1-11-1": {
     "InternalAddrs": [
       {"Public": [{"Addr": "10.1.0.1", "L4Port": 30097},
                   {"Addr": "2001:db8:a0b:1f::1", "L4Port": 30097}]},
       {"Public": [{"Addr": "10.1.0.2", "L4Port": 30097},
                   {"Addr": "2001:db8:a0b:1f::2", "L4Port": 30097}]}],
     "Interfaces": {
       "1": {"InternalAddrIdx": 0, "Overlay": "UDP/IPv4",
         "Bind": {"Addr": "10.0.0.1", "L4Port": 30090},
         "Public": {"Addr": "192.0.2.1", "L4Port": 44997},
         "Remote": {"Addr": "219.33.0.2", "L4Port": 1239},
         "Bandwidth": 500000000000, "ISD_AS": "1-12",
         "LinkType": "CORE", "MTU": 1472},
       "3": {"InternalAddrIdx": 0, "Overlay": "IPv6",
         "Public": {"Addr": "2001:db8:a0b:1f::1", "L4Port": 50000},
         "Remote": {"Addr": "2a00:14:4a:807::2e", "L4Port": 33997},
         "Bandwidth": 500000000000, "ISD_AS": "1-12",
         "LinkType": "CORE", "MTU": 4430},
       "8": { "InternalAddrIdx": 1, "Overlay": "IPv4",
         "Bind": {"Addr": "10.0.0.2", "L4Port": 40000},
         "Public": {"Addr": "192.0.2.2", "L4Port": 50000},
         "Remote": {"Addr": "156.3.22.37", "L4Port": 8112},
         "Bandwidth": 250000000000, "ISD_AS": "1-13",
         "LinkType": "CHILD", "MTU": 1480}}}},
 "ConsistencyService": {
   "1": {"Addr": "10.0.0.10", "L4Port": 2181},
    ... },
 "BeaconService": {
   "bs1-11-1": {
     "Bind": [{"Addr": "192.168.0.1", "L4Port": 30045},
              {"Addr": "2001:db8:a0b:1f::100", "L4Port": 30045}],
     "Public": [{"Addr": "10.1.0.100", "L4Port": 30045,
                "OverlayPort": 3004},
                {"Addr": "2001:db8:a0b:1f::100", "L4Port": 30045}]},
   "bs1-11-2": {
     "Public": [{"Addr": "10.1.0.101", "L4Port": 30053},
                {"Addr": "2001:db8:a0b:1f::101", "L4Port": 30053}]}},
 "CertificateService": {
    ... },
 "PathService": {
    ... },
 "SibraService": {
    ... },
 "RainsService": {
    ... }
}
```

Figure 16.4: An example of a discovery service configuration file.

17 Cryptographic Algorithms

ADRIAN PERRIG, PAWEL SZALACHOWSKI

In this chapter, we describe the algorithm agility property provided by SCION, and the cryptographic algorithms used in the SCION architecture. Algorithm selection was motivated by two main requirements, security and efficiency, and was based on standards related to cryptography, recommendations, best practices, and performance evaluations [95, 190, 221, 224].

Chapter Contents

17.1 Algorithm Agility

Algorithm agility is a property that allows a protocol to easily migrate from one algorithm to another one. It is especially important in the context of cryptographic algorithms, which become weaker over time. Since it is not possible to predict advances in cryptanalysis techniques, every future-proof protocol that employs cryptographic algorithms should provide a mechanism for algorithm agility.

In SCION, cryptographic algorithms are deployed extensively. However, in some cases algorithm selection is local to an AS and does not influence other parties. The design of the SCION elements that require coordinated lists of cryptographic algorithms follows the best current practice for cryptographic algorithm agility [112] and provides powerful mechanisms to enhance algorithm agility.

A. Perrig et al., *SCION: A Secure Internet Architecture*, Information Security and Cryptography, https://doi.org/10.1007/978-3-319-67080-5_17

17.1.1 Local Algorithms

Some cryptographic operations in SCION are performed locally, only within an AS's infrastructure. This is an ideal case for algorithm agility, as the algorithms used by an AS neither have to be synchronized with other parties, nor even known by other ASes or end hosts. In particular, these operations are:

- **Key generation:** An AS needs to generate its own keys. At least one symmetric (master key) and asymmetric key pair are required. To generate them, a strong pseudorandom number generator (PRNG) must be used.
- **Key derivation:** All symmetric keys used within an AS (e.g., keys used for hop-field authentication) are derived from the master key, which is shared among relevant infrastructure elements. ASes can freely select their favorite key derivation algorithm, as this decision influences only locally used keys.
- **Hop-field creation and verification:** Hop fields are created (see Section 7.1) with a MAC algorithm involved. However, hop fields are processed only locally (within an AS), so the AS does not have to even inform other parties which algorithm it is currently using. After hop fields are created, they are used in forwarding paths, where they are verified by border routers. Again, hop-field verification is local to the AS which has created the hop field. For verification, the same MAC scheme is used that has been used for hop-field creation.

As these operations use locally selected algorithms, ASes can change their PRNG, MAC, and key derivation algorithms at any time, with no synchronization with the rest of the network. It is only required that the AS's infrastructure elements use a consistent algorithm.

17.1.2 Mandatory-to-Implement Algorithms

In contrast to the locally used algorithms, some operations in SCION require a globally coordinated list of supported cryptographic algorithms. These operations are:

- **TRC creation and verification:** TRCs rely heavily on digital signatures. Each entity within a TRC is specified via its public-key certificate, which in turn specifies the digital signature algorithm used. The list of used algorithms has to be specified globally, as TRCs are cross-signed by other ISDs and verified by remote end hosts.
- **Beaconing:** Similarly to TRCs, beacons are protected with digital signatures. Also, AS-level certificates are used to identify the deployed algorithm. Note that each AS can use a different algorithm (as long as it is supported by other parties) to authenticate its AS entry. A beacon

is only forwarded if every AS that received it can verify all signatures. However, to support algorithm agility, SCION permits ASes to add multiple signatures to beacons. This can result in beacons (and path segments consequently) with different security properties.

- **Path-segment registration and lookup:** Every path segment is authenticated by a single AS, and similarly the algorithm used is determined from the AS certificate.
- **Path revocation:** Path revocation messages are authenticated using hash trees (see Section 7.3), which are built with cryptographic hash functions. As revocation messages can be verified by everyone, the algorithm used has to be standardized. However, each AS can pick (from the agreed list) its preferred hash function.
- **SCMP authentication:** SCMP packets are authenticated via the SCION packet security extension, which requires either a public-key encryption and a hash algorithm or a MAC scheme.
- **DRKey:** The DRKey protocol requires that a public-key encryption algorithm and a key derivation algorithm are specified.
- **Name authentication:** As name authentication in SCION is built upon RAINS, SCION relies on RAINS's algorithm suites [240].
- **End-entity authentication:** End-entity authentication relies on the TLS protocol, which provides algorithm agility and defines its own algorithm suites [68].

SCION defines a set of mandatory-to-implement cryptographic algorithms, and every allowed algorithm has a unique identifier assigned.

Authentication of TRCs, beacons, and path segments is based on AS-level certificates. (General information on AS-level certificates is provided in Section 4.2.3, while the detailed format of AS-level certificates is given in Section 16.2.) Every certificate contains the `algorithm` field, which specifies the digital signature algorithm used. An identifier of the digital signature algorithm implicitly determines the size of the private and public keys used, as well as the size of signatures produced. The certificates are short-lived, so an AS that wishes to change its digital signature scheme can do it with the next certificate re-issuance (up to a few days).

Path revocation messages are authenticated using a hash function. The hash algorithm used is specified in beacons and path segments whose lifetime is short, hence an AS can freely change the hash function used. An identifier of the hash function implicitly specifies its output length. More concretely, a hash function identifier is part of: (a) the `RevocationToken` field of every AS entry (see details in Section 15.3.3 on Page 359), and (b) each revocation message (see an example in Figure 15.18 on Page 362).

17.2 Symmetric Primitives

The only mandatory symmetric primitives in SCION are a cryptographic hash function and a MAC scheme. The other symmetric primitives are local to an AS, and do not have unique identifiers assigned. However, throughout this section we list the algorithms supported in our current specification.

17.2.1 Pseudorandom Number Generator

Pseudorandom number generators (PRNGs) are used to generate AS symmetric and asymmetric keys. As a PRNG's output is critical for the security of AS operations, it has to be selected carefully. The most common PRNGs are provided by underlying operating systems, and may be software or hardware components. By default, our implementation relies on the standard Linux PRNG [147].

17.2.2 Key Derivation Function

Key derivation functions are used by ASes to generate temporary symmetric keys used in production (e.g., for keys for hop-field generation). By default, our implementation deploys Password-Based Key Derivation Function 2 [126], with the following parameters: the output length is set to 16 bytes, the number of iterations is 1,000, and its internal pseudorandom function is set to HMAC-SHA256 [136].

17.2.3 Cryptographic Hash Function

Cryptographic hash functions are often elements of digital signature schemes, however, in SCION they are also used as stand-alone primitives in path revocation.

SCION supports two cryptographic hash functions. The first one is Secure Hash Algorithm 2 (SHA2) [185]. The supported variants of SHA2 are SHA-256, SHA-384, and SHA-512, which produce a hash value of 256, 384, and 512 bits, respectively.

The second algorithm supported is SHA3 (Keccak) [32, 203]. Similarly, the supported versions of SHA3 are SHA3-256, SHA3-384, and SHA3-512, which produce a hash value of 256, 384, and 512 bits, respectively.

17.2.4 Message Authentication Codes (MACs)

Although the MAC scheme used for hop-field authentication/verification can be freely selected by an AS, it is one of the most important cryptographic primitives in SCION. This function is executed by border routers for every packet forwarded. Hence, besides security, it has to provide outstanding performance.

In our implementation, we use an AES-based CMAC [226] as the default MAC scheme. This choice is motivated by efficiency. First, AES is commonly supported in hardware. Second, CMAC requires a minimal number of AES executions for authentication tag computation.

17.3 Asymmetric Primitives

The mandatory asymmetric primitives are digital signatures and public-key encryption (although other protocols built on top of SCION may require other asymmetric primitives). Throughout this section we list the algorithms supported in our current specification.

17.3.1 Digital Signatures

The requirements for digital signature schemes used in SCION are (a) security (as certificates, beacons, path segments, and TRCs are protected with them), (b) efficiency (as signature creation and verification are often executed), and (c) short signature and public key.

There are two digital signature algorithms used currently in SCION. As the default option the Edwards-curve Digital Signature Algorithm is used. More specifically, SCION deploys its instantiation called Ed25519 [30]. This signature scheme provides outstanding efficiency and signatures created are relatively short (64 bytes).

An alternative to Ed25519 is the Elliptic Curve Digital Signature Algorithm (ECDSA). Following the current best practice, we use the ECDSA-256 and ECDSA-384 algorithms. In comparison with Ed25519, the ECDSA algorithms are slower and produce longer signatures; however, when compared to standard digital signature algorithms (such as RSA), they provide higher efficiency and shorter signatures.

17.3.2 Public-Key Encryption

Public-key encryption in SCION is used only for key establishment in the DRKey protocol (see Section 12.5). The requirements for this primitive are similar to those for digital signatures, i.e., security (as the DRKey protocol protects SCMP messages), and efficiency (as key establishment should not incur significant overhead). In our current specification we use an Elliptic Curve Integrated Encryption Scheme (ECIES) based on the curve Curve25519 [29].

17.4 Post-Quantum Cryptography

With advances in quantum computers, some asymmetric-cryptography systems are threatened — as it becomes possible, for example, for a quantum computer to factor an RSA modulus with a lower computational overhead. Thanks to SCION's algorithm agility, new crypto systems can be introduced to provide a second layer of cryptography. This is relatively straightforward in the case of digital signatures, hash functions, or message authentication codes, as for example, a second signature can be computed with a different cryptographic scheme, the second signature can be added to the message, and the receiver can independently verify the signatures.

Algorithm agility is harder to achieve for encryption. A message that is encrypted with two different encryption algorithms needs to be decrypted with both algorithms; thus, if a receiver does not support an algorithm then it cannot decrypt the message. Separately encrypting a message with different encryption algorithms is in vain, as one achieves weakest-link security: the final message is only as secure as the weakest of the algorithms, and potentially even weaker if the encryption techniques compose unfavorably. In the case of digital signatures and message authentication codes described above, the combination of multiple schemes composes favorably, and the resulting security strength is at least that of the strongest algorithm, if implemented correctly.

We now discuss the quantum-resilient cryptographic algorithms that the basic SCION infrastructure can utilize.

Recently proposed symmetric-key cryptographic algorithms, such as PRFs for key derivation, hash functions, encryption, or MACs, resist quantum computers as long as the key length is sufficiently long.

For post-quantum key exchange mechanisms, new protocols that are based on hard problems in ideal lattices have recently been proposed [5, 40]. These approaches are based on the Ring Learning With Errors (R-LWE) problem.

One-time signatures provide a promising mechanism for the construction of post-quantum signature algorithms [31, 43]. These types of signatures are based on one-way functions, which can be efficiently instantiated with cryptographic hash functions.

The salient point here is that SCION is already prepared to accommodate and switch to any cryptographic algorithm, thus scaling with respect to cryptographic progress.

Bibliography

[1] Martín Abadi, Andrew Birrell, Ilya Mironov, Ted Wobber, and Yinglian Xie. Global authentication in an untrustworthy world. In *Proceedings of Workshop on Hot Topics in Operating Systems (HotOS)*, May 2013. ▻ Page 10.

[2] Bernhard Ager, Nikolaos Chatzis, Anja Feldmann, Nadi Sarrar, Steve Uhlig, and Walter Willinger. Anatomy of a large European IXP. In *Proceedings of the ACM SIGCOMM Conference*, 2012. ▻ Page 54.

[3] William Aiello, John Ioannidis, and Patrick McDaniel. Origin authentication in interdomain routing. In *Proceedings of the ACM Conference on Computer and Communications Security (CCS)*, October 2003. ▻ Page 307.

[4] Kahraman Akdemir, Martin Dixon, Wajdi Feghali, Patrick Fay, Vinodh Gopal, Jim Guilford, Erdinc Ozturk, Gil Wolrich, and Ronen Zohar. Breakthrough AES performance with Intel AES New Instructions. *White paper, June*, 2010. ▻ Page 11.

[5] Erdem Alkim, Léo Ducas, Thomas Pöppelmann, and Peter Schwabe. Post-quantum key exchange: A new hope. Technical Report 2015/1092, Cryptology ePrint Archive, March 2016. ▻ Page 386.

[6] American Registry for Internet Numbers (ARIN). Resource Public Key Infrastructure (RPKI). `https://www.arin.net/resources/rpki/`. ▻ Page 61.

[7] David G. Andersen, Hari Balakrishnan, Nick Feamster, Teemu Koponen, Daekyeong Moon, and Scott Shenker. Accountable Internet Protocol (AIP). In *Proceedings of the ACM SIGCOMM Conference*, 2008. ▻ Pages 25, 28, 94, 284, and 320.

[8] David G. Andersen, Hari Balakrishnan, M. Frans Kaashoek, and Robert Morris. Resilient overlay networks. In *Proceedings of ACM Symposium on Operating Systems Principles (SOSP)*, October 2001. ▻ Pages 9, 24, and 192.

[9] Tom Anderson, Ken Birman, Robert Broberg, Matthew Caesar, Douglas Comer, Chase Cotton, Michael J. Freedman, Andreas Haeberlen, Zachary G. Ives, Arvind Krishnamurthy, William Lehr, BoonThau Loo, David Mazieres, Antonio Nicolosi, Jonathan M. Smith, Ion Stoica, Robbert Renesse, Michael Walfish, Hakim Weatherspoon, and Christopher S. Yoo. The NEBULA future Internet architecture. In *The Future Internet*,

A. Perrig et al., *SCION: A Secure Internet Architecture*, Information Security and Cryptography, https://doi.org/10.1007/978-3-319-67080-5

Lecture Notes in Computer Science. Springer-Verlag, 2013. ▷ Pages 14 and 331.

[10] Tom Anderson, Timothy Roscoe, and David Wetherall. Preventing Internet denial-of-service with capabilities. *ACM SIGCOMM Computer Communication Review*, 2004. ▷ Pages 244 and 276.

[11] Apache. ZooKeeper. `http://zookeeper.apache.org`. ▷ Pages 147, 322, and 375.

[12] Maria Apostolaki, Aviv Zohar, and Laurent Vanbever. Hijacking Bitcoin: Routing attacks on cryptocurrencies. In *Proceedings of the IEEE Symposium on Security and Privacy (S&P)*, 2017. ▷ Pages 32 and 192.

[13] Roy Arends, Rob Austein, Matt Larson, Dan Massey, and Scott Rose. DNS security introduction and requirements. RFC 4033, March 2005. ▷ Pages 10 and 83.

[14] Katerina Argyraki and David R. Cheriton. Network capabilities: The good, the bad and the ugly. In *ACM HotNets*, 2005. ▷ Page 276.

[15] Brice Augustin, Xavier Cuvellier, Benjamin Orgogozo, Fabien Viger, Timur Friedman, Matthieu Latapy, Clémence Magnien, and Renata Teixeira. Avoiding traceroute anomalies with Paris traceroute. In *Proceedings of the ACM Internet Measurement Conference (IMC)*, 2006. ▷ Page 222.

[16] Jozef Babiarz, Kwok Chan, and Fred Baker. Configuration guidelines for DiffServ service classes. RFC 4594, August 2006. ▷ Page 244.

[17] Fred Baker, Carol Iturralde, Francois Le Faucheur, and Bruce Davie. Aggregation of RSVP for IPv4 and IPv6 reservations. RFC 3175, September 2001. ▷ Page 277.

[18] Fred Baker and Pekka Savola. Ingress filtering for multihomed networks. RFC 3704, March 2004. ▷ Page 320.

[19] Hitesh Ballani, Yatin Chawathe, Sylvia Ratnasamy, Timothy Roscoe, and Scott Shenker. Off by default! In *Proceedings of ACM Workshop on Hot Topics in Networks (HotNets)*, 2005. ▷ Page 30.

[20] David Barrera, Laurent Chuat, Adrian Perrig, Raphael M. Reischuk, and Pawel Szalachowski. The SCION Internet architecture. *Communications of the ACM*, 60(6), June 2017. ▷ Page 306.

[21] Cristina Basescu, Yue-Hsun Lin, Haoming Zhang, and Adrian Perrig. High-speed inter-domain fault localization. In *Proceedings of the IEEE Symposium on Security and Privacy (S&P)*, May 2016. ▷ Pages xvi and 281.

[22] Cristina Basescu, Raphael M. Reischuk, Pawel Szalachowski, Adrian Perrig, Yao Zhang, Hsu-Chun Hsiao, Ayumu Kubota, and Jumpei Urakawa. SIBRA: Scalable Internet bandwidth reservation architecture. In *Proceedings of the Symposium on Network and Distributed Systems Security (NDSS)*, February 2016. ▷ Pages xvi, 39, 243, 270, and 306.

[23] David Basin, Cas Cremers, Tiffany Hyun-Jin Kim, Adrian Perrig, Ralf Sasse, and Pawel Szalachowski. ARPKI: Attack resilient public-key infrastructure. In *Proceedings of the ACM Conference on Computer and Communications Security (CCS)*, November 2014. ▻ Pages xvi, 29, 40, 87, 89, and 302.

[24] David Basin, Cas Cremers, Tiffany Hyun-Jin Kim, Adrian Perrig, Ralf Sasse, and Pawel Szalachowski. Design, analysis, and implementation of ARPKI: an attack-resilient public-key infrastructure. *IEEE Transactions on Dependable and Secure Computing (TDSC)*, 2017. ▻ Pages 40, 87, 89, and 302.

[25] BBC. Asia communications hit by quake. `http://news.bbc.co.uk/2/hi/asia-pacific/6211451.stm`, December 2006. ▻ Pages 10 and 307.

[26] Stefan Bechtold and Adrian Perrig. Accountability in future Internet architectures: Can technical and legal aspects be happily intertwined? *Communications of the ACM*, 57(9):21–23, September 2014. ▻ Page xvi.

[27] Steven Bellovin, David Clark, Adrian Perrig, and Dawn Song. A clean-slate design for the next-generation secure Internet. Report for NSF Global Environment for Network Innovations (GENI) workshop, July 2005. ▻ Page xvii.

[28] Mark Berman, Jeffrey S. Chase, Lawrence Landweber, Akihiro Nakao, Max Ott, Dipankar Raychaudhuri, Robert Ricci, and Ivan Seskar. GENI: A federated testbed for innovative network experiments. *Computer Networks*, 61:5–23, 2014. ▻ Page 15.

[29] Daniel J. Bernstein. Cryptography in NaCl. *Networking and Cryptography library*, 3, 2009. ▻ Page 385.

[30] Daniel J Bernstein, Niels Duif, Tanja Lange, Peter Schwabe, and Bo-Yin Yang. High-speed high-security signatures. *Journal of Cryptographic Engineering*, 2(2):77–89, 2012. ▻ Pages 353 and 385.

[31] Daniel J. Bernstein, Daira Hopwood, Andreas Hülsing, Tanja Lange, Ruben Niederhagen, Louiza Papachristodoulou, Michael Schneider, Peter Schwabe, and Zooko Wilcox-O'Hearn. SPHINCS: practical stateless hash-based signatures. In *Advances in Cryptology - EUROCRYPT*, 2015. ▻ Page 386.

[32] Guido Bertoni, Joan Daemen, Michaël Peeters, and Gilles Van Assche. Keccak sponge function family main document. *Submission to NIST (Round 2)*, 3:30, 2009. ▻ Page 384.

[33] Robert Beverly, Ryan Koga, and KC Claffy. Initial longitudinal analysis of IP source spoofing capability on the Internet. *Internet Society*, 2013. ▻ Pages 313 and 320.

[34] Bobby Bhattacharjee, Ken Calvert, Jim Griffioen, Neil Spring, and James Sterbenz. Postmodern internetwork architecture. Technical Report ITTC

Technical Report ITTC-FY2006-TR-45030-01, University of Kansas, February 2006. ▻ Pages 13 and 40.

[35] Andrew Birrell, Butler Lampson, Roger Needham, and Michael Schroeder. A global authentication service without global trust. In *Proceedings of the IEEE Symposium on Security and Privacy (S&P)*, 1986. ▻ Pages 48 and 62.

[36] Burton H. Bloom. Space/time trade-offs in hash coding with allowable errors. *Communications of the ACM*, 13(7):422–426, 1970. ▻ Page 270.

[37] Rakesh Babu Bobba, Laurent Eschenauer, Virgil Gligor, and William Arbaugh. Bootstrapping security associations for routing in mobile ad-hoc networks. In *Proceedings of IEEE Global Telecommunications Conference (Globecom)*, 2003. ▻ Page 65.

[38] Raffaele Bolla, Roberto Bruschi, Franco Davoli, and Flavio Cucchietti. Energy efficiency in the future Internet: a survey of existing approaches and trends in energy-aware fixed network infrastructures. *IEEE Communications Surveys & Tutorials*, 13(2):223–244, 2011. ▻ Page 331.

[39] Carsten Bormann and Paul Hoffman. Concise binary object representation (CBOR). RFC 7049, October 2013. ▻ Page 114.

[40] Joppe Bos, Craig Costello, Léo Ducas, Ilya Mironov, Michael Naehrig, Valeria Nikolaenko, Ananth Raghunathan, and Douglas Stebila. Frodo: Take off the ring! Practical, quantum-secure key exchange from LWE. Technical Report 2016/659, Cryptology ePrint Archive, June 2016. ▻ Page 386.

[41] R.T. Braden, D.A. Borman, and C. Partridge. Computing the Internet checksum. RFC 1071, September 1988. ▻ Page 364.

[42] Tim Bray. The JavaScript Object Notation (JSON) Data Interchange Format. RFC 7159, March 2014. ▻ Page 369.

[43] Johannes Buchmann, Erik Dahmen, and Andreas Hülsing. XMSS - a practical forward secure signature scheme based on minimal security assumptions. Technical Report 2011/484, Cryptology ePrint Archive, November 2011. ▻ Page 386.

[44] Kevin Butler, Toni R. Farley, Patrick McDaniel, and Jennifer Rexford. A Survey of BGP Security Issues and Solutions. *Proceedings of the IEEE*, 98(1), 2010. ▻ Page 307.

[45] Matthew Caesar and Jennifer Rexford. BGP routing policies in ISP networks. *IEEE Network: The Magazine of Global Internetworking*, 2005. ▻ Page 5.

[46] Kenneth L. Calvert, James Griffioen, and Leonid Poutievski. Separating routing and forwarding: A clean-slate network layer design. In *Proceedings of International Conference on Broadband Communications, Networks and Systems (BROADNETS)*, 2007. ▻ Pages 13 and 40.

[47] Isidro Castineyra, Noel Chiappa, and Martha Steenstrup. The Nimrod routing architecture. RFC 1992, August 1996. ▻ Pages 13 and 40.

[48] Stephen Checkoway, Jacob Maskiewicz, Christina Garman, Joshua Fried, Shaanan Cohney, Matthew Green, Nadia Heninger, Ralf-Philipp Weinmann, Eric Rescorla, and Hovav Shacham. A systematic analysis of the Juniper Dual EC incident. In *Proceedings of the ACM Conference on Computer and Communications Security (CCS)*, October 2016. ▻ Page 305.

[49] Chen Chen, Daniele Enrico Asoni, David Barrera, George Danezis, and Adrian Perrig. HORNET: High-speed onion routing at the network layer. In *Proceedings of the ACM Conference on Computer and Communications Security (CCS)*, October 2015. ▻ Pages xvi, 35, 39, 306, and 319.

[50] Chen Chen, David Barrera, and Adrian Perrig. Modeling data-plane power consumption of future Internet architectures. In *Proceedings of the IEEE Conference on Collaboration and Internet Computing (CIC)*, November 2016. ▻ Pages xvi, 332, 333, and 334.

[51] Chen Chen and Adrian Perrig. PHI: Path-hidden lightweight anonymity protocol at network layer. In *Proceedings on Privacy Enhancing Technologies (PoPETs)*, July 2017. ▻ Page xvi.

[52] Laurent Chuat, Pawel Szalachowski, Adrian Perrig, Ben Laurie, and Eran Messeri. Efficient gossip protocols for verifying the consistency of certificate logs. In *Proceedings of the IEEE Conference on Communications and Network Security (CNS)*, 2015. ▻ Pages xvi and 92.

[53] Cisco. Requirements for next-generation core routing systems. `https://perma.cc/58AX-TAQJ`. ▻ Page 306.

[54] Cisco. Field notice: Endless BGP convergence problem in Cisco IOS software releases, 2001. ▻ Page 235.

[55] Cisco. IOS and IOS XE software cluster management protocol remote code execution vulnerability. `https://perma.cc/M7P8-FH26`, March 2017. ▻ Page 305.

[56] David Clark, Robert Braden, Aaron Falk, and Venkata Pingali. FARA: Reorganizing the addressing architecture. In *Proceedings of the ACM SIGCOMM Workshop on Future Directions in Network Architecture*, 2003. ▻ Pages 13 and 40.

[57] David Clark, Karen Sollins, John Wroclawski, Dina Katabi, Joanna Kulik, Xiaowei Yang, Robert Braden, Ted Faber, Aaron Falk, Venkata Pingali, Mark Handley, and Noel Chiappa. NewArch: Future generation Internet architecture. Technical report, Air Force Research Labs, 2004. ▻ Pages 13 and 40.

[58] The PlanetLab Consortium. PlanetLab, an open platform for developing, deploying, and accessing planetary-scale services. https://www.planet-lab.org/, 2016. ▷ Page 221.

[59] Danny Cooper, Ethan Heilman, Kyle Brogle, Leonid Reyzin, and Sharon Goldberg. On the risk of misbehaving RPKI authorities. In *Proceedings of the ACM Workshop on Hot Topics in Networks (HotNets)*, November 2013. ▷ Page 65.

[60] David Cooper, Stefan Santesson, Stephen Farrell, Sharon Boeyen, Russ Housley, and William Polk. Internet X.509 public key infrastructure certificate and certificate revocation list (CRL) profile. RFC 5280, May 2008. ▷ Pages 369 and 370.

[61] Miguel Medeiros Correia and Mustafa Tok. DNS-based Authentication of Named Entities (DANE). Technical report, Universidade do Porto, 2011–2012. ▷ Page 103.

[62] Court of Justice of the European Union (CURIA). Personal data: Protection of individuals with regard to the processing of such data. https://perma.cc/J2T3-VD67. ▷ Page 35.

[63] Jim Cowie. The new threat: Targeted Internet traffic misdirection. http://research.dyn.com/2013/11/mitm-internet-hijacking/, November 2013. ▷ Pages 7, 32, and 280.

[64] Stephen E. Deering and Robert Hinden. Internet protocol, version 6 (IPv6) specification. RFC 2460, December 1998. ▷ Page 4.

[65] Alan Demers, Srinivasan Keshav, and Scott Shenker. Analysis and simulation of a fair queueing algorithm. *ACM SIGCOMM Computer Communication Review*, 1989. ▷ Page 244.

[66] Amogh Dhamdhere and Constantine Dovrolis. Twelve years in the evolution of the Internet ecosystem. *IEEE/ACM Transactions on Networking*, 19(5):1420–1433, Sep 2011. ▷ Page 38.

[67] DICE Control Plane Working Group. Inter-Domain Controller (IDC) protocol specification. http://www.controlplane.net/, February 2010. Version 1.1. ▷ Page 271.

[68] Tim Dierks and Eric Rescorla. The Transport Layer Security (TLS) protocol version 1.2. RFC 5246, August 2008. ▷ Page 383.

[69] Daniel Eran Dilger. Oops: Microsoft leaks its golden key, unlocking Windows Secure Boot and exposing the danger of backdoors. https://perma.cc/444S-SPR6, 2016. ▷ Page 44.

[70] Wenxiu Ding, Zheng Yan, and Robert H. Deng. A survey on future Internet security architectures. *IEEE Access*, 4:4374–4393, July 2016. ▷ Pages 302 and 330.

[71] Mo Dong, Qingxi Li, Doron Zarchy, P. Brighten Godfrey, and Michael Schapira. PCC: Re-architecting congestion control for consistent high

performance. In *Proceedings of the USENIX Symposium on Networked Systems Design and Implementation (NSDI)*, 2015. ▻ Page 189.

[72] John R. Douceur. The Sybil attack. In *First International Workshop on Peer-to-Peer Systems (IPTPS '02)*, March 2002. ▻ Page 48.

[73] Jim Duffy. Cisco's IOS vs. Juniper's JUNOS. `https://perma.cc/FY59-WV9D`, April 2008. Network World. ▻ Page 306.

[74] Adam Dunkels. Design and implementation of the lwIP TCP/IP stack. *Swedish Institute of Computer Science*, 2:77, 2001. ▻ Page 187.

[75] Zakir Durumeric, James Kasten, David Adrian, J. Alex Halderman, Michael Bailey, Frank Li, Nicolas Weaver, Johanna Amann, Jethro Beekman, Mathias Payer, and Vern Paxson. The matter of Heartbleed. In *Proceedings of the ACM Internet Measurement Conference (IMC)*, 2014. ▻ Page 74.

[76] Morris J. Dworkin. SP 800-38D, Recommendation for block cipher modes of operation: Galois/Counter Mode (GCM) and GMAC, 2007. ▻ Page 353.

[77] Hurricane Electric. BGP peer report. `http://bgp.he.net/report/peers`, 2016. ▻ Page 146.

[78] Electronic Frontier Foundation. SSL Observatory. `https://www.eff.org/observatory`, 2010. ▻ Page 10.

[79] European Commission. FIRE: Future Internet research and experimentation. `https://www.ict-fire.eu`. ▻ Page 15.

[80] European Commission. FIWARE: Core platform of the future Internet. `https://www.fiware.org`. ▻ Page 15.

[81] European Commission. FORWARD: Managing emerging threats in ICT infrastructures. `http://www.ict-forward.eu`. ▻ Page 14.

[82] European Commission. SysSec: A European network of excellence in managing threats and vulnerabilities in the future Internet. `http://www.syssec-project.eu/`. ▻ Page 14.

[83] Dino Farinacci, Vince Fuller, David Meyer, and Darrel Lewis. The locator/ID separation protocol (LISP). RFC 6830, January 2013. ▻ Page 25.

[84] Adrian Farrel, Jean-Philippe Vasseur, and Jerry Ash. A path computation element (PCE)-based architecture. RFC 4655, August 2006. ▻ Page 277.

[85] Stephen Farrell and Hannes Tschofenig. Pervasive monitoring is an attack. RFC 7258, May 2014. ▻ Page xvii.

[86] Seyed Kaveh Fayazbakhsh, Yin Lin, Amin Tootoonchian, Ali Ghodsi, Teemu Koponen, Bruce Maggs, K.C. Ng, Vyas Sekar, and Scott Shenker. Less pain, most of the gain: incrementally deployable ICN. In *Proceedings of the ACM SIGCOMM Conference*, August 2013. ▻ Pages 12, 335, 336, and 337.

[87] Paul Ferguson and Daniel Senie. Network ingress filtering: Defeating denial of service attacks which employ IP source address spoofing. RFC 2827, May 2000. ▻ Page 320.

[88] Clarence Filsfils, Stefano Previdi, Bruno Decraene, Stephane Litkowski, and Rob Shakir. Segment routing architecture. Internet-draft, February 2017. ▻ Page 14.

[89] The Apache Software Foundation. Apache License Version 2.0, January 2004. `http://www.apache.org/licenses/LICENSE-2.0`, 2016. ▻ Page 306.

[90] Eva Galperin, Seth Schoen, and Peter Eckersley. A post mortem on the Iranian DigiNotar attack. `https://www.eff.org/deeplinks/2011/09/post-mortem-iranian-diginotar-attack`, October 2011. ▻ Page 7.

[91] Lixin Gao and Jennifer Rexford. Stable Internet routing without global coordination. *Networking, IEEE/ACM Trans on*, 9(6):681–692, 2001. ▻ Pages 234 and 328.

[92] GEANT. Bandwidth on demand. `http://geant3.archive.geant.net/service/BoD/pages/home.aspx`, 2015. ▻ Page 271.

[93] Yossi Gilad and Amir Herzberg. Plug-and-play IP security: Anonymity infrastructure instead of PKI. In *Proceedings of ESORICS*, 2013. ▻ Page 284.

[94] Phillipa Gill, Michael Schapira, and Sharon Goldberg. A survey of interdomain routing policies. *ACM SIGCOMM Computer Communication Review*, 44(1):28–34, 2013. ▻ Pages 233, 234, and 235.

[95] Damien Giry. Cryptographic key length recommendation. `http://www.keylength.com`, 2016. ▻ Page 381.

[96] James Glanz. Power, pollution and the Internet. `https://nyti.ms/2k5AX0q`, September 2012. The New York Times. ▻ Page 331.

[97] Virgil Gligor, Shyh-Wei Luan, and Joseph Pato. On inter-realm authentication in large distributed systems. In *Proceedings of the IEEE Symposium on Security and Privacy (S&P)*, 1992. ▻ Pages 48 and 62.

[98] P. Brighten Godfrey, Igor Ganichev, Scott Shenker, and Ion Stoica. Pathlet routing. In *Proceedings of the ACM SIGCOMM Conference*, 2009. ▻ Page 14.

[99] Fernando Gont. ICMP Attacks against TCP. RFC 5927, July 2010. ▻ Pages 7, 82, and 156.

[100] Geoffrey Goodell, William Aiello, Timothy Griffin, John Ioannidis, Patrick D. McDaniel, and Aviel D. Rubin. Working around BGP: An incremental approach to improving security and accuracy in interdomain routing. In *Proceedings of the Symposium on Network and Distributed Systems Security (NDSS)*, February 2003. ▻ Page 307.

[101] Google. Roughtime. https://roughtime.googlesource.com, 2016. ▷ Pages 160 and 289.

[102] Google. IPv6 adoption statistics. https://perma.cc/ZR8C-4CJG, April 2017. ▷ Page 4.

[103] Timothy G. Griffin, F. Bruce Shepherd, and Gordon Wilfong. The stable paths problem and interdomain routing. *IEEE/ACM Transactions on Networking (ToN)*, 10(2):232–243, 2002. ▷ Page 235.

[104] Ben Grubb. These graphs show the impact Netflix is having on the Australian Internet. http://www.smh.com.au/digital-life/digital-life-news/these-graphs-show-the-impact-netflix-is-having-on-the-australian-internet-20150401-1mdc1i, 2015. The Sydney Morning Herald. ▷ Pages 273 and 275.

[105] Ryan Hamilton, Janardhan Iyengar, Ian Swett, and Alyssa Wilk. QUIC: A UDP-based secure and reliable transport for HTTP/2. Internet-Draft, January 2016. ▷ Pages 179 and 188.

[106] Dongsu Han, Ashok Anand, Fahad Dogar, Boyan Li, Hyeontaek Lim, Michel Machado, Arvind Mukundan, Wenfei Wu, Aditya Akella, David G. Andersen, John W. Byers, Srinivasan Seshan, and Peter Steenkiste. XIA: Efficient support for evolvable internetworking. In *Proceedings of the USENIX Symposium on Networked Systems Design and Implementation (NSDI)*, 2012. ▷ Page 14.

[107] Garrett Hardin. The tragedy of the commons. *Science*, 1968. ▷ Page 245.

[108] Helion Technology. Giga AES cores. http://www.heliontech.com/aes_giga.htm. ▷ Page 334.

[109] Stephen Herzog. Revisiting the Estonian cyber attacks: Digital threats and multinational responses. *Journal of Strategic Security*, 4(2):49–60, 2011. ▷ Page 7.

[110] R. Hinden and B. Haberman. Unique local IPv6 unicast addresses. RFC 4193, October 1995. ▷ Page 114.

[111] Kerry Hinton, Jayant Baliga, Michael Feng, Robert Ayre, and Rodney S. Tucker. Power consumption and energy efficiency in the Internet. *IEEE Network*, 25(2):6–12, 2011. ▷ Page 331.

[112] Russ Housley. Guidelines for cryptographic algorithm agility and selecting mandatory-to-implement algorithms. RFC 7696, 2015. ▷ Page 381.

[113] Hsu-Chun Hsiao, Tiffany Hyun-Jin Kim, Adrian Perrig, Akira Yamada, Samuel C. Nelson, Marco Gruteser, and Wei Meng. LAP: Lightweight anonymity and privacy. In *Proceedings of the IEEE Symposium on Security and Privacy (S&P)*, 2012. ▷ Pages xvi and 39.

[114] Hsu-Chun Hsiao, Tiffany Hyun-Jin Kim, Sangjae Yoo, Xin Zhang, Soo Bum Lee, Virgil Gligor, and Adrian Perrig. STRIDE: Sanctuary trail – refuge from Internet DDoS entrapment. In *Proceedings of the ACM*

Asia Conference on Computer and Communications Security (AsiaCCS), 2013. ▻ Pages xvi, 276, and 277.

[115] Yih-Chun Hu, Adrian Perrig, and David Johnson. Wormhole attacks in wireless networks. *IEEE Journal on Selected Areas in Communications (JSAC)*, 24(2), February 2006. ▻ Pages 311 and 312.

[116] Yih-Chun Hu, Adrian Perrig, and Marvin Sirbu. SPV: Secure path vector routing for securing BGP. In *Proceedings of the ACM SIGCOMM Conference*, September 2004. ▻ Page 307.

[117] Geoff Huston. BGP in 2014. `http://www.potaroo.net/ispcol/2015-01/bgp2014.html`, January 2015. ▻ Pages 10 and 307.

[118] Red Hat Inc. Ansible, automation for everyone. `https://www.ansible.com`, 2016. ▻ Page 218.

[119] International Telecommunication Union. Information technology — ASN.1 encoding rules: Specification of basic encoding rules (BER), canonical encoding rules (CER), and distinguished encoding rules (DER). ITU-T Recommendation X.690, 2002. ▻ Page 370.

[120] Internet Assigned Numbers Authority (IANA). Protocol numbers. `https://perma.cc/FBE8-S2W5`. ▻ Pages 345 and 350.

[121] Internet Corporation for Assigned Names and Numbers (ICANN). Montevideo statement on the future of Internet cooperation. `https://www.icann.org/news/announcement-2013-10-07-en`, October 2013. ▻ Page 44.

[122] Internet Corporation for Assigned Names and Numbers (ICANN). Stewardship of IANA functions transitions to global Internet community as contract with U.S. government ends. `https://www.icann.org/news/announcement-2016-10-01-en`, October 2016. ▻ Page 44.

[123] Van Jacobson, Diana K. Smetters, James D. Thornton, Michael F. Plass, Nicholas H. Briggs, and Rebecca L. Braynard. Networking named content. In *Proceedings of the International Conference on Emerging Networking Experiments and Technologies (CoNEXT)*, 2009. ▻ Page 13.

[124] Sushant Jain, Alok Kumar, Subhasree Mandal, Joon Ong, Leon Poutievski, Arjun Singh, Subbaiah Venkata, Jim Wanderer, Junlan Zhou, Min Zhu, Jonathan Zolla, Urs Hölzle, Stephen Stuart, and Amin Vahdat. B4: Experience with a globally-deployed software defined WAN. In *Proceedings of the ACM SIGCOMM Conference*, August 2013. ▻ Page 33.

[125] Petri Jokela, András Zahemszky, Christian Esteve Rothenberg, Somaya Arianfar, and Pekka Nikander. LIPSIN: Line speed publish/subscribe inter-networking. In *Proceedings of the ACM SIGCOMM Conference*, 2009. ▻ Page 14.

[126] Burt Kaliski. PKCS #5: Password-based cryptography specification version 2.0. RFC 2898, 2000. ▻ Page 384.

[127] Farouk Kamoun and Leonard Kleinrock. Stochastic performance evaluation of hierarchical routing for large networks. *Computer Networks*, 3:337–353, November 1979. ▻ Pages 13 and 40.

[128] Min Suk Kang and Virgil D. Gligor. Routing bottlenecks in the Internet: Causes, exploits, and countermeasures. In *Proceedings of the ACM Conference on Computer and Communications Security (CCS)*, 2014. ▻ Page 322.

[129] Min Suk Kang, Soo Bum Lee, and Virgil D. Gligor. The Crossfire attack. In *Proceedings of the IEEE Symposium on Security and Privacy (S&P)*, May 2013. ▻ Pages 39, 245, 276, and 322.

[130] Josh Karlin, Stephanie Forrest, and Jennifer Rexford. Pretty Good BGP: improving BGP by cautiously adopting routes. In *Proceedings of the IEEE Conference on Network Protocols (ICNP)*, 2006. ▻ Page 328.

[131] Ethan Katz-Bassett, Colin Scott, David Choffnes, Italo Cunha, Vytautas Valancius, Nick Feamster, Harsha Madhyastha, Thomas Anderson, and Arvind Krishnamurthy. LIFEGUARD: Practical repair of persistent route failures. In *Proceedings of the ACM SIGCOMM Conference*, August 2012. ▻ Page 24.

[132] Tiffany Hyun-Jin Kim, Cristina Basescu, Limin Jia, Soo Bum Lee, Yih-Chun Hu, and Adrian Perrig. Lightweight source authentication and path validation. In *Proceedings of the ACM SIGCOMM Conference*, August 2014. ▻ Pages xvi, 279, 282, 291, 295, and 302.

[133] Tiffany Hyun-Jin Kim, Lin-Shung Huang, Adrian Perrig, Collin Jackson, and Virgil Gligor. Accountable key infrastructure (AKI): A proposal for a public-key validation infrastructure. In *Proceedings of the International World Wide Web Conference (WWW)*, 2013. ▻ Page xvi.

[134] Leonard Kleinrock and Farouk Kamoun. Hierarchical routing for large networks: Performance evaluation and optimization. *Computer Networks*, 1:155–174, 1977. ▻ Pages 13 and 40.

[135] Teemu Koponen, Scott Shenker, Hari Balakrishnan, Nick Feamster, Igor Ganichev, Ali Ghodsi, P. Brighten Godfrey, Nick McKeown, Guru Parulkar, Barath Raghavan, Jennifer Rexford, Somaya Arianfar, and Dmitriy Kuptsov. Architecting for innovation. *ACM SIGCOMM Computer Communication Review*, July 2011. ▻ Page 14.

[136] Hugo Krawczyk, Mihir Bellare, and Ran Canetti. HMAC: Keyed-hashing for message authentication. RFC 2104, 2000. ▻ Pages 353 and 384.

[137] Brian Krebs. DDoS on Dyn impacts Twitter, Spotify, Reddit. `https://krebsonsecurity.com/2016/10/ddos-on-dyn-impacts-twitter-spotify-reddit/`, 2016. ▻ Pages 7 and 30.

[138] Brian Krebs. Hacked cameras, DVRs powered today's massive Internet outage. `https://krebsonsecurity.com/2016/10/hacked-`

cameras-dvrs-powered-todays-massive-internet-outage/, 2016. ▷ Page 30.

[139] Brian Krebs. Israeli online attack service ‘vDOS’ earned $600,000 in two years. http://krebsonsecurity.com/2016/09/israeli-online-attack-service-vdos-earned-600000-in-two-years/, 2016. ▷ Page 320.

[140] Brian Krebs. KrebsOnSecurity Hit With Record DDoS. https://krebsonsecurity.com/2016/09/krebsonsecurity-hit-with-record-ddos/, 2016. ▷ Pages 30 and 244.

[141] Mirjam Kühne and Vasco Asturiano. Update on AS Path Lengths Over Time. https://labs.ripe.net/Members/mirjam/update-on-as-path-lengths-over-time, 2012. ▷ Page 38.

[142] Sanjeev Kumar. Smurf-based distributed denial of service (DDoS) attack amplification in Internet. In *Second International Conference on Internet Monitoring and Protection (ICIMP)*, July 2007. ▷ Page 7.

[143] Yi-Hsuan Kung, Tae-Ho Lee, Po-Ning Tseng, Hsu-Chun Hsiao, Tiffany Hyun-Jin Kim, Soo Bum Lee, Yue-Hsun Lin, and Adrian Perrig. A practical system for guaranteed access in the presence of DDoS attacks and flash crowds. In *Proceedings of the IEEE Conference on Network Protocols (ICNP)*, October 2015. ▷ Page xvi.

[144] Nate Kushman, Srikanth Kandula, and Dina Katabi. Can you hear me now?!: It must be BGP. *ACM SIGCOMM Computer Communication Review*, April 2007. ▷ Pages 5 and 24.

[145] Craig Labovitz, Abha Ahuja, Abhijit Bose, and Farnam Jahanian. Delayed Internet routing convergence. In *Proceedings of the ACM SIGCOMM Conference*, 2000. ▷ Page 5.

[146] Craig Labovitz, Scott Iekel-Johnson, Danny McPherson, Jon Oberheide, and Farnam Jahanian. Internet inter-domain traffic. *Proceedings of the ACM SIGCOMM Conference*, 2010. ▷ Page 273.

[147] Patrick Lacharme, Andrea Rock, Vincent Strubel, and Marion Videau. The Linux Pseudorandom Number Generator Revisited. https://hal.archives-ouvertes.fr/hal-01005441, 2012. ▷ Page 384.

[148] Leslie Lamport, Robert Shostak, and Marshall Pease. The Byzantine Generals Problem. *ACM Transactions on Programming Languages and Systems (TOPLAS)*, 4(3):382–401, July 1982. ▷ Page 71.

[149] Ben Laurie, Adam Langley, and Emilia Kasper. Certificate transparency. RFC 6962, June 2013. ▷ Page 87.

[150] Soo Bum Lee and Virgil D. Gligor. FLoc: Dependable link access for legitimate traffic in flooding attacks. In *IEEE ICDCS*, 2010. ▷ Pages 244 and 276.

[151] Soo Bum Lee, Min Suk Kang, and Virgil D. Gligor. CoDef: Collaborative defense against large-scale link-flooding attacks. In *Proceedings of the International Conference on Emerging Networking Experiments and Technologies (CoNEXT)*, 2013. ▻ Page 277.

[152] Tae-Ho Lee, Pawel Szalachowski, David Barrera, and Adrian Perrig. Bootstrapping real-world deployment of future Internet architectures. *arXiv:1508.02240*, August 2015. ▻ Pages 327 and 329.

[153] Taeho Lee, Christos Pappas, David Barrera, Pawel Szalachowski, and Adrian Perrig. Source accountability with domain-brokered privacy. In *Proceedings of the International Conference on Emerging Networking Experiments and Technologies (CoNEXT)*, December 2016. ▻ Pages xvi and 319.

[154] Taeho Lee, Christos Pappas, Cristina Basescu, Jun Han, Torsten Hoefler, and Adrian Perrig. Source-based path selection: The data plane perspective. In *Proceedings of ACM Conference on Future Internet Technologies (CFI)*, June 2015. ▻ Page xvi.

[155] Taeho Lee, Christos Pappas, Adrian Perrig, Virgil Gligor, and Yih-Chun Hu. The case for in-network replay suppression. In *Proceedings of the ACM Asia Conference on Computer and Communications Security (AsiaCCS)*, April 2017. ▻ Pages xvi and 323.

[156] Taeho Lee, Christos Pappas, Pawel Szalachowski, and Adrian Perrig. Communication based on per-packet one-time addresses. In *Proceedings of the IEEE Conference on Network Protocols (ICNP)*, November 2016. ▻ Pages xvi and 319.

[157] Matt Lepinski. BGPsec protocol specification. Internet-Draft, June 2016. ▻ Pages 33 and 307.

[158] Matt Lepinski and Sean Turner. An overview of BGPsec. Internet-Draft, June 2016. ▻ Pages 27, 33, and 307.

[159] SoundCloud Limited. Prometheus, from metrics to insight. `https://prometheus.io/`, 2016. ▻ Page 220.

[160] Pat Litke and Joe Stewart. BGP hijacking for cryptocurrency profit. `https://www.secureworks.com/research/bgp-hijacking-for-cryptocurrency-profit`, August 2014. ▻ Page 32.

[161] Bisheng Liu, Jerry T. Chiang, Jason J. Haas, and Yih-Chun Hu. Coward attacks in vehicular networks. *ACM SIGMOBILE Mobile Computing and Communications Review*, 14(3):34–36, 2010. ▻ Pages 280, 282, and 292.

[162] Doug Madory. Sprint, Windstream: Latest ISPs to hijack foreign networks. `http://research.dyn.com/2014/09/latest-isps-to-hijack/`, September 2014. ▻ Page 32.

[163] Georgios Mantas, Natalia Stakhanova, Hugo Gonzalez, Hossein Hadian Jazi, and Ali A. Ghorbani. Application-layer denial of service attacks: taxonomy and survey. *International Journal of Information and Computer Security*, 7(2-4):216–239, 2015. ▷ Page 320.

[164] Ronald Margolis, Leslie Derr, Michelle Dunn, Michael Huerta, Jennie Larkin, Jerry Sheehan, Mark Guyer, and Eric D. Green. The National Institutes of Health's Big Data to Knowledge (BD2K) initiative: capitalizing on biomedical big data. *Journal of the American Medical Informatics Association*, 2014. ▷ Page 273.

[165] Moxie Marlinspike. SSL and the future of authenticity. `http://blog.thoughtcrime.org/ssl-and-the-future-of-authenticity`, Apr 2011. ▷ Page 10.

[166] Stephanos Matsumoto and Raphael M. Reischuk. Certificates-as-an-Insurance: Incentivizing accountability in SSL/TLS. In *Workshop on Security of Emerging Networking Technologies (SENT)*, 2015. ▷ Pages 7 and 87.

[167] Stephanos Matsumoto and Raphael M. Reischuk. IKP: Turning a PKI around with decentralized automated incentives. In *Proceedings of the IEEE Symposium on Security and Privacy (S&P)*, 2017. ▷ Pages 7 and 87.

[168] Stephanos Matsumoto, Raphael M. Reischuk, Pawel Szalachowski, Tiffany Hyun-Jin Kim, and Adrian Perrig. Authentication Challenges in a Global Environment. *ACM Transactions on Privacy and Security (TOPS)*, 20(1), 2017. ▷ Page xvi.

[169] Stephanos Matsumoto, Samuel Steffen, and Adrian Perrig. CASTLE: CA signing in a touch-less environment. In *Proceedings of Annual Computer Security Applications Conference (ACSAC)*, December 2016. ▷ Pages xvi and 37.

[170] David Mazieres, Michael Kaminsky, M. Frans Kaashoek, and Emmett Witchel. Separating key mangement from file system security. In *Proceedings of SOSP*, 1999. ▷ Page 284.

[171] Nick McKeown, Tom Anderson, Hari Balakrishnan, Guru Parulkar, Larry Peterson, Jennifer Rexford, Scott Shenker, and Jonathan Turner. Openflow: enabling innovation in campus networks. *ACM SIGCOMM Computer Communication Review*, 38(2):69–74, March 2008. ▷ Page 15.

[172] Danny McPherson, Vijay Gill, Daniel Walton, and Alvaro Retana. Border Gateway Protocol (BGP) Persistent Route Oscillation Condition. RFC 3345, August 2002. ▷ Page 235.

[173] Tom Mendelsohn. Secure Boot snafu: Microsoft leaks backdoor key, firmware flung wide open [updated]. `http://arstechnica.co.uk/security/2016/08/microsoft-secure-boot-firmware-snafu-leaks-golden-key/`, 2016. ▷ Page 44.

[174] Ralph C. Merkle. A digital signature based on a conventional encryption function. In *Proceedings of Advances in Cryptology*, 1988. ▻ Pages 88 and 144.

[175] David Mills, Jim Martin, Jack Burbank, and William Kasch. Network Time Protocol version 4: Protocol and algorithms specification. RFC 5905, 2010. ▻ Pages 160 and 289.

[176] Stephen A. Misel. Wow, AS7007! `https://web.archive.org/web/20151027055050/http://merit.edu/mail.archives/nanog/1997-04/msg00340.html`, April 1997. ▻ Page 5.

[177] Mitre Corporation. CVE-2008-1447: "DNS insufficient socket entropy vulnerability" or "the Kaminsky bug". `https://cve.mitre.org/cgi-bin/cvename.cgi?name=CVE-2008-1447`, 2008. ▻ Page 134.

[178] Paul Mockapetris and Kevin J. Dunlap. Development of the Domain Name System. *ACM SIGCOMM Computer Communication Review*, 18(4):123–133, August 1988. ▻ Page 139.

[179] Robert Moskowitz and Pekka Nikander. Host Identity Protocol (HIP) Architecture. RFC 4423, May 2006. ▻ Page 284.

[180] Robert Moskowitz, Pekka Nikander, Petri Jokela, and Thomas R. Henderson. Host Identity Protocol (HIP). RFC 5201, 2008. ▻ Pages 28 and 94.

[181] John Nagle. On Packet Switches with Infinite Storage. RFC 970, December 1985. ▻ Page 244.

[182] Jad Naous, Michael Walfish, Antonio Nicolosi, David Mazieres, Michael Miller, and Arun Seehra. Verifying and enforcing network paths with ICING. In *Proceedings of the International Conference on Emerging Networking Experiments and Technologies (CoNEXT)*, 2011. ▻ Page 280.

[183] Maitreya Natu and Jelena Mirkovic. Fine-grained capabilities for flooding DDoS defense using client reputations. In *ACM LSAD*, 2007. ▻ Pages 244 and 276.

[184] NDN. Named Data Networking (NDN) - A Future Internet Architecture. `http://www.named-data.net/`, June 2015. ▻ Pages 13 and 331.

[185] NIST. FIPS PUB 180-2, Secure Hash Standard (SHS), 2008. ▻ Page 384.

[186] Erik Nordstrom, David Shue, Prem Gopalan, Rob Kiefer, Matvey Arye, Steven Ko, Jennifer Rexford, and Michael J. Freedman. Serval: An end-host stack for service-centric networking. In *Proceedings of the USENIX Symposium on Networked Systems Design and Implementation (NSDI)*, 2012. ▻ Pages 14 and 153.

[187] North American Network Operators' Group (NANOG). Mailing list and archives. `https://www.nanog.org/list/archives`. ▻ Page 244.

[188] Octave Klaba / Oles. 1156 Gbps DDoS. `https://twitter.com/olesovhcom/status/778019962036314112`, 2016. ▻ Page 244.

[189] Open Networking Foundation. OpenFlow. https://www.opennetworking.org/sdn-resources/openflow. ▻ Page 15.

[190] Hilarie Orman and Paul Hoffman. Determining strengths for public keys used for exchanging symmetric keys. RFC 3766, April 2004. ▻ Page 381.

[191] Charlie Osborne. Microsoft Secure Boot key debacle causes security panic. http://www.zdnet.com/article/microsoft-secure-boot-key-debacle-causes-security-panic/, 2016. ▻ Page 44.

[192] Palo Alto Research Center (PARC). Project CCNx. http://blogs.parc.com/ccnx. ▻ Page 13.

[193] Rong Pan, Balaji Prabhakar, and Konstantinos Psounis. CHOKe – a stateless active queue management scheme for approximating fair bandwidth allocation. In *IEEE INFOCOM*, 2000. ▻ Page 277.

[194] Christos Pappas, Katerina Argyraki, Stefan Bechtold, and Adrian Perrig. Transparency instead of neutrality. In *Proceedings of the ACM Workshop on Hot Topics in Networks (HotNets)*, 2015. ▻ Pages xvi and 35.

[195] Bryan Parno, Adrian Perrig, and David Andersen. SNAPP: Stateless network-authenticated path pinning. In *Proceedings of the ACM Asia Conference on Computer and Communications Security (AsiaCCS)*, March 2008. ▻ Page 14.

[196] Bryan Parno, Dan Wendlandt, Elaine Shi, Adrian Perrig, Bruce Maggs, and Yih-Chun Hu. Portcullis: Protecting connection setup from denial-of-capability attacks. In *Proceedings of the ACM SIGCOMM Conference*, 2007. ▻ Pages 244 and 276.

[197] Diego Perino and Matteo Varvello. A reality check for content centric networking. In *ACM SIGCOMM Workshop on Information-centric Networking*, 2011. ▻ Page 335.

[198] Nicole Perlroth. Hackers used new weapons to disrupt major websites across U.S. https://nyti.ms/2oPyvhB, October 2016. ▻ Page 7.

[199] Simon Peter, Umar Javed, Qiao Zhang, Doug Woos, Thomas Anderson, and Arvind Krishnamurthy. One tunnel is (often) enough. In *Proceedings of the ACM SIGCOMM Conference*, 2014. ▻ Pages 194 and 416.

[200] Nathaniel Popper. How China took center stage in Bitcoin's civil war. https://nyti.ms/2k7L5lg, June 2016. ▻ Page 94.

[201] Jon Postel. Internet Protocol. RFC 791, September 1981. ▻ Page 4.

[202] Matthew Prince. Technical details behind a 400Gbps NTP amplification DDoS attack. https://perma.cc/9ZNP-TBHV, February 2014. ▻ Page 244.

[203] Penny Pritzker and Patrick D. Gallagher. SHA-3 standard: Permutation-based hash and extendable-output functions. *Information Tech Laboratory National Institute of Standards and Technology*, pages 1–35, 2014. ▻ Page 384.

[204] Barath Raghavan and Alex C. Snoeren. A system for authenticated policy-compliant routing. In *Proceedings of the ACM SIGCOMM Conference*, 2004. ▷ Page 14.

[205] Barath Raghavan, Patric Verkaik, and Alex C. Snoeren. Secure and policy-compliant source routing. *IEEE/ACM Transactions on Networking*, 17(3), 2009. ▷ Page 14.

[206] Costin Raiciu, Sebastien Barre, Christopher Pluntke, Adam Greenhalgh, Damon Wischik, and Mark Handley. Improving datacenter performance and robustness with Multipath TCP. In *Proceedings of the ACM SIGCOMM Conference*, 2011. ▷ Pages 189 and 223.

[207] Costin Raiciu, Christoph Paasch, Sebastien Barre, Alan Ford, Michio Honda, Fabien Duchene, Olivier Bonaventure, and Mark Handley. How hard can it be? Designing and implementing a deployable multipath TCP. In *Proceedings of the USENIX Symposium on Networked Systems Design and Implementation (NSDI)*, 2012. ▷ Pages 188, 189, and 223.

[208] Dipankar Raychaudhuri, Kiran Nagaraja, and Arun Venkataramani. MobilityFirst: A robust and trustworthy mobility-centric architecture for the future Internet. *ACM SIGMOBILE Mobile Computing and Communications Review*, July 2012. ▷ Page 14.

[209] Yakov Rekhter, Tony Li, and Susan Hares. A Border Gateway Protocol 4 (BGP-4). RFC 4271, January 2006. ▷ Page 4.

[210] Yakov Rekhter, Bob Moskowitz, Daniel Karrenberg, Geert Jan de Groot, and Eliot Lear. Address allocation for private internets. RFC 1918, February 1996. ▷ Pages 114, 202, 232, 296, and 345.

[211] Réseaux IP Européens Network Coordination Centre (RIPE NCC). YouTube hijacking: A RIPE NCC RIS case study. `https://perma.cc/4DK6-FKR3`, March 2008. ▷ Page 6.

[212] Ronald Rivest. Can we eliminate certificate revocation lists? In *Financial Cryptography*, 1998. ▷ Pages 75 and 76.

[213] Scott Rose. DNS Security (DNSSEC) DNSKEY Algorithm IANA Registry Updates. RFC 6725, August 2012. ▷ Page 370.

[214] Christian Rossow. Amplification hell: revisiting network protocols for DDoS abuse. In *Proceedings of the Symposium on Network and Distributed Systems Security (NDSS)*, February 2014. ▷ Page 31.

[215] Benjamin Rothenberger, Daniele E. Asoni, David Barrera, and Adrian Perrig. Internet kill switches demystified. In *Proceedings of EuroSec*, April 2017. ▷ Page 325.

[216] Amit Sahoo, Krishna Kant, and Prasant Mohapatra. BGP convergence delay under large-scale failures: Characterization and solutions. *Computer Communications*, 32(7), May 2009. ▷ Pages 10 and 307.

[217] Jerome H. Saltzer, David P. Reed, and David D. Clark. End-to-end arguments in system design. *ACM Transactions on Computer Systems*, 2(4), November 1984. ▻ Page 11.

[218] Sandstorm. Cap'n Proto data interchange format and RPC system. `https://capnproto.org/`. ▻ Pages 212 and 356.

[219] Max Schuchard, Eugene Y. Vasserman, Abdelaziz Mohaisen, Denis Foo Kune, Nicholas Hopper, and Yongdae Kim. Losing control of the Internet: Using the data plane to attack the control plane. In *Proceedings of the Symposium on Network and Distributed Systems Security (NDSS)*, February 2011. ▻ Pages 5, 11, 274, 291, and 307.

[220] Abhigyan Sharma, Xiaozheng Tie, Hardeep Uppal, Arun Venkataramani, David Westbrook, and Aditya Yadav. A global name service for a highly mobile internetwork. In *Proceedings of the ACM SIGCOMM Conference*, August 2014. ▻ Pages 14 and 34.

[221] Yaron Sheffer, Ralph Holz, and Peter Saint-Andre. Recommendations for Secure Use of Transport Layer Security (TLS) and Datagram Transport Layer Security (DTLS). RFC 7525, May 2015. ▻ Page 381.

[222] Madhavapeddi Shreedhar and George Varghese. Efficient fair queuing using deficit round-robin. *IEEE/ACM Transactions on Networking*, 1996. ▻ Page 277.

[223] Karen E. Sirois and Stephen T. Kent. Securing the Nimrod routing architecture. In *Proceedings of the Symposium on Network and Distributed Systems Security (NDSS)*, February 1997. ▻ Page 13.

[224] Nigel P. Smart, Vincent Rijmen, Bogdan Warinschi, Gaven Watson, and Rodica Tirtea. Algorithms, key size and parameters report. Technical report, European Union Agency for Network and Information Security Agency (ENISA), November 2014. ▻ Page 381.

[225] Christopher Soghoian and Sid Stamm. Certified lies: Detecting and defeating government interception attacks against SSL. In *Financial Cryptography and Data Security*. Springer, 2012. ▻ Page 87.

[226] Junhyuk Song, Radha Poovendran, Jicheol Lee, and Tetsu Iwata. The AES-CMAC algorithm. RFC 4493, June 2006. ▻ Pages 353 and 385.

[227] Neil Spring, Ratul Mahajan, and David Wetherall. Measuring ISP topologies with Rocketfuel. *ACM SIGCOMM Computer Communication Review*, 32(4):133–145, 2002. ▻ Page 335.

[228] Toby Sterling. Second firm warns of concern after Dutch hack. `https://perma.cc/MUU3-W996`, 2011. ▻ Page 7.

[229] Ion Stoica, Daniel Adkins, Shelley Zhuang, Scott Shenker, and Sonesh Surana. Internet indirection infrastructure. *IEEE/ACM Transactions on Networking*, April 2004. ▻ Page 14.

[230] Ion Stoica, Scott Shenker, and Hui Zhang. Core-stateless fair queueing: a scalable architecture to approximate fair bandwidth allocations in high-speed networks. *IEEE/ACM Trans. Netw.*, 11(1):33–46, February 2003. ▻ Pages 268 and 277.

[231] Ahren Studer and Adrian Perrig. The Coremelt attack. In *Proceedings of the European Symposium on Research in Computer Security (ESORICS)*, September 2009. ▻ Pages 39, 244, 245, 274, 276, and 322.

[232] Lakshminarayanan Subramanian, Matthew Caesar, Cheng Tien Ee, Mark Handley, Morley Mao, Scott Shenker, and Ion Stoica. HLP: A next generation inter-domain routing protocol. In *Proceedings of the ACM SIGCOMM Conference*, 2005. ▻ Pages 13 and 40.

[233] Pawel Szalachowski, Laurent Chuat, Taeho Lee, and Adrian Perrig. RITM: Revocation in the middle. In *Proceedings of IEEE International Conference on Distributed Computing Systems (ICDCS)*, June 2016. ▻ Page xvi.

[234] Pawel Szalachowski, Laurent Chuat, and Adrian Perrig. PKI safety net (PKISN): Addressing the too-big-to-be-revoked problem of the TLS ecosystem. In *Proceedings of the IEEE European Symposium on Security and Privacy (Euro S&P)*, April 2016. ▻ Pages xvi and 92.

[235] Pawel Szalachowski, Stephanos Matsumoto, and Adrian Perrig. PoliCert: Secure and flexible TLS certificate management. In *Proceedings of the ACM Conference on Computer and Communications Security (CCS)*, November 2014. ▻ Pages xvi, 29, 40, 87, and 89.

[236] Oliver Tamm, Christian Hermsmeyer, and Allen M. Rush. Eco-sustainable system and network architectures for future transport networks. *Bell Labs Technical Journal*, 14(4):311–327, 2010. ▻ Page 333.

[237] Sasu Tarkoma, Mark Ain, and Kari Visala. The Publish/Subscribe Internet Routing Paradigm (PSIRP): Designing the future Internet architecture. In *Future Internet Assembly*, 2009. ▻ Page 13.

[238] Andree Toonk. Massive route leak causes Internet slowdown. `http://www.bgpmon.net/massive-route-leak-cause-internet-slowdown/`, June 2015. ▻ Page 31.

[239] Brian Trammell. Properties of an ideal naming service. Internet-Draft, September 2016. ▻ Page 102.

[240] Brian Trammell. RAINS (another Internet naming service) protocol specification. Internet-Draft, September 2016. ▻ Pages 104, 106, 114, and 383.

[241] Brian Trammell and Dominik Schatzmann. On flow concurrency in the Internet and its implications for capacity sharing. In *ACM CSWS*, 2012. ▻ Page 273.

[242] Paul F. Tsuchiya. The landmark hierarchy: a new hierarchy for routing in very large networks. In *Proceedings of the ACM SIGCOMM Conference*, 1988. ▷ Pages 13 and 40.

[243] Kannan Varadhan, Ramesh Govindan, and Deborah Estrin. Persistent route oscillations in inter-domain routing. *Computer networks*, 32(1):1–16, 2000. ▷ Page 235.

[244] Jean-Philippe Vasseur and Jean-Louis Le Roux. Path computation element (PCE) communication protocol (PCEP). RFC 5440, March 2009. ▷ Page 277.

[245] Willem Vereecken, Ward Van Heddeghem, Didier Colle, Mario Pickavet, and Piet Demeester. Overall ICT footprint and green communication technologies. In *International Symposium on Communications, Control and Signal Processing*, 2010. ▷ Page 331.

[246] Jai Vijayan. Juniper discovers unauthorized code in its firewall OS. `http://www.darkreading.com/vulnerabilities---threats/juniper-discovers-unauthorized-code-in-its-firewall-os-/d/d-id/1323622`, 2015. ▷ Page 305.

[247] Michael Walfish, Jeremy Stribling, Maxwell Krohn, Hari Balakrishnan, Robert Morris, and Scott Shenker. Middleboxes no longer considered harmful. In *Proceedings of OSDI*, 2004. ▷ Page 284.

[248] Tao Wan, Evangelos Kranakis, and Paul C. van Oorschot. Pretty secure BGP, psBGP. In *Proceedings of the Symposium on Network and Distributed Systems Security (NDSS)*, 2005. ▷ Page 307.

[249] Yuefeng Wang, Flavio Esposito, Ibrahim Matta, and John Day. RINA: An architecture for policy-based dynamic service management. Technical Report BUCS-TR-2013-014, November 2013. ▷ Page 15.

[250] Dan Wendlandt, David G. Andersen, and Adrian Perrig. Perspectives: Improving SSH-style Host Authentication with Multi-Path Probing. In *Proceedings of USENIX Annual Technical Conference*, June 2008. ▷ Page 6.

[251] WikiLeaks. Vault 7: Working with MikroTik RouterOS 6.X. `https://wikileaks.org/ciav7p1/cms/page_44957707.html`, 2017. ▷ Page 305.

[252] Chris Williams. Bungling Microsoft singlehandedly proves that golden backdoor keys are a terrible idea. `http://www.theregister.co.uk/2016/08/10/microsoft_secure_boot_ms16_100/`, 2016. ▷ Page 44.

[253] Tilman Wolf, James Griffioen, Kenneth L. Calvert, Rudra Dutta, George N. Rouskas, Ilia Baldine, and Anna Nagurney. ChoiceNet: toward an economy plane for the Internet. *ACM SIGCOMM Computer Communication Review*, 44(3):58–65, July 2014. ▷ Page 14.

[254] John Wroclawski. The use of RSVP with IETF integrated services. RFC 2210, September 1997. ▻ Page 244.

[255] Hao Wu, Hsu-Chun Hsiao, and Yih-Chun Hu. Efficient large flow detection over arbitrary windows: An algorithm exact outside an ambiguity region. In *Proceedings of the ACM Internet Measurement Conference (IMC)*, 2014. ▻ Page 269.

[256] Qinghua Wu, Zhenyu Li, Jianer Zhou, Heng Jiang, Zhiyang Hu, Yunjie Liu, and Gaogang Xie. Sofia: toward service-oriented information centric networking. *IEEE Network*, 28(3):12–18, 2014. ▻ Page 153.

[257] Xilinx. Virtex-7 power estimator. `https://www.xilinx.com/products/technology/power/xpe.html`. ▻ Page 334.

[258] Abraham Yaar, Adrian Perrig, and Dawn Song. SIFF: A stateless Internet flow filter to mitigate DDoS flooding attacks. In *Proceedings of the IEEE Symposium on Security and Privacy (S&P)*, 2004. ▻ Pages 244 and 276.

[259] Xiaowei Yang, David Clark, and Arthur W. Berger. NIRA: A new inter-domain routing architecture. *IEEE/ACM Transactions on Networking*, 2007. ▻ Pages 13, 14, and 40.

[260] Xiaowei Yang, David Wetherall, and Thomas Anderson. A DoS-limiting network architecture. *ACM SIGCOMM Computer Communication Review*, 2005. ▻ Pages 244 and 276.

[261] Kim Zetter. Secret code found in Juniper's firewalls shows risk of government backdoors. `https://www.wired.com/2015/12/juniper-networks-hidden-backdoors-show-the-risk-of-government-backdoors/`, 2015. ▻ Page 305.

[262] Kim Zetter. New discovery around Juniper backdoor raises more questions about the company. `https://www.wired.com/2016/01/new-discovery-around-juniper-backdoor-raises-more-questions-about-the-company/`, 2016. ▻ Page 305.

[263] Fuyuan Zhang, Limin Jia, Cristina Basescu, Tiffany Hyun-Jin Kim, Yih-Chun Hu, and Adrian Perrig. Mechanized network origin and path authenticity proofs. In *Proceedings of the ACM Conference on Computer and Communications Security (CCS)*, November 2014. ▻ Pages xvi, 279, 282, and 291.

[264] Lixia Zhang, Stephen Deering, Deborah Estrin, Scott Shenker, and Daniel Zappala. RSVP: A new resource reservation protocol. *IEEE Network*, 1993. ▻ Page 277.

[265] Xin Zhang. *Secure and efficient network fault localization*. PhD thesis, Carnegie Mellon University, 2012. ▻ Page 281.

[266] Xin Zhang, Hsu-Chun Hsiao, Geoffrey Hasker, Haowen Chan, Adrian Perrig, and David Andersen. SCION: Scalability, Control, and Isolation

On Next-Generation Networks. In *Proceedings of the IEEE Symposium on Security and Privacy (S&P)*, May 2011. ▷ Pages 3 and 306.

[267] Philip Zimmerman. PGP user's guide. `http://www.pa.msu.edu/reference/pgpdoc1.html`, October 1994. ▷ Page 40.

Frequently Asked Questions

In this section, we answer frequently asked questions regarding the design, deployment, and operation of SCION. We start off with general questions, then we address more technical and deployment-related questions.

1. General Questions

Do you use countries as ISDs? Doesn't this cause a lot of problems?

We are currently looking into the best way to partition the Internet into ISDs, so using countries as ISDs is only one possible option. Countries have the advantage of providing a uniform legal environment, allowing misbehavior in an ISD to be handled according to the legal framework of that ISD.

Thanks to the possibility of overlapping ISDs, ASes in SCION can obtain additional guarantees by joining ISDs with functional characteristics (e.g., an ISD for financial transactions). For more information, we refer to Chapter 3, in particular to Section 3.6 on Page 56 for overlapping ISDs, and to Section 3.5 on Page 51 for ISD governance models.

Doesn't SCION create opportunities for government intervention and censorship?

No, SCION makes it decidedly harder for a government to censor networks. We are not aware of an attack that a government could perform on SCION that could not be performed on today's Internet. On the other hand, many attacks that are possible in today's Internet do not work in SCION, for example route-hijacking attacks — a SCION path cannot be diverted due to the separation of control and data planes, and packet-carried forwarding state. Moreover, the transparency properties of SCION reveal censorship actions, and the path control properties enable end hosts to avoid certain ASes. Finally, peering links that traverse to an AS in another ISD can allow communication to different ISDs without traversing any core ASes that may censor communications.

A. Perrig et al., *SCION: A Secure Internet Architecture*, Information Security and Cryptography, https://doi.org/10.1007/978-3-319-67080-5

Why partition the Internet into ISDs at all? Shouldn't the Internet be a globally connected entity?

Even with SCION's ISDs, the Internet remains globally connected — entities anywhere in the world can communicate with one another using SCION. ISDs provide isolation guarantees with respect to this communication, such that if two entities in the same ISD communicate, no traffic between them will ever exit the ISD. Additionally, ISDs guarantee that their internal routing decisions for external destinations cannot affect the rest of the Internet, which would, for instance, prevent an ISP in Pakistan from globally hijacking traffic to YouTube (as happened in February 2008).

To learn more about the role of isolation, see Section 3.1 on Page 43.

How is SCION different from source routing?

In source routing, an end host knows the network topology and selects a path through that topology to reach the destination. The path is embedded in the packet header. Source routing does not scale to the size of the Internet because the source would need to know the entire network topology to determine paths. Source routing would not allow the receiver to control the path, nor does it enable path control at the ISPs.

In SCION, an end host cannot freely choose the path, but merely select it from a set of offered paths. Thus, the end host does not need to know the network topology, although the offered paths provide a limited amount of network topology information. So in contrast to source routing, source nodes in SCION can combine up to three path segments (an up-segment, a core-segment, and a down-segment) without requiring knowledge of the entire network topology. Moreover, SCION allows ISPs, sources, and destinations to control the end-to-end path. This approach is fundamentally different from source routing. See Section 8.2 on Page 164 for more details on how to construct paths.

Does SCION need flow state to be set up on intermediate routers before communication can occur?

No, SCION routers do not have any per-flow state. Senders can simply place hop fields from a path in a packet header and send the packet to the destination without any required setup. See Section 8.2 on Page 164 for more details on how to construct paths.

How does SCION overcome network configuration errors?

Configuration errors can affect the control plane or the data plane. In the control plane of today's Internet, a misconfiguration of BGP can result in a route hijack, and traffic not intended for the misconfigured AS takes a detour and is sent to that AS. SCION's secure routing architecture prevents such misconfigurations from affecting other ASes; they simply reject incorrect announcements due to the absence of correct cryptographic authenticators. Another misconfiguration in today's Internet is an incorrect BGP route filter, which can result in the hijack of the internal IP address space. SCION's control-plane structure also makes such misconfigurations impossible.

In the data plane of today's Internet, a misconfiguration can result in incorrect forwarding tables. In SCION, ASes require correct hop fields to forward traffic, which contain a cryptographic token that each AS verifies. So without the correct hop fields, a packet cannot traverse the network and is dropped when it enters the network. Since both an AS's ingress and egress points are contained in the hop field and are verified, packets can by design not be forwarded along an incorrect path.

SCION's PCBs reveal information about how ASes are willing to route traffic (or some fraction of that information since only a subset of all paths are shared at a given time). Are ASes willing to share such information broadly?

ASes in SCION can decide which links they want to announce to which downstream customers. However, SCION provides more information about connectivity of an AS than BGP currently reveals. Nothing about an AS's internal connectivity is revealed in SCION; and even in today's Internet, an analysis of BGP messages over a time period shows how an AS is connected to its neighbors.

How does multipath communication conceal link failures from applications? A loss of some packets on some path will still introduce delays and retransmissions.

A SCION multipath socket will notice a lost packet after an RTT delay, based on the lack of a (positive or negative) acknowledgment. The multipath socket can then re-send the packet on a different working path, thus transparently masking a link failure.

Moreover, SCION's secure link revocation system (see Section 7.3) enables rapid failover to a working path.

How is SCION different from OSPF, SDN, or MPLS?

SCION is used for *inter*-domain communication, to scale to the size of the Internet. For *intra*-domain routing, each domain can use any protocol, such as

OSPF, SDN, or MPLS-based forwarding. SCION's hop fields may resemble MPLS tags, but they have authentication and integrity protection.

How is SCION different from SDN?

Most current work in SDN is primarily applicable to *intra*-domain communication, while SCION is primarily concerned with improving *inter*-domain communication, with minimal changes to the network within each domain. SCION can make use of SDN to provide intra-domain communication. Therefore, the two approaches complement each other.

While some emerging SDN projects attempt to provide inter-domain properties within the existing Internet architecture, such as explicit multipath support, they lack most of the security and scalability properties offered by SCION.

How is SCION different from Asynchronous Transfer Mode (ATM)?

Of the many differences, we will describe some of the main ones. First, ATM faces inherent scalability problems due to fundamental design decisions. Forwarding and label rewriting can lead to per-channel state at ATM switches. The ATM cell headers can identify up to 2^{24} channels at an ingress switch (the UNI interface) and up to 2^{28} channels at a core switch (the NNI interface). Increasing the length of the corresponding header fields would enable support for more channels, but does not address the root of the problem: per-channel state is not a scalable solution, especially for core switches. Furthermore, enabling multipath communication would further harm scalability, since one pair of hosts would set up multiple virtual channels. In SCION, the required forwarding state is embedded into packets, which in practice makes switches stateless for forwarding. This design decision makes SCION extremely scalable with respect to the end-point population. Furthermore, ATM's design decision of having short cells (instead of larger packets) in order to support real-time voice traffic is obsolete in modern networks; full-length 1,500-byte packets do not incur significant processing delays in practice.

Second, ATM was not designed with security in mind, which raises multiple problems in an inter-domain deployment scenario. Quality of Service (QoS), which is one of the most attractive properties of ATM, cannot be guaranteed in an inter-domain environment where a channel traverses multiple domains with conflicting interests. For example, channel setup signals could be modified en route without being detected. Furthermore, a host has no control over the selected path and cannot avoid potentially malicious domains. However, the lack of path control is also a feature of other architectures and protocols (e.g., IP). SCION treats security as an integral design principle, offering strong security guarantees such as enforceable path control.

Would ISPs be happy with letting clients control their communication paths?

A detailed discussion of this aspect is in Section 10.9, which describes how ISPs can define and enforce their path policies. SCION ISPs can also control the amount of bandwidth they let flow over a given link. With such bandwidth control they can steer the traffic, and clients will automatically be diverted toward paths with more bandwidth. Especially with the SIBRA extension (see Chapter 11 on Page 243), the ISP has fine-grained and explicit control over how much bandwidth is provided on each link.

2. Questions Regarding the Operation of SCION

What does a SCION address look like?

A SCION address is a 3-tuple of the form $\langle$ `I,A,e` $\rangle$, where `I` identifies the ISD, `A` identifies the AS in ISD `I`, and `e` identifies the end host inside AS `A`. For more details see Section 15.1.2 on Page 345. We note that, in contrast to a public IP address, a public SCION address of a host `H` does not enable a host outside AS `A` to communicate with `H`. Any host (in particular network adversaries) needs to have at least one valid SCION path to the host's AS in order to send packets to `H`.

Does SCION require public IP addresses?

A customer of a SCION-supporting ISP would not need a public IP address, as it can communicate with other SCION ASes via SCION paths. At the current stage of SCION deployment, however, an AS without a SCION-supporting provider ISP needs to have a public IP address in order to connect to other SCION ASes using the current Internet as an underlay network.

Does SCION work for hosts behind NATs?

Yes. Section 10.3 on Page 201 illustrates how subnets behind NATs are connected in SCION. Section 10.8 on Page 223 shows the packet headers for an end-to-end example in which two hosts communicate over SCION.

Which ports does SCION run on?

The SCION protocol currently requires ports `30041` and `50000` to be open for UDP packets. Port numbers may change in the future.

What is the operating system that SCION runs on?

SCION currently runs on Ubuntu 16.04. For more information, take a look at Section 10.4 on Page 211.

3. Questions Regarding the Deployment of SCION

Does the entire world need to switch to SCION at the same time, or can SCION be incrementally deployed?

SCION can be incrementally deployed at ISPs, who can make use of the SCION features to route traffic through the backbone of the Internet. This is done by encapsulating SCION packets inside IP packets (transparently to end users). At the edges, the encapsulation is removed and packets are routed as usual. If a SCION path exists between two or more ISPs, encapsulation is not needed.

We discuss the deployment of SCION in Chapter 10 on Page 191. In particular, we illustrate various deployment scenarios in Section 10.1.1, talk about deployment incentives in Section 2.5, and discuss incremental deployment strategies in Section 10.1.2.

Can SCION run on existing network hardware?

Since SCION re-uses communication within an AS, no changes to the internal network infrastructure are needed. The border routers, however, do need to support SCION. We currently use standard PC hardware-based border routers, but we believe that hardware routers can be upgraded to efficiently process SCION packets. Sections 10.1 and 10.2 describe more details of SCION deployment for ISPs and end domains.

Assume my ISP has deployed SCION. Do I have to use TCP/SCION in my communication?

As part of SCION's incremental deployment plan, you can still send legacy traffic in terms of standard TCP/IP packets. The *SCION-IP gateway* will automatically encapsulate your traffic if needed. The details are described in Section 10.3 on Page 201.

I'd like to download and use SCION, what can I do?

Fetch the SCION codebase and its documentation from our GitHub repository hosted at `https://github.com/netsec-ethz/scion`. For more information, take a look at Section 10.4 on Page 211.

What are the benefits to an ISP of deploying SCION?

An ISP can offer new services to its customers, for example: (a) high availability through multipath communication; (b) secure paths that cannot be hijacked; (c) client path control, ensuring that a packet will only traverse ASes that were specified in the forwarding path (this can help with communication compliance), and that sensitive traffic did not leave a jurisdiction; and (d) guaranteed communication despite DDoS attacks. We discuss deployment incentives in detail in Section 2.5 on Page 34.

My company would like to try out SCION. What are typical use cases and how can we get started?

In the initial stages of SCION deployment, we believe that the early use cases will encompass corporations wanting to secure a point-to-point connection (where they control both endpoints) to achieve properties similar to a leased line, but without the costs and time delays associated with setting up and running a leased line. In this context, here are a few concrete use cases:

- **Highly available communication:** Many aspects of SCION provide higher availability, especially when multipath communication is used. SCION is immune to attacks such as prefix hijacking.
- **Client path control, for example for compliance purposes:** SCION can guarantee what path each packet takes; in particular, which ISPs are *not* traversed.
- **Secret paths that can only be used by selected communication partners:** The cryptographic path protection enables path hiding even if an attacker knows the network topology, thus making the path impossible to DDoS.
- **VPN link protection:** A VPN link can be provided by the SCION network, providing all the properties listed above for the end-to-end VPN tunnel.

Companies in Switzerland can experience the benefits of a native SCION deployment if they are a customer of a SCION-deploying ISP (e.g., Swisscom or SWITCH at the moment) — in essence, one can communicate without any reliance on BGP (therefore avoiding BGP vulnerabilities or weaknesses). As additional ISPs deploy SCION, these properties will also become available to customers of those ISPs.

For more information, take a look at Section 10.4 on Page 211.

What properties can SCION offer if some of its links connecting deploying ISPs are over the traditional Internet?

The properties that SCION achieves are weaker if some of the traffic traverses the current Internet, but major benefits can still be achieved.

As our simulations reported in Section 13.9 (and independent research by Peter et al. [199]) show, paths composed of shorter tunnels are much more resilient to BGP hijacking attacks than the longer end-to-end paths.

Another major benefit is that SCION's source-based path selection and multipath routing enables relatively fine-grained selection of paths. For instance, for traffic from Switzerland to Australia, the sender can select paths that go eastwards via Singapore or westwards via the US.

Many of the properties continue to hold, though in weakened form depending on the deployment density and network topology. With steadily increasing deployment density, the properties will continue to strengthen as well, further increasing deployment incentives in a virtuous feedback cycle.

Glossary

Autonomous System (AS). An autonomous system is a locally connected network under a common administrative control, e.g., the network at ETH Zurich constitutes an AS. If an organizational entity operates multiple networks that are not directly connected through a local area network, then the different networks are considered different ASes in SCION.

Beacon Server. Beacon servers enable SCION's path exploration mechanism. Core beacon servers start beaconing (i.e., propagating *path-segment construction beacons (PCBs)*⋆) to allow the construction of path segments from core ASes to leaf ASes. Upon receiving a PCB, a beacon server can choose to register the learned path with a core path server and a local path server. The beacon server then propagates the PCB to its downstream and peering ASes and appends a corresponding routing decision for each of these ASes, known as a hop field (HF). A beacon server located in a leaf AS only registers paths with the local path server and the core path server, since PCBs are never propagated upstream.

Certificate Server. SCION's certificate servers keep cached copies of *trust root configurations (TRCs)*⋆ and AS certificates. Certificate servers are queried by beacon servers when validating the authenticity of PCBs.

Certification Authority (CA). A certification authority is an entity issuing digital certificates that bind information (most often a domain name) to a public key. CAs have the responsibility of ensuring that the information-to-public-key binding is correct. They also possess one or more key pairs used to sign and verify certificates. The term *certificate authority* is used interchangeably.

Control Plane. The control plane is responsible for the discovery of network paths, i.e., for the exchange of routing information between network nodes. The control plane thus makes decisions about where traffic is sent and deals with questions such as how routes are established, which paths exist, what quality individual links offer, etc. After all routing-related tasks are completed, data packets are forwarded in the *data plane*⋆.

A. Perrig et al., *SCION: A Secure Internet Architecture*, Information Security and Cryptography, https://doi.org/10.1007/978-3-319-67080-5

Core AS. The ASes that manage an ISD are referred to as core ASes. They are typically operated by large ISPs that in the current Internet are the main operators in a given region.

Data Plane. The data plane (sometimes also referred to as the *forwarding plane*) is responsible for forwarding data packets that end hosts have injected into the network. After routes have been established in the *control plane**, packets are forwarded according to these routes in the data plane.

End Domain. See *leaf AS**.

Fastpath. A router's fastpath handles packet processing and forwarding on the line card, and is thus performance-critical. See also *slowpath**.

Forwarding Path. A forwarding path is a complete end-to-end path between two SCION hosts. A forwarding path is used to transmit packets in the data plane and can be created with a combination of up to three *path segments** (an up-segment, a core-segment, and a down-segment). § 8.2 (Page 164)

IP Prefix. In an IP address, the *prefix* or *network number* is the group of most significant bits that identifies the network or subnetwork. The remaining bits in the IP address are used to identify hosts. For instance, in CIDR notation, the IPv4 prefix `10.0.0.0/8` denotes that only the first 8 bits are used to identify the network, and the remaining 24 bits are used to address hosts within the network. IPv6 prefixes are analogous.

ISD Core. The ISD core is the set of ASes within an *isolation domain (ISD)** responsible for handling inter-ISD routing, quality of service within an ISD, and accountability management (e.g., disabling connectivity for misbehaving ASes). § 3.2 (Page 47)

Isolation Domain (ISD). An isolation domain (ISD) is a hierarchical grouping of networks under a common organizational domain. Networks within an ISD should share a common jurisdiction. Each ISD designates a set of *core ASes** that provide service to other ASes. § 3 (Page 43)

Leaf AS. Autonomous systems (ASes) that are not providers to any other ASes are referred to as leaf ASes or end domains.

Log Server. A log server records the operations (most commonly certificate issuance and/or revocation) of *certification authorities (CAs)*⋆ in order to make any misbehavior publicly visible. Operations are most commonly logged using an append-only *Merkle hash tree*⋆.

Merkle Hash Tree (MHT). Merkle hash trees, or just hash trees, are data structures in which leaf nodes hold data objects and non-leaf nodes hold the hash of their child nodes. One can prove that a data object is present in the hash tree using a logarithmic number of nodes, and if the data is ordered in the leaf nodes, one can also prove that a data object is absent from the tree.

Multihoming. An AS with more than one provider is referred to as a multi-homed AS. Multihoming is thus the action of obtaining Internet connectivity through multiple providers.

Name Server. Name servers in SCION are similar to DNS servers in today's Internet; they translate a human-understandable name into an address. End-to-end paths can be looked up and created based on the (ISD, AS) tuple in the *SCION address*⋆ returned by the name server. The end-host address and end-to-end path are then placed in the SCION packet header to enable delivery to a given destination.

Packet-Carried Forwarding State (PCFS). Packet forwarding information is contained in the packet — more precisely in its header — and may, for example, specify the next hop on a path.

Path Segment. Path segments are derived from *path-segment construction beacons (PCBs)*⋆ and registered at path servers. A path segment can be any of the following:

- up-segment (path between a non-core AS and a core AS in the same ISD)
- down-segment (same as an up-segment, but in the opposite direction)
- core-segment (path between core ASes)

Up to three path segments can be used to create an end-to-end *forwarding path*⋆.

Path Server. End hosts register their up- and down-segments at path servers within their ISD. End hosts can also query path servers in order to obtain path segments to a destination. Path servers located in a core AS are called *core path servers*. Beacon servers registers down-segments with core path servers. The core path server, when it receives a SCION destination address as input from a local path server, returns down-segments to the destination AS. If the destination AS resides within a different ISD, the core path server requests the down-segments from the remote (destination) ISD's core path server before returning these paths to the local path server. All core path servers within a single ISD run a consistency protocol to ensure a consistent view of intra-ISD paths. Path servers located in non-core ASes are called *local path servers*. Local path servers, when they receive a SCION destination address as input (from a local client), return path segments between the AS where the local path server resides and the destination's AS. The local path server, as it cannot independently resolve a full path across ISDs, may send queries to core path servers.

Path-Segment Construction Beacon (PCB). Each core AS generates inter-domain and intra-domain PCBs to explore inter- and intra-domain paths, respectively. Each AS further propagates selected PCBs to its neighboring ASes. As a PCB traverses the network, it assembles a path segment, which can subsequently be used as a component for traffic forwarding.

SCION Address. Network-level address of a device using SCION. A SCION address is a 3-tuple of the form (`I`, `A`, `e`), where `I` identifies the ISD, `A` identifies the AS in ISD `I`, and `e` identifies the end host inside AS `A`. The end-host may be identified using an IPv4 or IPv6 address, for example. § 15.1.2 (Page 345)

Slowpath. A router's slowpath runs routing protocols and performs the network management and flow setup. Slowpath operations are typically executed on the router's main CPU and are thus less performance-critical than *fastpath*⋆ operations.

Ternary Content-Addressable Memory (TCAM). A content-addressable memory (CAM) is a special storage device that stores the address together with each data entry. When retrieving the data entry, all the addresses are searched and the matching data entry is returned. In contrast, a regular memory stores a contiguous address range. A TCAM is a special CAM where the address supports a third state for each retrieval address bit, where the third state indicates "don't care" — so an entry will match if all the stored address bits match the query address, excluding the "don't care" bits. The advantage of a TCAM is that maximum-prefix-matching route table lookups can be performed efficiently. The disadvantage of TCAMs is their very high energy consumption.

Trust Root Configuration (TRC). The trust root configuration (TRC) of an ISD defines the roots of trust (i.e., public keys) for verification of control-plane certificates, name resolution certificates, and end-entity certificates. The TRC also contains the policy on how the TRC can be updated.

Abbreviations

AES	Advanced Encryption Standard
ARPKI	Attack-Resilient Public-Key Infrastructure
AS	Autonomous System
ASE	AS Entry
BGP	Border Gateway Protocol
BGPsec	BGP Security Extension
BR	Border Router
CA	Certification Authority
CDN	Content Delivery Network
CIDR	Classless Inter-domain Routing
CT	Certificate Transparency
DDoS	Distributed Denial of Service
DILL	Dynamic Inter-domain Leased Line
DNS	Domain Name System
DNSSEC	DNS Security Extensions
DoS	Denial of Service
DRKey	Dynamically Recreatable Key
FIA	Future Internet Architecture
HE	Hop Entry
HF	Hop Field
IAC	IP Allocation Configuration
IANA	Internet Assigned Numbers Authority
ICANN	Internet Corporation for Assigned Names and Numbers
ICMP	Internet Control Message Protocol
IETF	Internet Engineering Task Force
INF	Info Field
IP	Internet Protocol
ISD	Isolation Domain
ISP	Internet Service Provider
IsSP	Isolation Service Provider
ITU	International Telecommunication Union
IXP	Internet Exchange Point
LIR	Local Internet Registry
MAC	Message Authentication Code

A. Perrig et al., *SCION: A Secure Internet Architecture*, Information Security and Cryptography, https://doi.org/10.1007/978-3-319-67080-5

MHT	Merkle Hash Tree
MITM	Man in the Middle
MPLS	Multiprotocol Label Switching
MSC	Multi-Signature Certificate
MTU	Maximum Transmission Unit
NAT	Network Address Translation
NCO	Naming Consistency Observer
OPT	Origin and Path Trace
OSPF	Open Shortest Path First
PCB	Path-Segment Construction Beacon
PCFS	Packet-Carried Forwarding State
PKI	Public-Key Infrastructure
PoP	Point of Presence
PRF	Pseudorandom Function
PRNG	Pseudorandom Number Generator
PVF	Path Verification Field
QoS	Quality of Service
RAINS	RAINS, Another Internet Naming Service
RIR	Regional Internet Registry
RLD	Registrant-Level Domain
RPC	Remote Procedure Call
RPKI	Resource Public Key Infrastructure
RTT	Round-Trip Time
RZK	Root Zone Key
SCP	Subject Certificate Policy
SDN	Software-Defined Networking
SIBRA	Scalable Internet Bandwidth Reservation Architecture
SIG	SCION-IP Gateway
SSP	SCION Stream Protocol
TCAM	Ternary Content-Addressable Memory
TCP	Transmission Control Protocol
TLD	Top-Level Domain
TLS	Transport Layer Security
TOFU	Trust on First Use
TRC	Trust Root Configuration
TTL	Time to Live
UDP	User Datagram Protocol
VPN	Virtual Private Network
ZK	Zone Key

Index

A. Perrig et al., *SCION: A Secure Internet Architecture*, Information Security and Cryptography, https://doi.org/10.1007/978-3-319-67080-5

B

C

D

$\mathcal{E}$

$\mathcal{F}$

$\mathcal{G}$

$\mathcal{H}$

$\mathcal{I}$

T

U

W

Z

Printed in the United States
By Bookmasters